DATE DUE

The Illustrated Flora of Illinois

The Illustrated Flora of Illinois
Robert H. Mohlenbrock, *General Editor*

The Illustrated Flora of Illinois

Flowering Plants

Pokeweeds, Four-o'clocks, Carpetweeds, Cacti, Purslanes, Goosefoots, Pigweeds, and Pinks

Robert H. Mohlenbrock

Illustrated by Paul W. Nelson

Maps by Wendy Preece

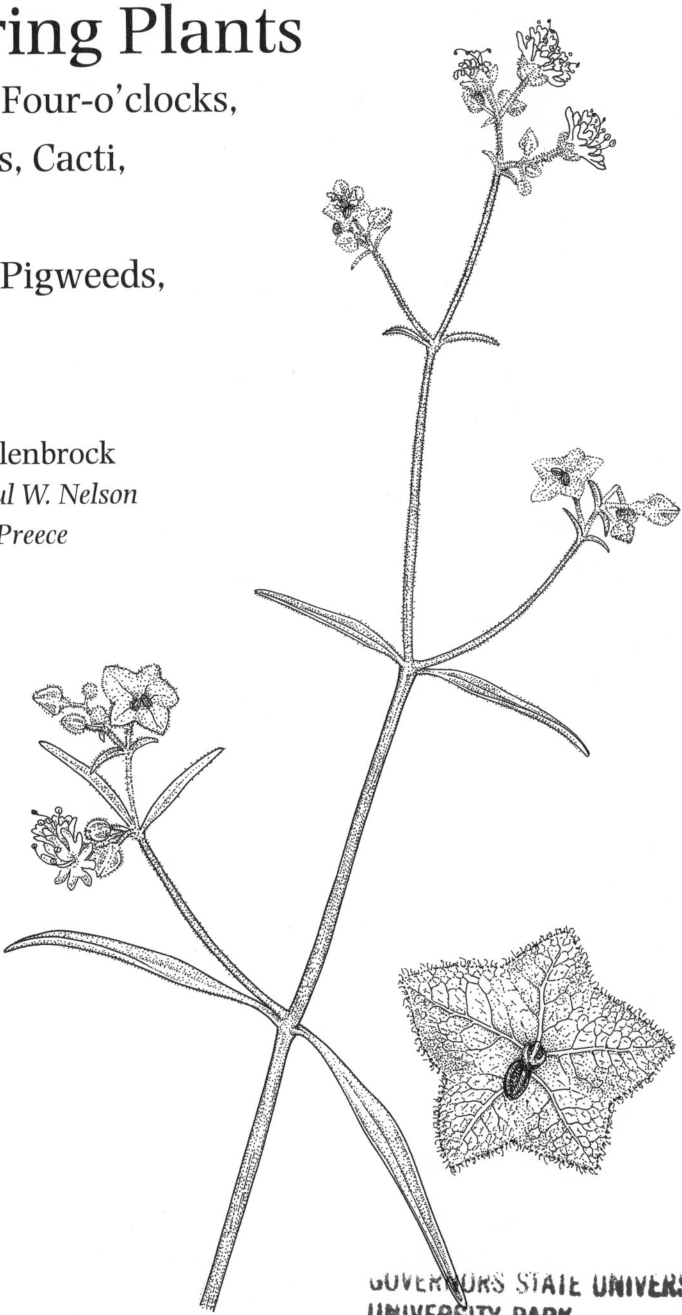

Southern
Illinois
University Press

Carbondale and Edwardsville

Printed in the United States of America
04 03 02 01 4 3 2 1

Library of Congress
Cataloging-in-Publication Data
Mohlenbrock, Robert H., 1931–
Flowering plants : pokeweeds, four-o'clocks, carpetweeds, cacti, purslanes, goosefoots, pigweeds, and pinks / Robert H. Mohlenbrock ; illustrated by Paul W. Nelson.
p. cm.—(The illustrated flora of Illinois)
Includes bibliographical references (p.) and index.
1. Angiosperms—Illinois. 2. Angiosperms—Illinois—Pictorial works. 3. Plants—Identification. I. Title. II. Series.
QK157 .M6224 2001
581.9773—dc21 00-063517
ISBN 0-8093-2380-X

The paper used in this publication meets the minimum requirements of American National Standard for Information Sciences—Permanence of Paper for Printed Library Materials, ANSI Z39.48-1992. ♾

This book is dedicated to Rolla M. and Alice Tryon,

friends, colleagues, and mentors.

Contents

Illustrations

Preface

Several volumes in the Illustrated Flora of Illinois series will be devoted to dicotyledonous flowering plants; this volume is the seventh one. It follows publication of one on ferns, six on monocotyledonous plants, and six on other dicots.

The concept of the Illustrated Flora of Illinois is to produce a multivolumed flora of the plants of the state of Illinois, which will include algae, fungi, mosses, liverworts, lichens, ferns, and seed plants. For each kind of plant known to occur in Illinois, complete descriptions, illustrations showing diagnostic features, distribution maps, and ecological notes will be provided. Keys to aid in identification of the plants will be presented.

An advisory board was created in 1964 to criticize, evaluate, and make suggestions for each volume of the Illustrated Flora during its preparation. The board consists of botanists eminent in their area of specialty—Dr. Robert F. Thorne, Rancho Santa Ana Botanical Garden (flowering plants) and Dr. Rolla M. Tryon Jr., University of South Florida (ferns).

There is no definite sequence for publication of the Illustrated Flora of Illinois. Volumes will appear as they are completed.

Herbaria from which specimens have been studied for this volume are located at Eastern Illinois University, the Field Museum of Natural History, the Gray Herbarium of Harvard University, the Illinois Natural History Survey, the Illinois State Museum, Knox College, the Missouri Botanical Garden, the Morton Arboretum, the New York Botanical Garden, Southern Illinois University Carbondale, the United States National Herbarium, University of Illinois, and Western Illinois University. In addition, some private collections have been examined. The author is indebted to the curators and staffs of these herbaria for the courtesies extended.

I am deeply grateful to the Gaylord and Dorothy Donnelley Foundation for their generous support that made this volume possible.

The illustrations for each species in this volume, depicting the habit and the distinguishing features, were prepared by Paul W. Nelson of the Missouri Department of Natural Resources. My daughter Wendy Preece prepared the maps and also prepared Mr. Nelson's illustrations for publication. My wife, Beverly, assisted me in several of the herbaria and typed all the drafts of the manuscript. Without the help of all those individuals and organizations mentioned above, this book would not have been possible.

The Illustrated Flora of Illinois

County Map of Illinois

Introduction

Flowering plants that form two "seed leaves," or cotyledons, when the seed germinates are called dicotyledons, or dicots. These far exceed the number of species of monocots, or flowering plants that produce a single "seed leaf" upon germination. This is the seventh volume of the Illustrated Flora of Illinois to be devoted to the dicots of Illinois.

The system of classification adopted for the Illustrated Flora of Illinois was proposed by Thorne in 1968. This system is a marked departure from the more familiar system of Engler and Prantl. This latter system, which is still followed in many regional floras, is out-of-date and does not reflect the vast information recently gained from the study of cytology, biochemistry, anatomy, and embryology.

Since the arrangement of orders and families proposed by Thorne is unfamiliar to many, an outline of the orders and families of flowering plants known to occur in Illinois is presented. In the following list, the names in italics have been described in previous volumes of the Illustrated Flora of Illinois. Those in boldface are described in this volume of the Illustrated Flora of Illinois.

Order *Annonales*
Family *Magnoliaceae*
Family *Annonaceae*
Family *Calycanthaceae*
Family *Aristolochiaceae*
Family *Lauraceae*
Family *Saururaceae*
Order *Berberidales*
Family *Menispermaceae*
Family *Ranunculaceae*
Family *Berberidaceae*
Family *Papaveraceae*
Order *Nymphaeales*
Family *Nymphaeaceae*
Family *Ceratophyllaceae*
Order *Sarraceniales*
Family *Sarraceniaceae*
Order *Theales*
Family *Aquifoliaceae*
Family *Hypericaceae* (called Clusiaceae by Thorne in 1968)
Family *Elatinaceae*
Family *Ericaceae*
Order *Ebenales*
Family *Ebenaceae*
Family *Styracaceae*
Family *Sapotaceae*
Order *Primulales*
Family *Primulaceae*
Order *Cistales*
Family *Violaceae*
Family *Cistaceae*
Family *Passifloraceae*
Family *Cucurbitaceae*
Family *Loasaceae*
Order *Salicales*
Family *Salicaceae*
Order *Tamaricales*
Family *Tamaricaceae*
Order *Capparidales*
Family *Capparidaceae*
Family *Resedaceae*
Family *Brassicaceae*
Order *Malvales*
Family *Sterculiaceae*
Family *Tiliaceae*
Family *Malvaceae*
Order *Urticales*

Family *Ulmaceae*
Family *Moraceae*
Family *Urticaceae*
Order *Rhamnales*
Family *Rhamnaceae*
Family *Elaeagnaceae*
Order *Euphorbiales*
Family *Thymelaeaceae*
Family *Euphorbiaceae*
Order *Solanales*
Family *Solanaceae*
Family *Convolvulaceae*
Family *Polemoniaceae*
Order *Campanulales*
Family *Campanulaceae*
Order *Santalales*
Family *Celastraceae*
Family *Santalaceae*
Family *Loranthaceae*
Order Oleales
Family Oleaceae
Order Geraniales
Family Linaceae
Family Zygophyllaceae
Family Oxalidaceae
Family Geraniaceae
Family Balsaminaceae
Family Polygalaceae
Order Rutales
Family Rutaceae
Family Simaroubaceae
Family Anacardiaceae
Family Sapindaceae
Family Aceraceae
Family Hippocastanaceae
Family Juglandacese
Order Myricales
Family Myricaceae
Order **Chenopodiales**
Family **Phytolaccaceae**
Family **Nyctaginaceae**
Family **Molluginaceae** (included in Aizoaceae by Thorne in 1968)
Family **Cactaceae**
Family **Portulacaceae**
Family **Chenopodiaceae**
Family **Amaranthaceae**
Family **Caryophyllaceae**
Family *Polygonaceae*
Order *Hamamelidales*
Family *Hamamelidaceae*
Family *Platanaceae*
Order *Fagales*
Family *Fagaceae*
Family *Betulaceae*
Family *Corylaceae*
Order Rosales
Family Rosaceae
Family Fabaceae
Family Crassulaceae
Family Saxifragaceae
Family Droseraceae
Family Staphyleaceae
Order Myrtales
Family Lythraceae
Family Melastomaceae
Family Onagraceae
Order Gentianales
Family Loganiaceae
Family Rubiaceae
Family Apocynaceae
Family Asclepiadaceae (included in Apocynaceae by Thorne in 1968)
Family Gentianaceae
Family Menyanthaceae
Order Bignoniales
Family Bignoniaceae
Family Martyniaceae
Family Scrophulariaceae
Family Plantaginaceae
Family Orobanchaceae
Family Lentibulariacee
Family Acanthaceae
Order Cornales
Family Vitaceae
Family Nyssaceae
Family Cornaceae
Family Haloragidaceae
Family Hippuridaceae
Family Araliaceae

Family Apiaceae (included in Araliaceae by Thorne in 1968)
Order Dipsacales
Family Caprifoliaceae
Family Adoxaceae
Family Valerianaceae
Family Dipsacaceae
Order Lamiales
Family Hydrophyllaceae
Family Boraginaceae
Family Verbenaceae
Family Phrymaceae (included in Verbenaceae by Thorne in 1968)
Family Callitrichaceae
Family Lamiaceae
Order Asterales
Family Asteraceae

Included in this volume is one order, the Chenopodiales, and eight families. A ninth family in the order, Polygonaceae, appeared in a previous volume in this series.

Since only a small number of dicot families are treated in this book, no general key to the dicot families has been provided. The reader is invited to use my companion books, *Guide to the Vascular Flora of Illinois* (revised and enlarged edition, 1986) and *Vascular Flora of Illinois* (forthcoming), for keys to all families of flowering plants in Illinois.

The nomenclature for the species and lesser taxa used in this volume has been arrived at after lengthy study of recent floras and monographs. Synonyms, with complete author citation, that have applied to species in Illinois are given under each species. A description, while not necessarily intended to be complete, covers the more important features of the species.

The common names are the ones used locally in Illinois.

The habitat designation is not always the habitat throughout the range of the species but only for the species in Illinois. The overall range for each species is given from the northeastern to the northwestern extremities, south to the southwestern limit, then eastward to the southeastern limit. The range has been compiled from various sources, including examination of herbarium material and field studies. A general statement is given concerning the range of each species in Illinois. Dot maps showing county distribution for each taxon are provided. Each dot represents a voucher specimen deposited in some herbarium. There has been no attempt to locate each dot with reference to the actual locality within each county.

The distribution has been compiled from extensive field study as well as herbarium study at Eastern Illinois University, the Field Museum of Natural History, the Gray Herbarium of Harvard University, the Illinois Natural History Survey, the Illinois State Museum, Knox College, the Missouri Botanical Garden, the Morton Arboretum, the New York Botanical Garden, Southern Illinois University Carbondale, the United States National Herbarium, University of Illinois, Western Illinois University, and some private collections.

Descriptions and Illustrations

Order **Chenopodiales**

All of the families in this order except the Cactaceae were included in Engler's Order Centrospermae (1897), while Bessey's system of classification (1915) included them in the Caryophyllales. The major characteristics of the order are the usually coiled or curved embryo and usually unilocular ovary. Wettstein's classification (1935) was the first major one to include the Cactaceae in the same order as the other families in the Centrospermae.

Within the Chenopodiales, the only families with an inferior ovary are the Cactaceae and part of the Nyctaginaceae and Portulacaceae. Several species in the order are apetalous.

Phytolaccaceae—Pokeweed Family

Herbs (in Illinois), shrubs, vines, or trees; leaves alternate, simple, without stipules; flowers perfect (in Illinois) or unisexual; calyx 4- or 5-parted, sometimes petaloid; petals absent; stamens 4–30; ovary superior, several-locular; fruit dry to fleshy, with several seeds.

Agdestis, a genus that does not occur in Illinois and that has an inferior ovary, is often included in the Phytolaccaceae, but I place it in its own family, the Agdestidaceae.

As considered by this author, there are about one hundred species of Phytolaccaceae distributed among twelve genera, most of them in the tropics.

Only the following genus occurs in Illinois.

1. **Phytolacca** L.—Pokeweed

Robust, aromatic, poisonous (when mature), perennial herbs; leaves alternate, simple, entire to undulate; flowers perfect, borne in terminal and lateral racemes; sepals 5, free, petaloid; petals absent; stamens 5–30; ovary superior, consisting of a ring of up to 12 carpels; fruit a berry.

Phytolacca is comprised of about thirty-five species, all but ours being found in warmer parts of the World.

Only the following species occurs in Illinois.

1. **Phytolacca americana** L. Sp. Pl. 441. 1753. Fig. 1.
Phytolacca dodecandra L. Sp. Pl. ed. 2, 631. 1762.

Robust perennial herb from a stout rootstock; stems very stout, erect, to 3.5 m tall, branched, glabrous, becoming bright purple-red with age; leaves alternate, simple, lanceolate to ovate, acute to acuminate at the apex, rounded or cuneate at the base, to 30 cm long, to 10 cm broad, entire or somewhat undulate, glabrous, aromatic, on petioles less than half as long as the blades; flowers perfect, up to 6 mm across, borne in terminal and lateral racemes, each lateral raceme opposite a leaf, the racemes pendulous at maturity, up to 20 cm long, on peduncles up to 20 cm

1. **Phytolacca americana**
a. Stem node showing pith
b. Leaf
c. Flower
d. Raceme of fruits

long; sepals 5, free, greenish white to pinkish, suborbicular, 2–3 mm long; petals absent; stamens usually 10, rarely 5; styles usually 10; pistils usually 10, green, borne in a ring; berries depressed-globose, dark purple, up to 1 cm in diameter; seeds several.

Common Name: Pokeweed; Pokeberry.
Habitat: Fields, woods, disturbed soil.
Range: Quebec to Minnesota, south to Texas and Florida.
Illinois Distribution: Common throughout the state.

This is one of the most robust herbs in Illinois, sometimes attaining heights of more than three meters with a stem diameter in excess of three centimeters. The usually ten carpels arranged in a ring are distinctive for this plant.

When young and the stems and leaves are totally green, they may be gathered and cooked as a vegetable. However, the stem and leaves become poisonous to eat at the time that they begin to take on a purplish red hue.

Pokeweed flowers from June to September.

Nyctaginaceae—Four-o'clock Family

Annual, biennial, or perennial (in Illinois) herbs, shrubs, vines, or trees; leaves usually opposite, simple, without stipules; flowers perfect (in Illinois) or unisexual, variously arranged, subtended by bracts or an involucre of bracts; sepals united, forming a campanulate or funnelform calyx persistent in fruit, often petaloid, 3- to 5-toothed or -lobed; petals absent; stamens 1–5; ovary inferior (in Illinois) or superior, 1-locular; style 1; fruit a 1-seeded utricle enclosed by the calyx tube.

There are about thirty genera and three hundred species in this family. Most of them are in the American tropics. Several genera have ornamental value, including *Myrabilis,* the four-o'clock, and *Bougainvillea,* a tropical woody vine or shrub.

1. **Mirabilis** L.—Four-o'clock

There is often an involucre of bracts that is calyxlike in appearance and that subtends or encloses one or more flowers. Those plants that have the greatly enlarged involucre are sometimes segregated into the genus *Oxybaphus.* The fruit, which includes the 1-seeded utricle enclosed by the calyx tube, is referred to as the anthocarp.

All of our species except *M. jalapa* were originally described in the genus *Allionia,* but that genus has winged anthocarps.

Key to the Species of **Mirabilis** in Illinois

1. Anthocarp (fruit) smooth or 5-angled; involucre not papery or membranaceous, scarcely or not enlarged in fruit . . . . 1. *M. jalapa*
1. Anthocarp (fruit) prominently 5-ribbed; involucre paper or membranaceous, greatly enlarged in fruit.
   2. Leaves petiolate, ovate, rounded or cordate at the base . . . . 2. *M. nyctaginea*
   2. Leaves sessile or nearly so, linear to lanceolate to oblong, not rounded or cordate at the base.
      3. Leaves linear to linear-lanceolate, 1–5 mm wide . . . . 3. *M. linearis*
      3. Leaves lanceolate to oblong, some or all of them more than 5 mm wide.

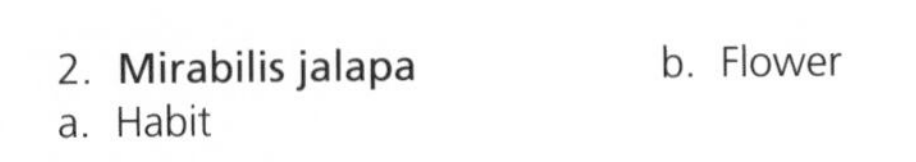

2. **Mirabilis jalapa**
a. Habit
b. Flower
c. Fruit
d. Seed

4. Stems densely hirsute, at least at the base and at the nodes .... 4. *M. hirsuta*
4. Stems glabrous or puberulent in lines .... 5. *M. albida*

1. **Mirabilis jalapa** L. Sp. Pl. 177. 1753. Fig. 2.

Perennial herb from woody or fleshy roots; stems erect, to 1 m tall, much branched, glabrous to sparsely pubescent; leaves opposite, ovate to deltate, acuminate at the tip, cordate or rounded at the base, to 15 cm long, to 8 cm broad, entire, glabrous to sparsely pubescent, occasionally viscid, with petioles up to half as long as the blades, or the uppermost leaves nearly sessile; flowers in glomerules at the ends of the branches, the glomerules on peduncles 1–2 mm long; bracts united, campanulate, up to 15 mm long, the lobes linear-lanceolate to lance-ovate, ciliolate, bristle-tipped, about twice as long as the tube, the involucre of bracts subtending 1 flower; calyx trumpet-shaped, 4–6 cm long, the limb 2.0–3.5 cm across, variously colored, notched around the edges, glabrous to sparsely pubescent on the outer surface; stamens 5, as long as or slightly longer than the calyx; staminodia 3 or 6; anthocarp oval to ovoid, 8–10 mm long, 5-angled, rugose or verrucose, glabrous or puberulent, dark brown to black.

Common Name: Garden Four-o'clock.
Habitat: Disturbed soil, sometimes on refuse heaps.
Range: Adventive in much of the United States; apparently native to tropical America.
Illinois Distribution: The only collection of this plant is from Grundy County.

This is the only species of *Mirabilis* in Illinois that does not have a greatly enlarged, membranaceous involucre of bracts. The anthocarp is 5-angled but not conspicuously 5-ribbed.

Mirabilis jalapa, the garden four-o'clock, is sometimes planted as a garden ornamental but not as often today as in the past.

This species flowers from June to October.

2. **Mirabilis nyctaginea** (Michx.) MacM. Metasp. Minn. 217. 1892. Fig. 3.
Allionia nyctaginea Michx. Fl. Bor. 1:100. 1803.
Oxybaphus nyctagineus (Michx.) Sweet, Hort. Brit. 429. 1830.
Oxybaphus floribundus Choisy in DC. Prodr. 13 (2):433. 1849.

Perennial herb from a thickened taproot; stems erect, to 1 m tall, branched, glabrous to sparsely pubescent; leaves opposite, lance-ovate to ovate, acute to acuminate at the apex, rounded to cordate at the base, to 12 cm long, to 7 cm broad, entire, glabrous, on petioles up to half as long as the blades, or the uppermost leaves nearly sessile; flowers perfect, 3–5 together subtended by an involucre of bracts; involucre up to 6 mm long in flower, up to 15 mm long in fruit, 5-lobed, the lobes obtuse to subacute, pilose, becoming enlarged, veiny, and often pinkish purple in fruit; calyx 5-lobed, pink or purple; stamens 3–5, exserted; anthocarp hardened at maturity,

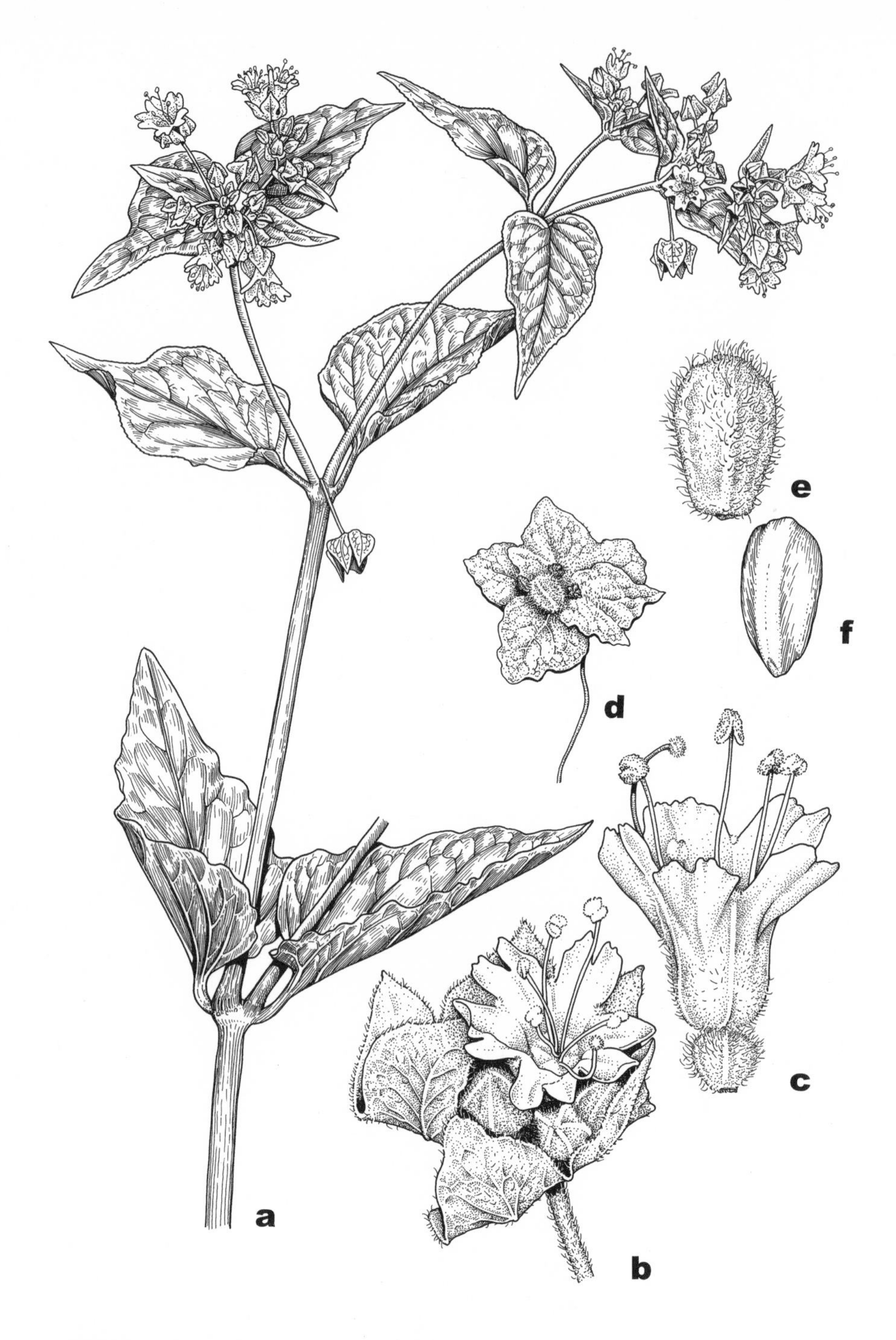

3. **Mirabilis nyctaginea**
a. Habit
b. Flower with involucre
c. Flower
d. Fruit with involucre
e. Fruit
f. Seed

narrowly obovoid, 4–6 mm long, densely pilose, rugose or verrucose, prominently 5-ribbed, grayish brown; seed obovoid, 2.5–3.0 mm long, light brown.

Common Name: Wild Four-o'clock; Umbrella-wort.
Habitat: Disturbed soil; particularly common along railroads.
Range: Manitoba to Montana, south to Colorado, Texas, and Louisiana; adventive in much the rest of the United States, including Illinois.
Illinois Distribution: Common throughout the state.

The wild four-o'clock is a common plant along railroad rights-of-way.

When in fruit, the involucre of bracts often becomes rather handsomely pinkish purple. This species differs from the other species in Illinois that have an enlarged, membranaceous involucre in its petiolate, cordate leaves.

Mirabilis nyctaginea flowers from May to September.

3. **Mirabilis linearis** (Pursh) Heimerl, Ann. Conserv. Jard. Bot. Geneve 5:186. 1901. Fig. 4.
Allionia linearis Pursh, Fl. Am. Sept. 728. 1814.
Oxybaphus linearis (Pursh) B.L. Robins. Rhodora 10:31. 1908.

Perennial herb from an elongated taproot; stems decumbent to ascending to erect, to 1 m tall, glabrous or puberulent in the lower half, short-hairy to viscid in the upper half, glaucous; leaves opposite, linear to linear-lanceolate, acute at the tip, cuneate at the base, to 10 cm long, up to 5 mm broad, entire, glabrous or viscid-pubescent, often glaucous on the lower surface, sessile or nearly so; flowers perfect, 3 together subtended by an involucre of bracts; involucre up to 4 mm long in flower, up to 2 cm long in fruit, 5-lobed, glandular-pubescent; calyx deeply 5-lobed, the lobes retuse, pink to bright purple, to 10 mm long, sparsely pilose; stamens 3–5, usually slightly exserted; anthocarp obovoid, 4–5 mm long, pubescent, rugose, 5-ribbed, olive-brown; seed obovoid, 2.5–3.0 mm long, light yellow-brown.

Common Name: Linear-leaved Wild Four-o'clock.
Habitat: Disturbed sandy soil and in railroad ballast.
Range: North Dakota to Montana, south to Arizona and Texas; Mexico; sparingly adventive in the eastern United States.
Illinois Distribution: Known from Cook, Madison, St. Clair, and Will counties.

The only collection of this plant made during the twentieth century was by Julian Neill in 1947 in St. Clair County.

This species is readily distinguished by its linear leaves.

Mirabilis linearis flowers from June to August.

4. **Mirabilis hirsuta** (Pursh) MacM. Metasp. Minn. 217. 1892. Fig. 5.
Allionia hirsuta Pursh, Fl. Am. Sept. 728. 1814.
Oxybaphus hirsutus (Pursh) Choisy in DC. Prodr. 13 (2):433. 1849.

Perennial herb from an elongated taproot; stems decumbent to ascending to erect, to 1 m tall, sparingly branched or unbranched, hirsute on the lower half of the stem

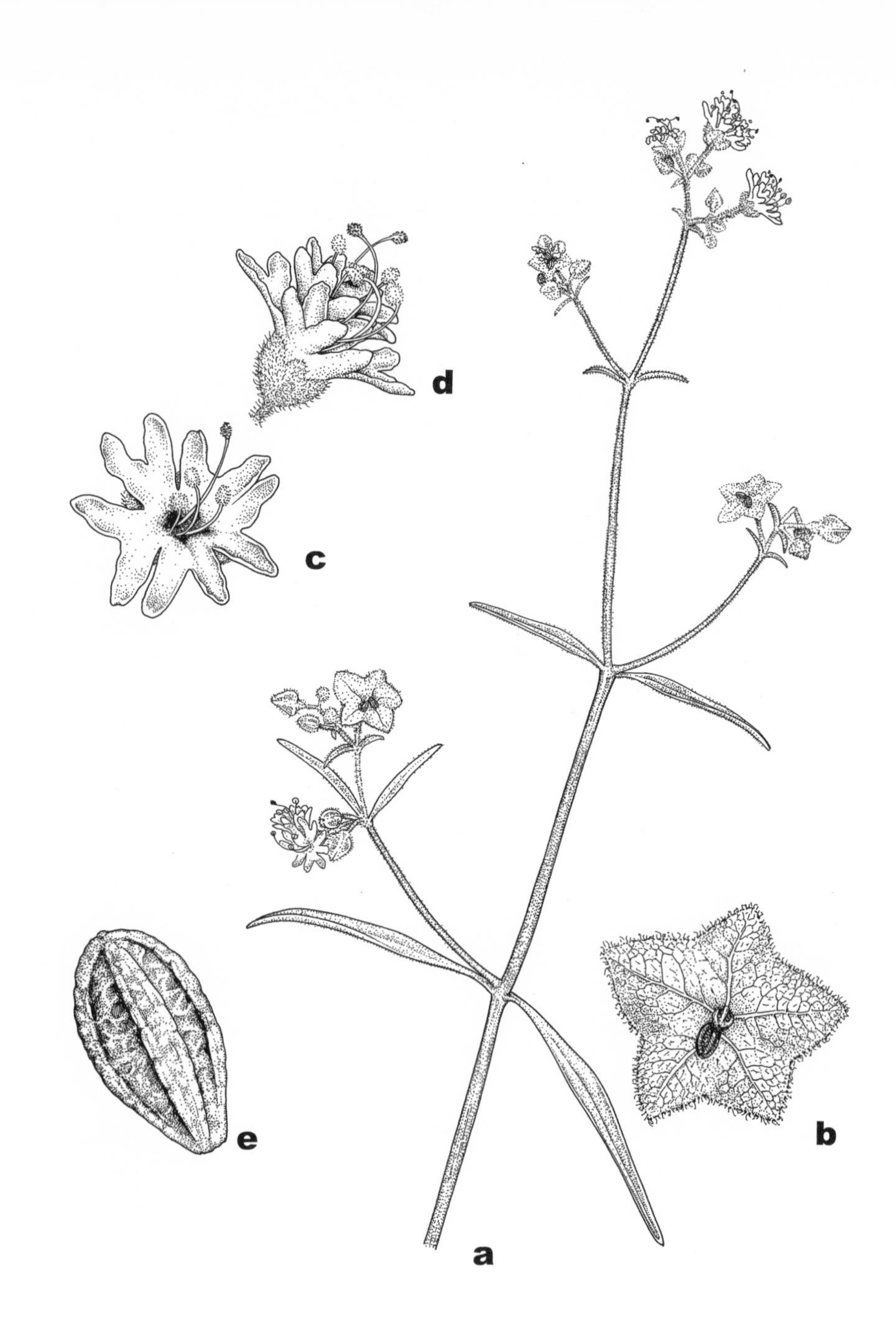

4. **Mirabilils linearis**
a. Habit
b. Fruit with involucre
c. Flower
d. Flower
e. Seed

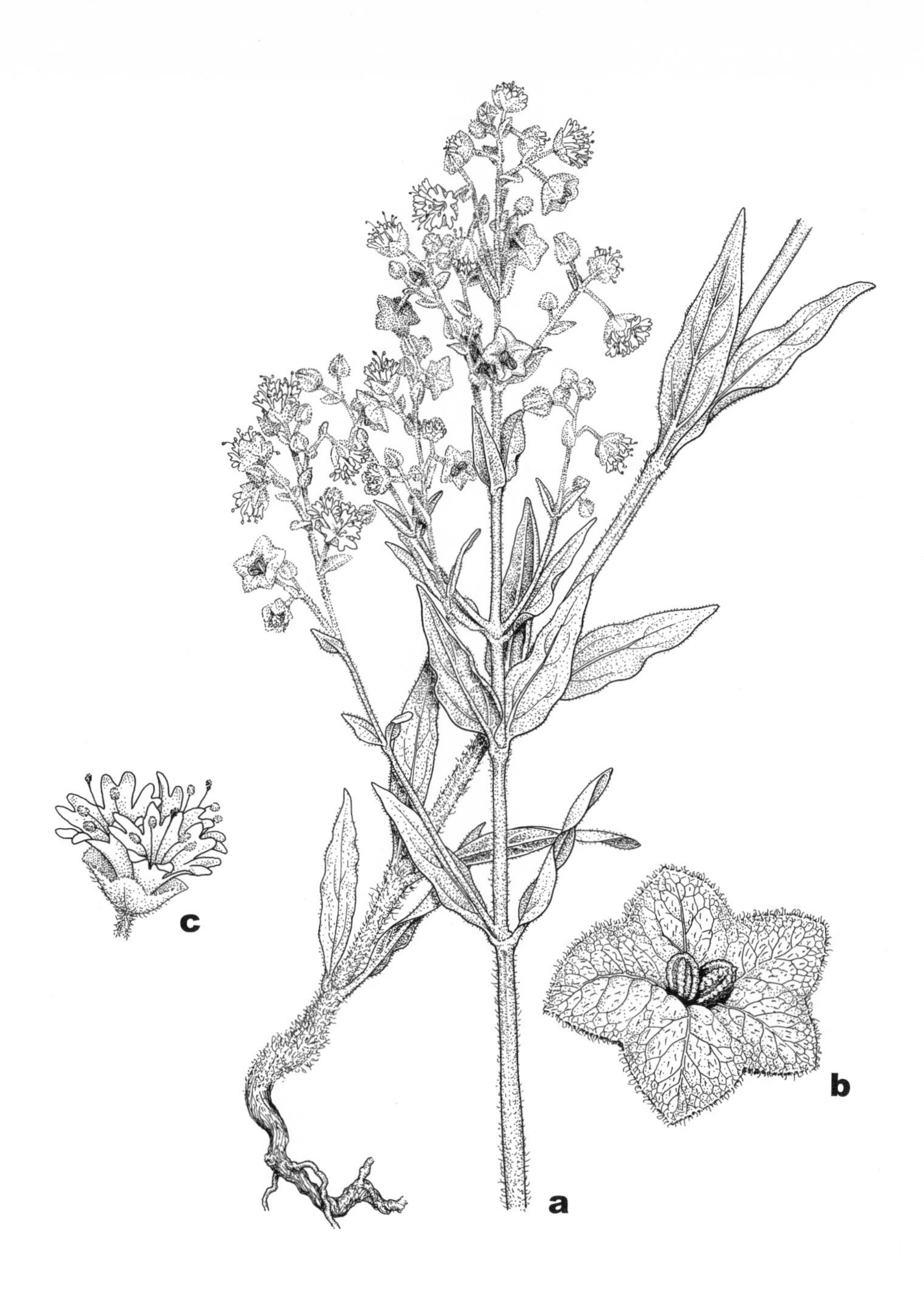

5. **Mirabilis hirsuta**
a. Habit
b. Fruit with involucre
c. Flowers

and at the nodes; leaves opposite, linear-lanceolate to lance-ovate, obtuse to acute at the apex, subcordate or cuneate at the base, to 10 cm long, 1–2 cm broad, entire, densely hirsute to viscid-pubescent, sessile or on petioles up to 5 mm long; flowers perfect, 3 together subtended by an involucre of bracts; involucre 4–5 mm long in flower, up to 2 cm long in fruit, 5-lobed, glandular-pubescent; calyx 5-lobed, pink, to 10 mm long, sparsely pilose; stamens 3–5, usually slightly exserted; anthocarp obovoid, 4–5 mm long, pubescent, rugose, 5-ribbed, deep olive; seed obovoid, 2.5–3.0 mm long, pale brown.

Common Name: Hairy Wild Four-o'clock.
Habitat: Disturbed soil, including railroad ballast; in hill prairies.
Range: Wisconsin to Saskatchewan, south to Arizona, Texas, Missouri, and Illinois; sparingly adventive elsewhere in the United States.
Illinois Distribution: Known from Cook, DuPage, JoDaviess, St. Clair, Tazewell, and Will counties.

The collection from JoDaviess County appears to be from a native population.

This species is distinguished by its hirsute stems and nodes.

Mirabilis hirsuta flowers during July and August.

5. **Mirabilis albida** (Walt.) Heimerl, Ann. Conserv. Jard. Bot. Geneve 5:182. 1901. Fig. 6.
Allionia albida Walt. Fl. Car. 84. 1788.
Oxybaphus albidus (Walt.) Sweet, Hort. Brit. 2:429. 1827.

Perennial herb from an elongated taproot; stems decumbent to ascending to erect, to 1 m tall, glabrous except for the viscid-pubescent upper part as well as just below each node, branched or unbranched; leaves opposite, linear-lanceolate to ovate, obtuse to acute at the apex, cuneate at the base, to 12 cm long, to 25 mm broad, entire, glabrous, glaucous on the lower surface, sessile or on petioles to 5 mm long; flowers perfect, 1–3 together subtended by an involucre of bracts; involucre 3–4 mm long in flower, up to 15 mm long in fruit, 5-lobed, the lobes deltate, viscid-pubescent; calyx shallowly 5-lobed, pink or white, 7–10 mm long, sparsely pubescent; stamens 3–5, exserted; anthocarp obovoid, 4–5 mm long, with tufts of silvery hairs, tuberculate, 5-ribbed, deep olive; seed obovoid, 3.0–3.5 mm long, pale brown.

Common Name: White Wild Four-o'clock.
Habitat: Disturbed soil.
Range: South Carolina to Kansas, south to Texas and Louisiana; sparingly adventive elsewhere in the United States.
Illinois Distribution: Known from Cook, Grundy, Logan, and Sangamon counties.

The distinguishing features of this species are the nearly glabrous stems and leaves and the tufts of silvery hairs on the anthocarp.

Mirabilis albida flowers during July and August.

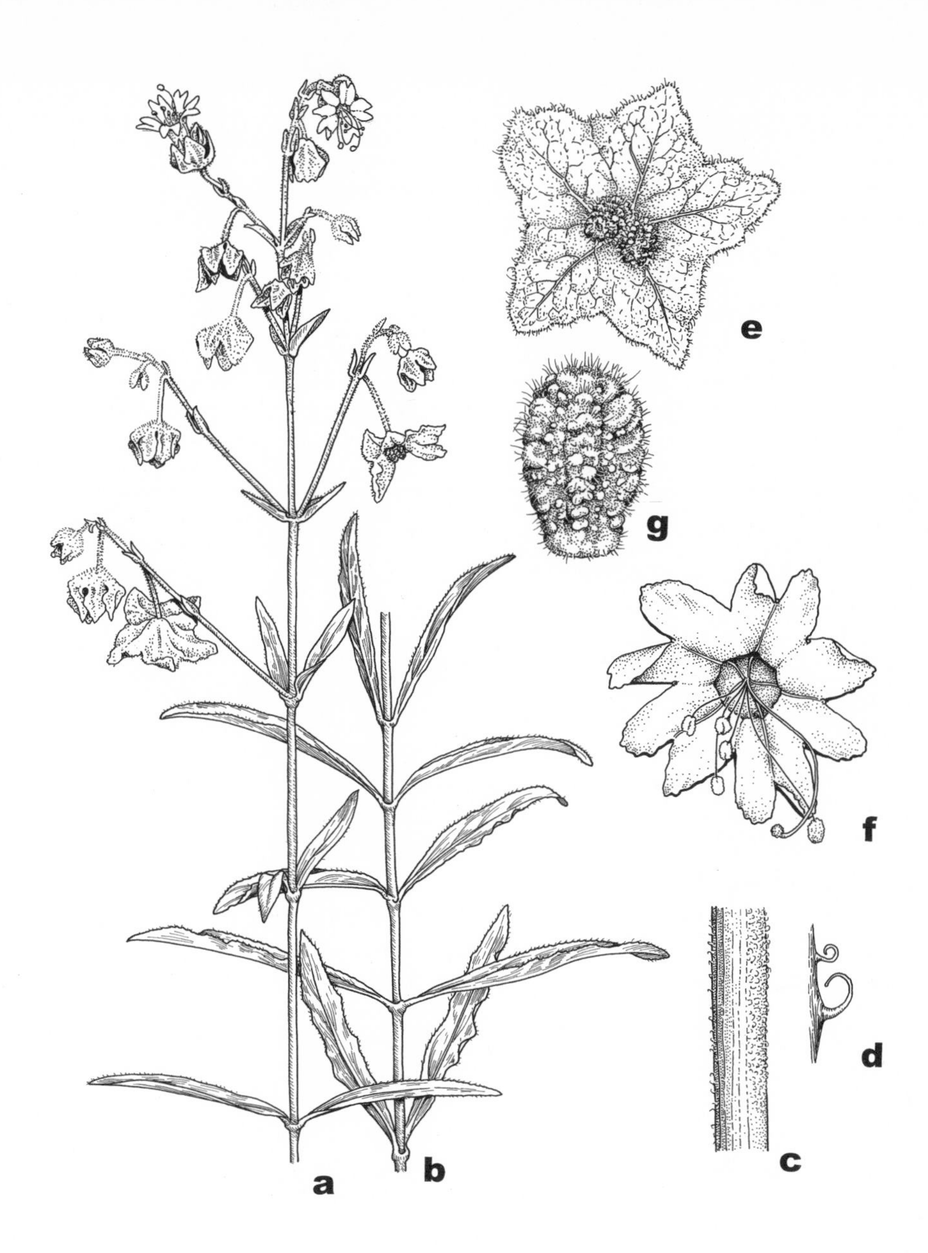

6. **Mirabilis albida**
a. Habit
b. Stems and leaves
c. Stem with pubescence
d. Curved hairs
e. Fruit with bracts
f. Flower
g. Fruit

Molluginaceae—Carpetweed Family

Annuals (in Illinois), perennials, or shrubs; leaves simple, whorled (in Illinois) or opposite, with or without stipules; flowers perfect, axillary, in cymes or umbels or solitary; sepals 4–5, free, usually persistent on the fruit; petals usually absent; stamens 5 or 10; ovary superior, 1- to 10-locular; styles (1–) 3–5; fruit a capsule, splitting vertically.

Many botanists in the past have merged this family within the Aizoaceae. The Aizoaceae differ in their circumscissile capsule, their unwhorled leaves, and their usually fleshy or succulent stems and leaves. As considered here, the Molluginaceae consist of fourteen genera and nearly one hundred species, mostly in the tropics.

Only the following genus occurs in Illinois.

1. **Mollugo** L.—Carpetweed

Annual herb; leaves whorled, simple, forming a basal rosette, without stipules; flowers perfect, axillary, solitary or in cymes; sepals 5, free, petaloid; petals absent; stamens 3 (in Illinois) or 5; ovary superior, 5-locular; styles 3–5; fruit a capsule, dehiscing vertically, enclosed in the persistent calyx; seeds numerous.

Mollugo consists of about twenty species, mostly in the tropics and subtropics.

Only the following species occurs in Illinois.

1. **Mollugo verticillata** L. Sp. Pl. 89. 1753. Fig. 7.

Annual herb from a slender taproot; stems prostrate, matted, repeatedly branched, up to 25 cm long, glabrous; leaves whorled, linear to narrowly oblanceolate, obtuse at the apex, cuneate at the base, to 3 cm long, to 1 cm broad, entire, glabrous, on petioles up to 3 mm long; flowers perfect, 2–5 at each node, on slender pedicels up to 15 mm long; sepals 5, free, petaloid, white or pale green, oblong to elliptic, 2.0–2.5 mm long; petals absent; stamens 3 (–4); styles 3 (–5); fruit a capsule, ovoid to ellipsoid, glabrous, 2.5–3.0 mm long, a little longer than the sepals, dehiscing vertically; seeds many, 0.5 mm long, reniform, more or less ridged down the back and sides, smooth, shiny.

Common Name: Carpetweed.
Habitat: Disturbed soil, often in sand or moist soil.
Range: Native to tropical America; adventive throughout the United States.
Illinois Distribution: Common throughout the state.

This species is readily distinguished by its whorled leaves and its small flowers that have five white sepals, no petals, and three stamens.

Most plants sprawl on the ground, radiating symmetrically from a central point, forming a mat. The mat may sometimes be up to 50 cm in diameter.

Mollugo verticillata flowers from June to October.

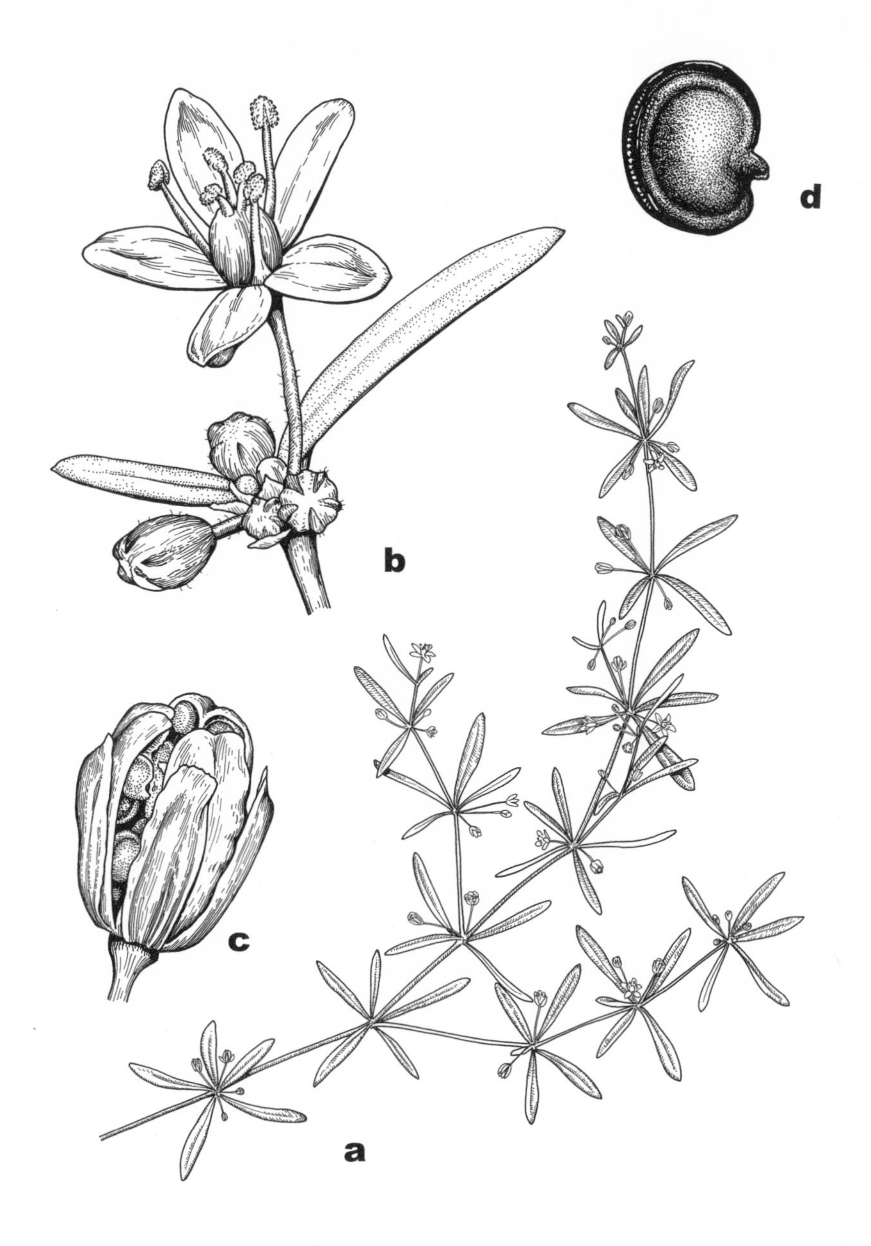

7. **Mollugo verticillata**
a. Habit
b. Flower and buds
c. Fruit
d. Seed

Cactaceae—Cactus Family

Perennial succulent herbs (in Illinois), shrubs, or trees; stems green, very succulent, flat and jointed or cylindrical; leaves reduced to sharp spines; flowers perfect, actinomorphic, usually showy, solitary, sessile; sepals numerous, free, in several rows, sometimes merging into the petals and united at the base with them to form a hypanthium; petals numerous, free, in several rows; stamens numerous, attached to the hypanthium; ovary inferior, 1-locular; style 1; fruit a dry or fleshy berry; seeds numerous.

The cactus family probably consists of about one hundred genera and two thousand species, but the delimitations of the genera are controversial among some botanists. The entire family is native to the New World, although the genus *Rhipsalis* is found along the coast of Africa where it is probably adventive. With very few exceptions, the leaves in this family are reduced to spines to eliminate the loss of water from leaf surfaces.

Only the following genus occurs in Illinois.

1. **Opuntia** Mill.—Prickly Pear

Perennial succulent herbs (in Illinois), shrubs, or trees; stems green, very succulent, flat or cylindrical, jointed; leaves reduced to spines located at areoles, with 1–10 spines per areole, with barbed bristles (glochidia present at each areole as well); flower solitary at each spine-bearing areole, perfect, actinomorphic, showy; sepals numerous; petals numerous, often waxy; stamens numerous; ovary inferior, 1-locular; style 1; fruit dry or fleshy; seeds numerous.

Illinois species of *Opuntia* have flat or somewhat cylindrical jointed stems or "pads." The fruits are sweet and edible, but one must be sure that all glochidia have been removed before eating the fruits.

Key to the Species of **Opuntia** in Illinois

1. Areoles with 3–9 spines; fruit dry, with spines and glochidia; stems not very flat .... 1. *O. fragilis*
1. Areoles with 1–2 (–6) spines; fruit fleshy, without spines but with glochidia; stems very flat.
   2. Joints of stem green; main roots fibrous; spines usually less than 3 cm long; margin of seed smooth, about 0.5 mm broad .... 2. *O. humifusa*
   2. Joints of stem more or less glaucous; main roots tuberous-thickened; spines usually more than 3 cm long; margin of seed irregularly erose, about 1 mm broad .... 3. *O. macrorhiza*

1. Opuntia fragilis (Nutt.) Haw. Suppl. Pl. Succ. 82. 1819. Fig. 8.
Cactus fragilis Nutt. Gen. 1:296. 1818.

Mound-forming cactus up to 50 cm across, from fibrous roots; stems prostrate to spreading; joints of stem turgid, somewhat flattened, obovoid to ellipsoid, to 7.5 cm long, to 2.5 cm broad, more or less glaucous; areoles with (2–) 3–9 spines, the spines white to gray, up to 2.5 cm long; flower solitary, showy, up to 5 cm across; perianth petaloid, greenish or yellowish; fruit obovoid, dry, tan, 1.2–1.5 cm long, spiny and with glochidia.

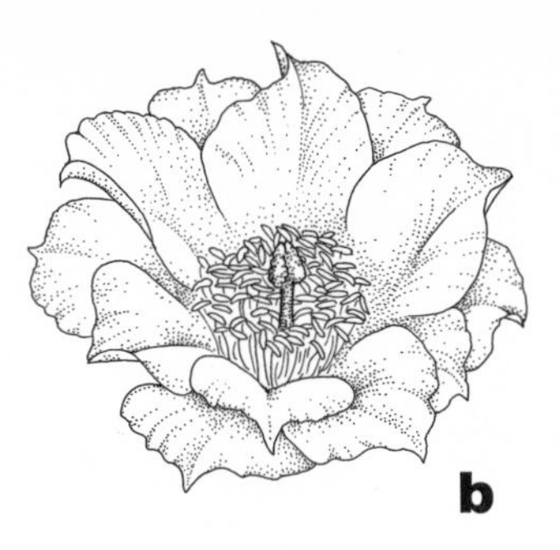

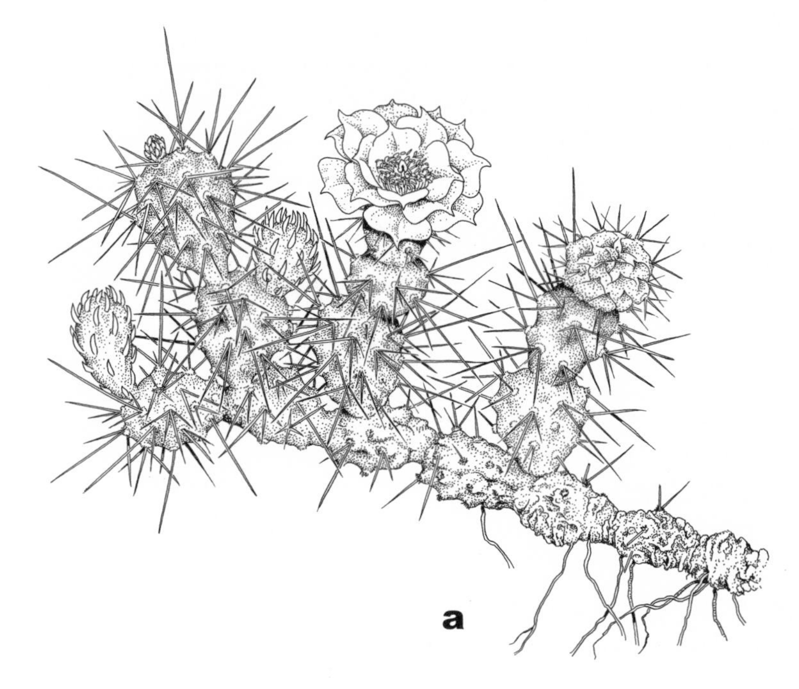

8. **Opuntia fragilis** a. Habit b. Flower

Common Name: Little Prickly Pear.
Habitat: Sandy soil.
Range: Michigan to British Columbia, south to California and Texas; northwestern Illinois.
Illinois Distribution: Known only from JoDaviess County.

This species was observed by Henry A. Gleason in 1910 near Hanover, JoDaviess County, but no specimen was collected. It was, however, rediscovered in JoDaviess County more than a half century later by John Schwegman and Randy Nyboer.

Opuntia fragilis differs from the other species of *Opuntia* in Illinois by its slightly smaller flowers and fruits, its more elongated stem joints that are scarcely flat, and the presence of spines as well as glochidia on the fruits.

The segments of the stem of *Opuntia fragilis* are easily disjointed, hence the specific epithet *fragilis.*

This species flowers in June and July.

2. **Opuntia humifusa** (Raf.) Raf. Med. Bot. 2:247. 1830. Fig. 9.
Cactus compressus Salisb. Prodr. 348. 1796, misapplied.
Cactus humifusus Raf. Ann. Nat. 15. 1820.
Opuntia rafinesquii Engelm. Pacif. R.R. Rep. 4:41. 1856.
Opuntia compressa (Salisb.) Macbr. Contr. Gray Herb. 65:41. 1922.

Clump-forming cactus from fibrous roots; stems prostrate to spreading; joints of stem oblong to orbicular, flat, 5–10 cm long, nearly as broad, green or becoming purplish in winter; areoles with 1 spine, the spine gray or brownish, up to 3 (–5) cm long; flower solitary, showy, up to 6 cm across; perianth petaloid, yellow, sometimes with a reddish center; fruit obovoid, fleshy, purplish, 2.5–3.8 cm long, not spiny but with glochidia; margin of seed smooth, about 0.5 mm broad.

Common Name: Prickly Pear.
Habitat: Dry soil; in sand; on exposed cliffs.
Range: Massachusetts to Ontario and Minnesota, south to Texas and Georgia.
Illinois Distribution: Scattered throughout the state.

This is the only species of cactus that is widespread in the northeastern United States.

Opuntia humifusa differs from the similar *O. macrorhiza,* the other species in Illinois with very flat joints, in its fibrous roots and its single, usually shorter spine at each areole.

This species with its waxy yellow petals flowers from May to July.

3. **Opuntia macrorhiza** Engelm. Bost. Journ. Nat. Hist. 6:206. 1850. Fig. 10.

Clump-forming cactus from tuberous-thickened roots; stems prostrate to spreading; joints of stem obovate to orbicular, flat, 5–10 cm long, not as broad, more or less

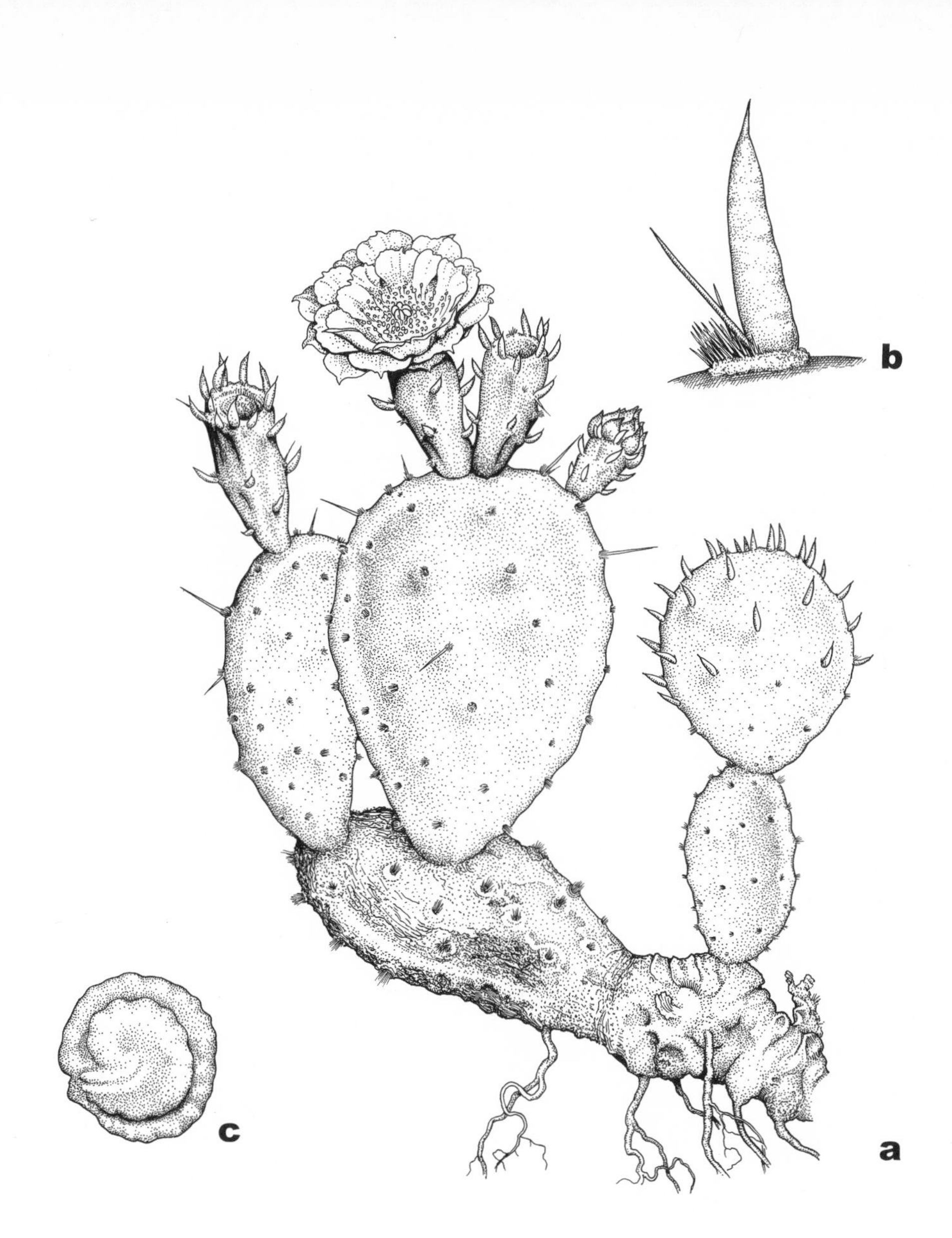

9. **Opuntia humifusa**
a. Habit
b. Aeriole
c. Seed

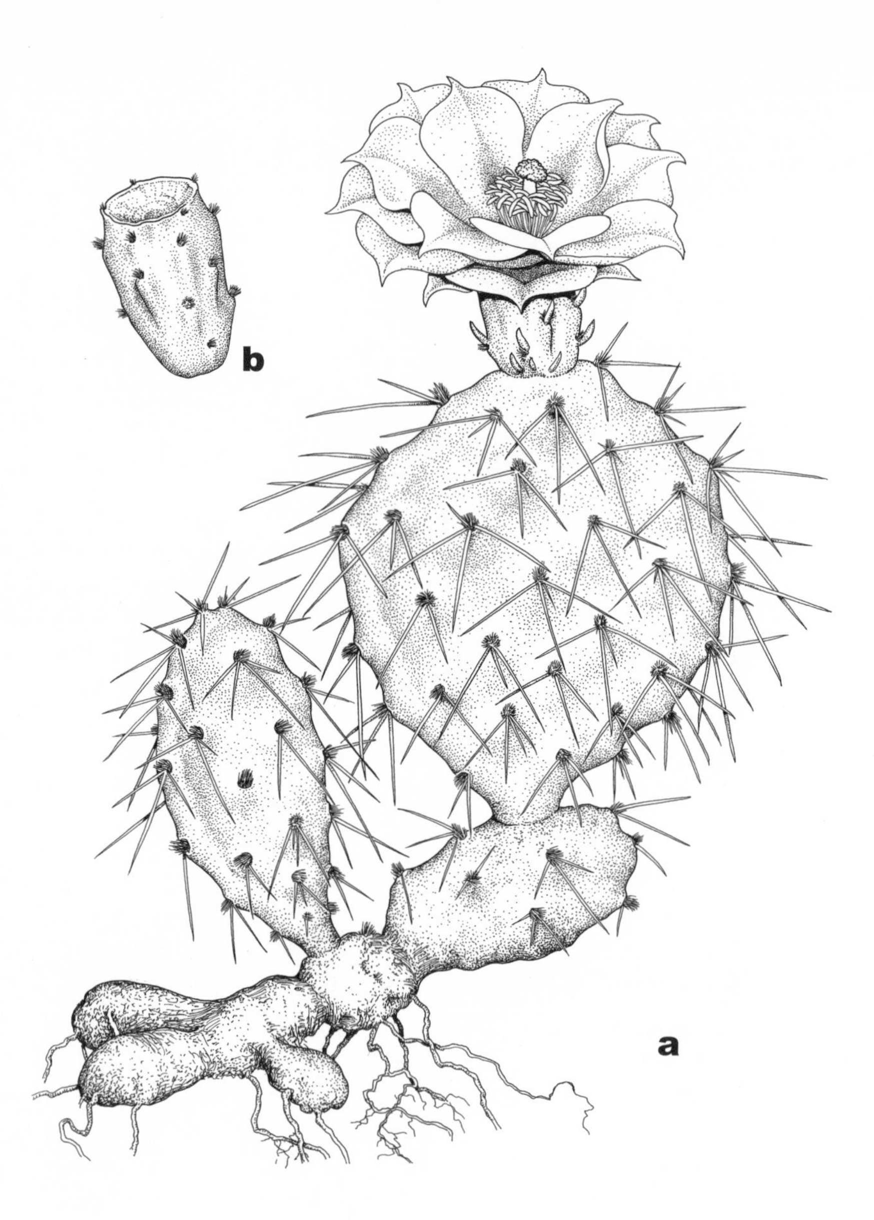

10. **Opuntia macrorhiza** a. Habit b. Portion of stem

glaucous; areoles usually with 2 (–6) spines, the spines white to yellow-brown, 3.5–6.0 cm long; flower solitary, showy, up to 6 cm across; perianth petaloid, yellow, usually with a reddish center; fruit obovoid, fleshy, purplish, 2.52–3.8 cm long, not spiny but with glochidia; margin of seed irregular, at least 1 mm broad.

Common Name: Plains Prickly Pear.
Habitat: Edge of cliffs.
Range: Southern Michigan to California, east to Texas and Louisiana.
Illinois Distribution: Confined to a few cliffs in the western side of the state.

This species looks very much like *O. humifusa,* but it differs in its tuberous-thickened roots and usually longer, yellow-brown spines that number two or more at each areole.

This is a western species that reaches its eastern limit in Illinois. *Opuntia macrorhiza* flowers during June and July.

Portulacaceae—Purslane Family

Annual or perennial herbs, often succulent; leaves alternate or opposite, simple, entire, usually with stipules; flowers perfect, actinomorphic, borne in cymes, racemes, glomerules, or solitary; sepals usually 2, free or united; petals usually 5, free or united at the base; stamens 4–many; ovary superior or inferior, 1-locular; styles 2–9; fruit a capsule; seeds 1–many.

Almost all members of this family have a horseshoe-shaped embryo surrounded by the endosperm and central or basal placentation. Some botanists have interpreted the sepals as bracts and the petals as sepals, with the flower lacking petals, a view not followed here.

The family consists of about twenty genera and four hundred species distributed in temperate parts of the world. Some species of *Portulaca, Lewisia,* and *Calandrinia* have showy flowers and are grown as ornamentals.

Key to the Genera of **Portulacaceae** in Illinois

1. Ovary inferior or partly so; capsule circumscissile; cauline leaves numerous . . . 1. *Portulaca*
1. Ovary superior; capsule dehiscing vertically; cauline leaves 2 or absent.
   2. Cauline leaves absent; all leaves basal or nearly so, terete, very fleshy; inflorescence cymose; flowers pink or rose . . . . 2. *Talinum*
   2. Cauline leaves 2; all leaves flat, not particularly fleshy; inflorescence racemose; flowers white, rarely faint pink . . . . 3. *Claytonia*

1. **Portulaca** L.—Purslane

Annual or perennial, more or less succulent herbs; leaves alternate (in Illinois), sometimes appearing whorled beneath the flower, flat or terete, entire, without stipules or with caducous stipules; flowers perfect, actinomorphic, solitary or clustered, variously colored; sepals 2, usually united at the base; petals 5–many, free; stamens 6–40; ovary inferior or at least partly so, 1-locular; styles 4–9; fruit a circumscissile capsule, many-seeded; seeds smooth to tuberculate or granular.

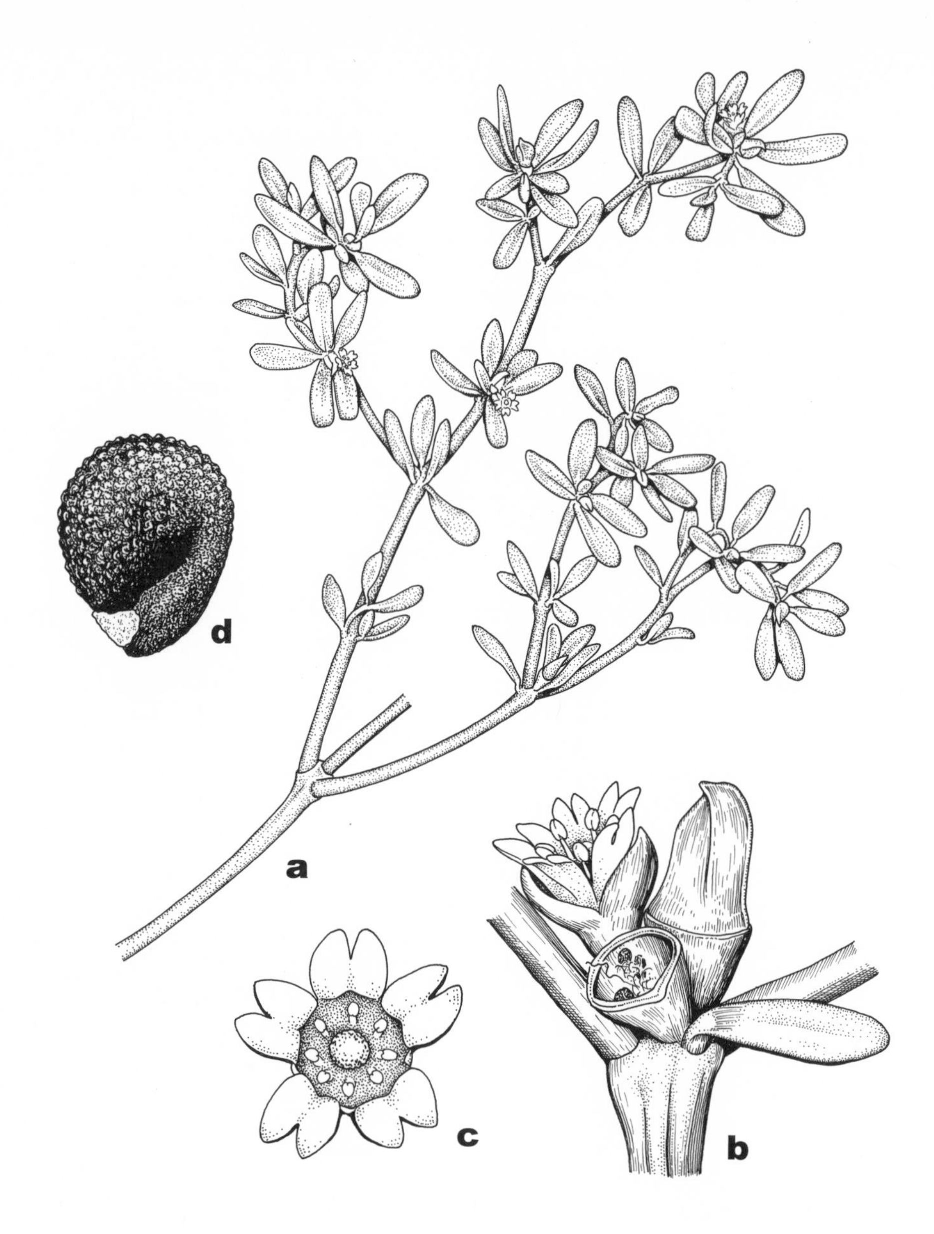

11. **Portulaca oleracea**
a. Habit
b. Flower and opened capsule
c. Flower
d. Seed

This genus differs from others in the family in its circumscissile capsules and its inferior or partly inferior ovary.

There are about one hundred species of *Portulaca* in the warmer parts of the World.

Key to the Species of **Portulaca** in Illinois

1. Leaves flat, spatulate to ovate; stems glabrous; flower up to 1 cm across ...... 1. *P. oleracea*
1. Leaves terete, linear; stems pubescent at the nodes; flowers 2–4 cm across .... 2. *P. grandiflora*

1. Portulaca oleracea L. Sp. Pl. 445. 1753. Fig. 11.

Annual, somewhat succulent herb from a taproot; stems prostrate to ascending, to 30 cm tall, frequently purplish red, glabrous; leaves alternate (rarely opposite), crowded and appearing whorled beneath the flowers, flat, spatulate to obovate, broadly rounded at the apex, cuneate at the base, to 30 mm long, to 15 mm broad, entire, glabrous, sessile; flowers solitary or clustered, sessile, up to 1 cm across; sepals 2, orbicular to broadly ovate, keeled, 3.0–4.5 mm long; petals 5, free, yellow, 3.0–4.5 mm long; stamens 6–10; styles 4–6; capsules ovoid, 5–9 mm long; seeds many, reniform, 0.7–1.0 mm long, black, granular or tuberculate.

Common Name: Common Purslane.
Habitat: Disturbed soil; often in cracks of sidewalks.
Range: Native of Europe; widely adventive in the United States.
Illinois Distribution: Common throughout the state.

This rather fleshy species is common in disturbed soil throughout Illinois. It often appears as a weed in gardens.

Portulaca oleracea differs from other members of the family in its flat, fleshy, spatulate leaves and its yellow flowers that are 5–10 mm across.

There is considerable variation within this species, and many of the varieties have been given names. I am unable to differentiate satisfactorily between the variants and am not recognizing them here. For a discussion of these, see Danin, Baker, and Baker (1978).

When young, this plant may be cooked and eaten as a vegetable.

Portulaca oleracea flowers from June to October.

2. Portulaca grandiflora Hook. Bot. Mag. pl. 2885. 1829. Fig. 12.

Annual succulent herb from a small taproot; stems ascending to spreading, bisexual, to 35 cm tall, pubescent at least at the nodes; leaves alternate, simple, often crowded, terete, to 3 cm long, 1.5–2.5 mm broad, entire, glabrous; flower often solitary, showy, variously colored, sometimes double-flowered, 2–4 cm across; sepals 2, free or united at the base, obtuse, scarious along the margins, 3–5 mm long; petals 5 (many in double-flowered forms), free, entire or erose, 1–2 cm long; stamens about 40; styles 5–9; capsule ovoid, 4–8 mm long; seeds many, reniform, 0.5–1.5 mm long, gray, shiny.

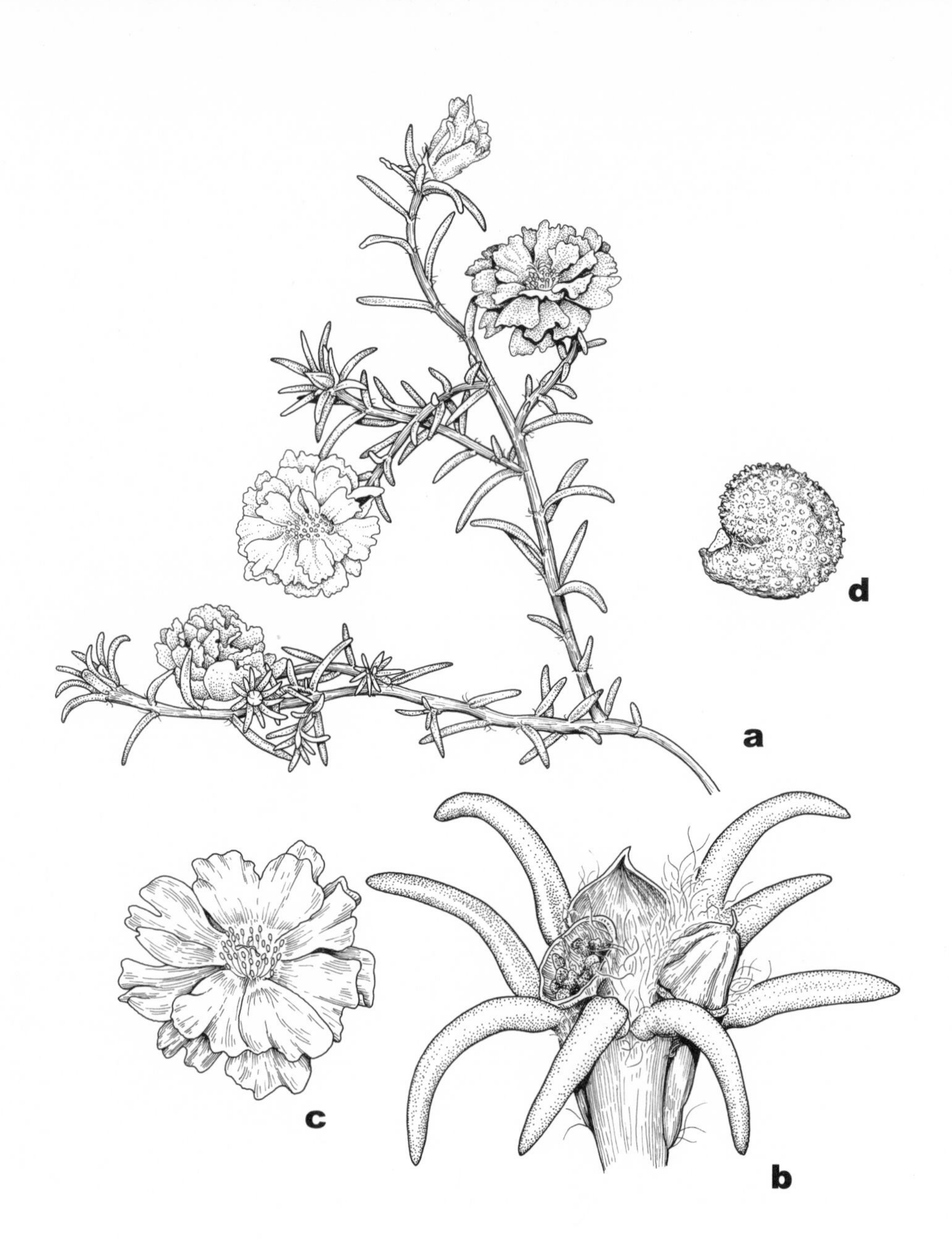

12. **Portulaca grandiflora**
a. Habit
b. Open capsule
c. Flower
d. Seed

Common Name: Rose Moss; Garden Purslane.
Habitat: Disturbed soil.
Range: Native of South America; sparingly escaped from cultivation in the United States.
Illinois Distribution: Known from Cass, Champaign, Jackson, and Mason counties.

This is a bright-flowered plant of gardens that rarely escapes from cultivation. The flowers, which are variously colored, open only in bright sunlight.

Portulaca oleracea flowers from June to October.

2. **Talinum** Adans.—Flower-of-an-hour; Fame-flower

Perennial, succulent, glabrous herbs; leaves basal or nearly so, terete, entire; flowers perfect, actinomorphic, solitary or in cymes, bracteate, usually on long peduncles; sepals 2, free or nearly so, caducous; petals 5, free, lasting only a short time; stamens 4–45; ovary superior, 1-locular; styles 3, more or less united; fruit a capsule dehiscing vertically, many-seeded; seeds flattened, more or less reniform, minutely roughened.

This genus of small succulent species has the terete leaves all or nearly all basal. The flowers last for one to three hours, then wither, accounting for the common name flower-of-an-hour. Each flower opens at a particular precise time every day.

About fifty species comprise this genus, usually found in warmer parts of the eastern and western hemispheres.

Key to the Species of **Talinum** in Illinois

1. Stamens 4–8; flowers pale pink .... 1. *T. parviflorum*
1. Stamens 10–45; flowers bright pink to rose.
    2. Petals 6–8 mm long; stamens 10–25; capsules 4–5 mm long .... 2. *T. rugospermum*
    2. Petals 12–16 mm long; stamens 30–45; capsules 6–8 mm long .... 3. *T. calycinum*

1. Talinum parviflorum Nutt. in Torr. & Gray, Fl. N. Am. 1:197. 1838. Fig. 13.

Perennial succulent herb from a thickened, fleshy taproot; stems scapose, to 20 cm tall, glabrous; leaves basal or nearly so, terete, acute at the apex, scarcely tapering to the broadened base, up to 5 cm long, 1.0–2.5 mm thick, entire, glabrous; flowers perfect, borne solitary or in a cyme 8–15 cm long, on slender peduncles up to 15 cm long; bracts deltate, 2–5 mm long, reflexed; sepals 2, oval to ovate, 2.5–4.0 mm long; petals 5, usually pale pink, 5–7 mm long; stamens 4–8; capsules ellipsoid, 3–5 mm long; seeds many, 0.8–0.9 mm long, smooth, dark gray.

Common Name: Small-flowered Flower-of-an-hour.
Habitat: Exposed sandstone cliffs.
Range: Minnesota to Colorado, south to Texas and Arkansas; Mexico.
Illinois Distribution: Known from Calhoun, Johnson, Pope, and Union counties.

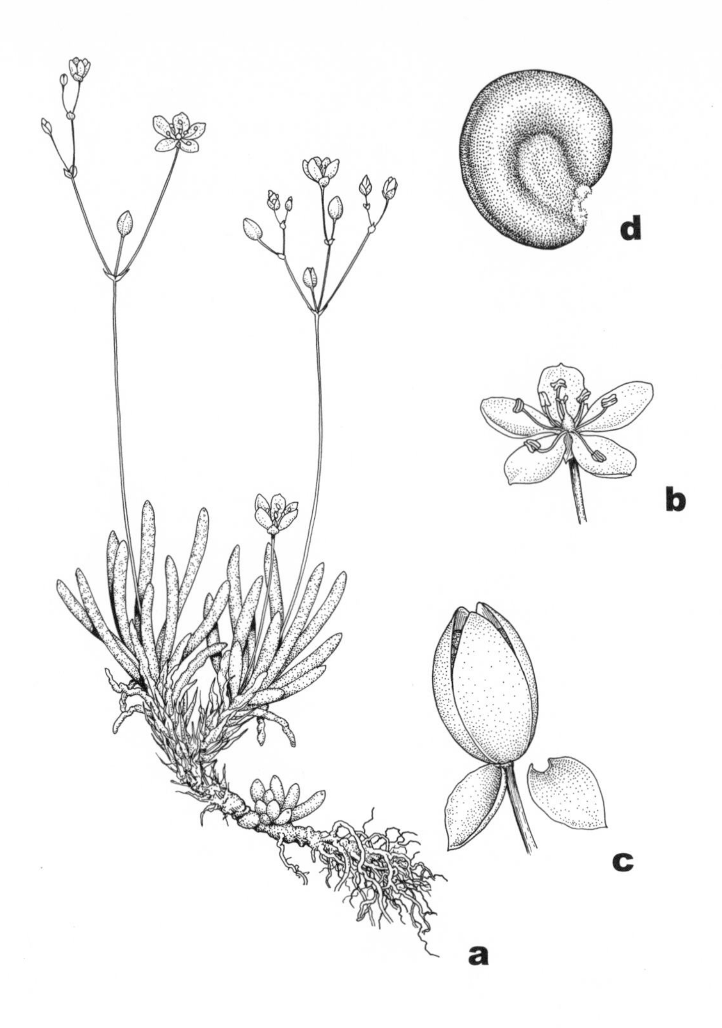

13. **Talinum parviflorum**
a. Habit
b. Flower
c. Fruit
d. Seed

This species occurs on the driest, most exposed sandstone cliffs in Illinois. Its succulent, terete leaves store water and permit it to survive under such xeric conditions.

Talinum parviflorum is smaller in all respects than the other two species of the genus in Illinois. It also has paler flowers.

This species flowers during the morning in June and July.

2. Talinum rugospermum Holz. Asa Gray Bull. 7:117. 1899. Fig. 14.

Perennial, succulent herb from a thickened, fleshy taproot; stems scapose, to 25 cm tall, glabrous; leaves basal or nearly so, terete, acute at the apex, tapering to the slightly narrower base, to 4.5 cm long, 1.5–2.5 mm thick, entire, glabrous; flowers perfect, borne solitary or in a cyme 10–20 cm long, on slender pedicels up to 15 cm long; bracts deltate, 2–5 mm long, reflexed; sepals 2, ovate to obovate, 3.0–4.5 mm long; petals 5, bright pink to rose, 6–8 mm long; stamens 10–25; capsules globose, 4–5 mm long; seeds many, 0.9–1.0 mm long, minutely roughened and wrinkled, shiny.

Common Name: Wrinkle-seeded Flower-of-an-hour.
Habitat: Sandy savannas; sandstone cliffs.
Range: Wisconsin and Minnesota, south to northeastern Iowa, northern Illinois, and northwestern Indiana.
Illinois Distribution: Scattered and generally not common in the northern half of the state.

This species has a very narrow range, being known only from Minnesota, Wisconsin, northeastern Iowa, northern Illinois, and northwestern Indiana.

It differs from *T. parviflorum* in its brighter flowers, more numerous stamens, and wrinkled seeds. It differs from *T. calycinum* in its smaller flowers, fewer stamens, smaller capsules, and wrinkled seeds.

For a long time this plant was called *T. teretifolium* by Illinois botanists, but that species of the southeastern United States differs in its nearly smooth seeds and its slightly larger capsules.

Talinum rugospermum flowers from June to April, opening late in the afternoon.

3. Talinum calycinum Engelm. in Wisliz. Rep. 88. 1848. Fig. 15.

Perennial succulent herb from a thickened, fleshy taproot; stems scapose, to 10 cm tall, glabrous; leaves basal or nearly so, terete, acute at the apex, triangular-broadened at the base, to 7 cm long, 1–2 mm thick, entire, glabrous; flowers perfect, borne solitary on in a cyme 10–30 cm long, on slender pedicels up to 20 cm long; bracts deltate, 2–5 mm long, reflexed; sepals 2, ovate to nearly orbicular, 4.5–8.0 mm long, not caducous and sometimes even persistent on the capsule; petals 5, rose, 12–16 mm long; stamens 30–45; capsules globose-ovoid, 6–8 mm long; seeds many, 1.0–1.1 mm long, smooth, shiny.

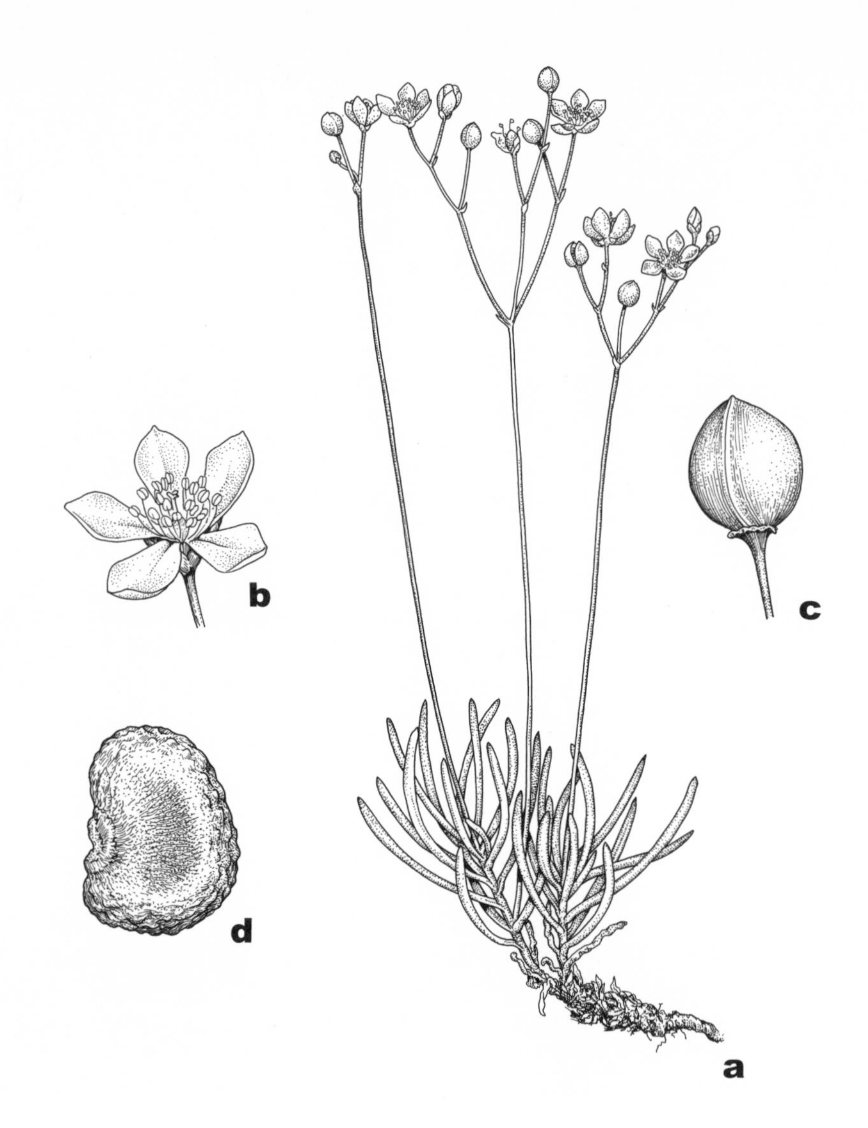

14. **Talinum rugospermum**
a. Habit
b. Flower
c. Fruit
d. Seed

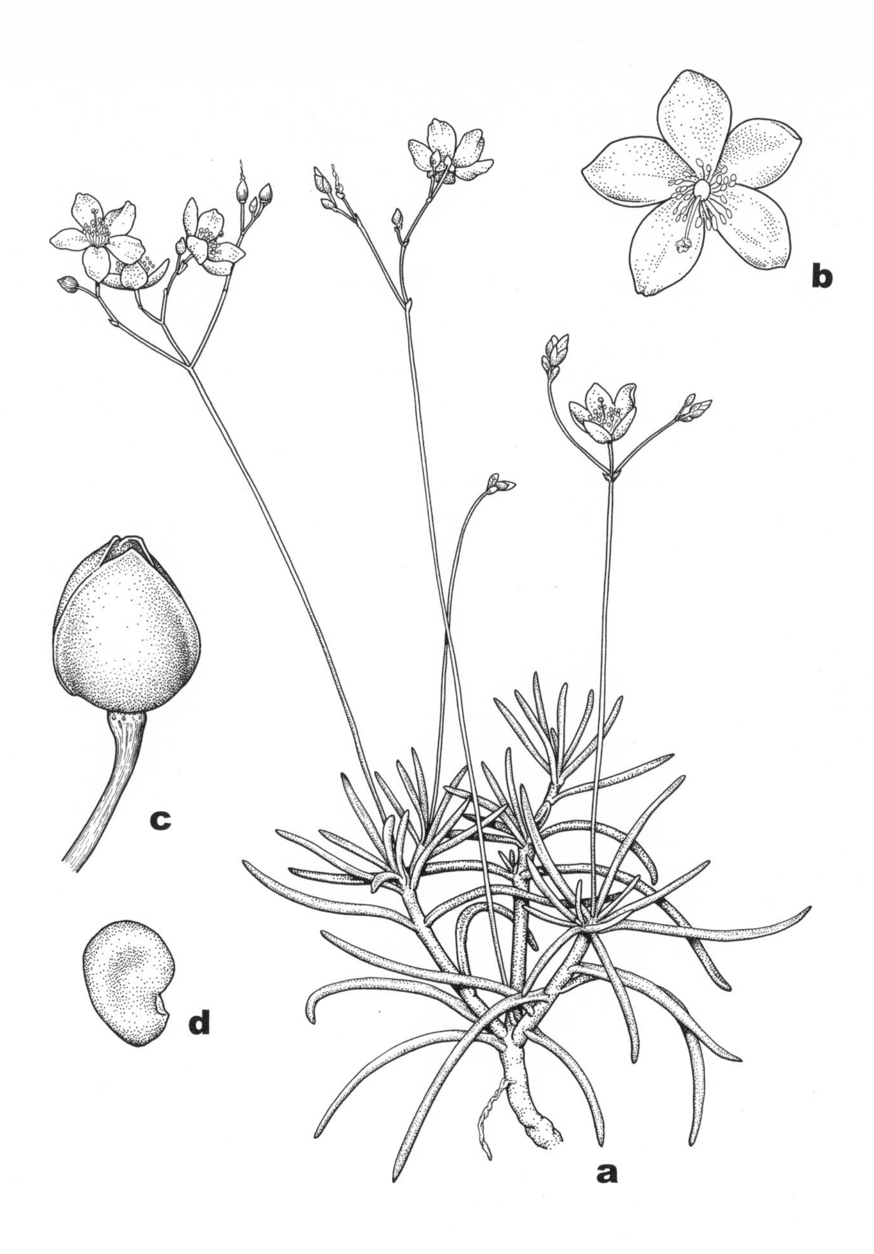

15. **Talinum calycinum** b. Flower c. Fruit
a. Habit d. Seed

Common Name: Large-flowered Flower-of-an-hour.
Habitat: Exposed edge of sandstone cliffs.
Range: Southwestern Illinois to Nebraska, south to Texas and Arkansas.
Illinois Distribution: Known only from Randolph County.

This handsome succulent was discovered in Illinois by the author on September 25, 1954. It was growing on an exposed sandstone ledge at Castle Rock near the site of the extinct village of Leanderville. Subsequently it was found on a nearby ledge. Although heavily encroached upon by aggressive weedy vegetation, this species still persists in Randolph County.

Talinum calycinum differs from other species of this genus in Illinois in its larger flowers, larger capsules, and more numerous stamens.

This species flowers early in the afternoon from June to September.

3. **Claytonia** L.—Spring Beauty

Perennial herbs from corms; leaves basal as well as a single opposite pair, entire, sometimes slightly fleshy; flowers perfect, actinomorphic, borne in a loose raceme, bracteate; sepals 2, free, persistent in fruit; petals 5, free; stamens 5, attached to the base of the petals; ovary superior, 1-locular; style 3-cleft; fruit a vertically dehiscent capsule, bearing 3–6 seeds; seeds orbicular.

Some botanists interpret the sepals in this genus to be bracts and the petals to be sepals, with no true petals present. I do not follow this view.

Claytonia is a genus of about twenty species, all native to the northern hemisphere. Davis (1966) has made a study of the genus.

Only the following species occurs in Illinois.

1. **Claytonia virginica** L. Sp. Pl. 204. 1753. Fig. 16.
Claytonia simsii Sweet, Brit. Fl. Gard. pl. 216. 1827.
Claytonia multicaulis A. Nels. var. *robusta* Somes, Iowa Nat. 2:67. 1910.
Claytonia robusta (Somes) Rydb. Brittonia 1:87. 1931.
Claytonia virginica L. f. *robusta* (Somes) Palmer & Steyerm. Ann. Mo. Bot. Gard. 22: 537. 1935.
Claytonia virginica L. var. *simsii* (Sweet) R.J. Davis, Brittonia 18:296. 1966.

Perennial herb from a globose corm; stems 1–several, erect to ascending, unbranched, to 30 cm tall, glabrous, somewhat fleshy; leaves linear to lanceolate, acute at the apex, cuneate at the base, to 20 cm long, to 1.5 (–2.5) cm broad, entire, glabrous, petiolate; flowers perfect, 6–15 in a terminal raceme, subtended by bracts up to 10 mm long, up to 6 mm broad; sepals 2, free, oval to ovate, obtuse at the apex, 5–7 mm long, persistent and becoming reddish in fruit; petals 5, free, oval, to 14 mm long, white or pale pink, pink- or purple-veined; stamens 5; styles 3, united to near the apex; capsules 3–5 mm long, ovoid, with 3–6 seeds; seeds 1.8–2.0 mm in diameter, orbicular, brownish black, shiny.

16. **Claytonia virginica**
a. Habit in flower
b. Flower
c. Flower from below
d. Habit in fruit
e. Fruit
f. Seed

Common Name: Spring Beauty.
Habitat: Woods, both rich and degraded; lawns.
Range: Quebec to Ontario and Minnesota, south to Texas and Georgia.
Illinois Distribution: Common throughout the state.

This species is readily distinguished by its single pair of cauline leaves and its flowers with two sepals, five free petals, and five stamens that are adnate to the base of the petals.

There is considerable variation, particularly with regard to leaf width. Specimens with leaves broader than one centimeter may be called f. *robusta.* These plants, which are scattered throughout the state, have sometimes been erroneously identified as *C. caroliniana. Claytonia caroliniana,* which has not been found in Illinois, has much smaller, scarious bracts. Rare plants with very narrow leaves up to 2 mm wide may be called var. *simsii.*

Spring beauty, which is probably the most common, widely distributed spring wildflower in Illinois, blooms from March to June.

Chenopodiaceae—Goosefoot Family

Herbs (in Illinois), shrubs, or rarely trees, sometimes monoecious or dioecious, often white-mealy or scaly; stems sometimes fleshy; leaves usually alternate, occasionally opposite, simple, sometimes fleshy; flowers solitary or in clusters, sometimes in spikes or panicles, with or without bracts; calyx with 1–5 sepals, sometimes absent in pistillate flowers; petals absent; stamens 1–5; ovary superior, 1-locular; styles 1–3; fruit a 1-seeded utricle; seeds vertical or horizontal.

This family consists of approximately one hundred genera and thirteen hundred species found worldwide. Many species occur in deserts or occupy saline or alkaline soils. Of commercial importance is *Beta,* the garden beet, and *Spinacia,* spinach.

Key to the Genera of Nonflowering, Nonfruiting or Immature Plants

1. Stems very fleshy, jointed; leaves reduced to scales . . . . 1. *Salicornia*
1. Stems not fleshy or only slightly fleshy, not jointed; leaves present.
    2. Leaves all opposite, crowded to appear fasciculate, threadlike to filiform . . . . 2. *Polycnemum*
    2. Some or all the leaves alternate, never appearing fasciculate, linear or broader.
        3. Leaves spine-tipped . . . . 3. *Salsola*
        3. Leaves not spine-tipped.
            4. Stems, leaves, and sometimes the calyx either glandular-pubescent, cobwebby-hairy, white-mealy, scurfy-scaly, or silvery-scaly.
                5. Stems and leaves glandular-pubescent . . . . 5. *Chenopodium*
                5. Stems and leaves not glandular-pubescent.
                    6. Stem and leaves and often the calyx white-mealy.
                        7. Leaves not triangular or, if triangular, usually with coarse teeth . . . . 5. *Chenopodium*
                        7. Some of the leaves triangular, entire or sparsely toothed . . . . 6. *Monolepis*

6. Stem and leaves not white-mealy.
   8. Stem and leaves with cobwebby hairs .... 7. *Cycloloma*
   8. Stem and leaves lacking cobwebby hairs but either scurfy-scaly or silvery-scaly .... 8. *Atriplex*

4. Stem and leaves neither glandular-pubescent, cobwebby-hairy, white-mealy, scurfy-scaly, or silvery-scaly.
   9. Some or all the leaves more than 5 mm wide.
      10. Leaves and stems often short-pubescent, or the leaves at least ciliate at base .... 4. *Kochia*
      10. Leaves and stems not pubescent, or leaves not ciliate at base ..... 5. *Chenopodium*
   9. None of the leaves more than 5 mm wide.
      11. Leaves ciliate at base .... 4. *Kochia*
      11. Leaves not ciliate at base.
         12. Leaves more or less terete, glaucous .... 9. *Suaeda*
         12. Leaves flat, not glaucous .... 10. *Corispermum*

Key to the Genera of Flowering and Fruiting Plants

1. Flowers sunken in the hollow of the uppermost joints of the very succulent stem; leaves reduced to scales .... 1. *Salicornia*
1. Flowers not sunken in the hollow of the uppermost joints of the stem; stems not very succulent nor jointed; leaves present.
   2. All flowers unisexual; stems and leaves either scurfy-scaly or silvery-scaly .... 8. *Atriplex*
   2. Some or all the flowers perfect; stems with various kinds of indument or glabrous but not scurfy-scaly or silvery-scaly.
      3. Sepal 1, or calyx 3-cleft.
         4. Sepal 1; styles not persistent on the fruit; leaves triangular to lanceolate, at least sparsely white-mealy .... 6. *Monolepis*
         4. Calyx usually 3-cleft, rarely 2 or 1; styles persistent on the fruit; leaves linear, not white-mealy .... 10. *Corispermum*
      3. Calyx 5-cleft.
         5. Lobes of calyx tuberculate at the tip .... 4. *Kochia*
         5. Lobes of calyx not tuberculate at the tip.
            6. Seeds narrowly or broadly winged.
               7. Seeds horizontal; wing of fruit more than 0.5 mm wide.
                  8. Fruits horizontally winged, flat; leaves not spine-tipped; bracts absent .... 7. *Cycloloma*
                  8. Fruits transversely winged, not flat; leaves spine-tipped; bracts present .... 3. *Salsola*
               7. Seeds vertical; wing of fruit up to 0.5 mm wide ..... 10. *Corispermum*
            6. Seeds unwinged.
               9. Stamens 3; styles 3; leaves opposite, filiform .... 2. *Polycnemum*
               9. Stamens 1–5, rarely 3; styles usually 2; leaves alternate.
                  10. Calyx lobes unequal; flowers perfect and pistillate on the same plant; plants not white-mealy .... 9. *Suaeda*
                  10. Calyx lobes equal; flowers all perfect; plants often white-mealy.
                     11. Leaves linear, 1-nerved; plants never white-mealy .... 10. *Corispermum*
                     11. Leaves rarely linear, usually more than 1-nerved; plants often white-mealy .... 5. *Chenopodium*

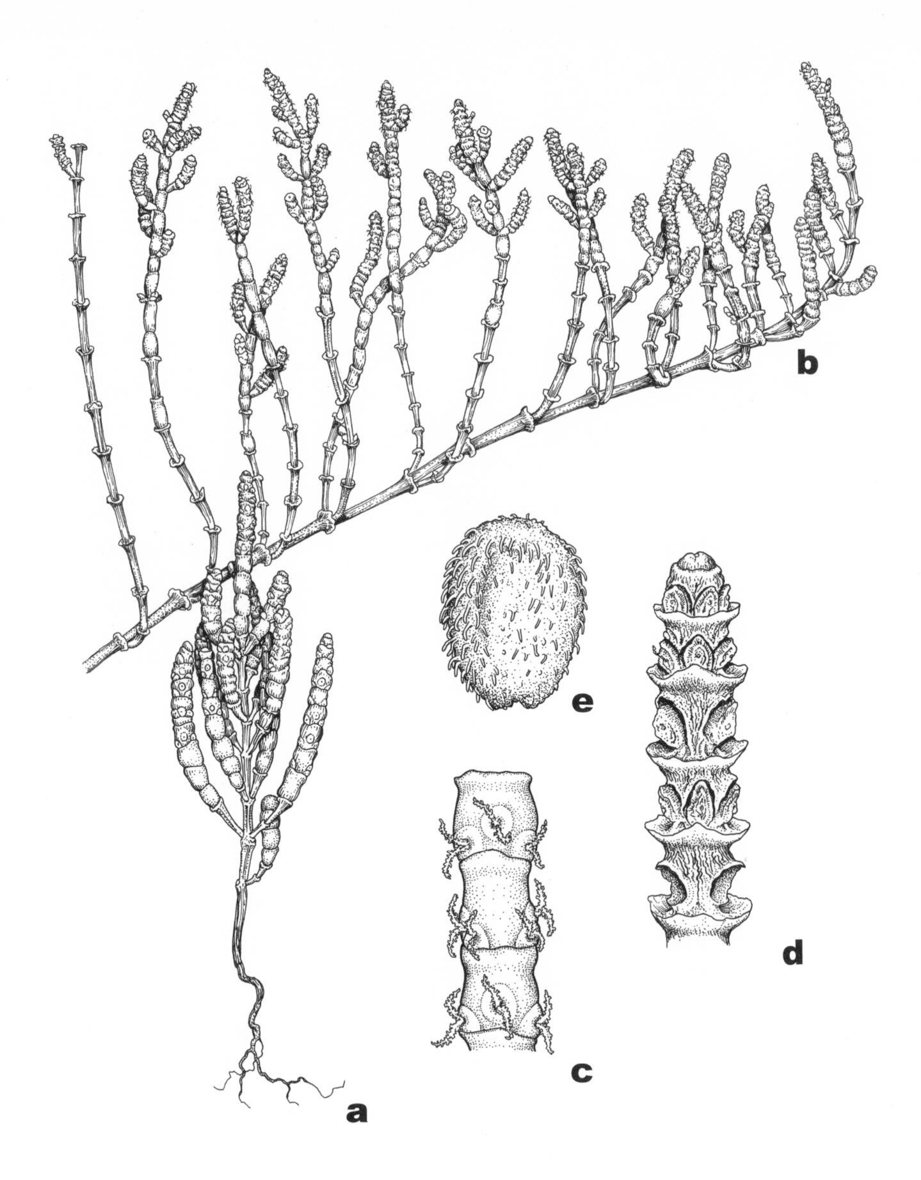

17. **Salicornia europaea**
a. Habit
b. Habit
c. Flowering stem
d. Fruiting stem
e. Seed

1. **Salicornia** L.—Glasswort

Annual or perennial herbs; stems succulent, jointed, with opposite branching; leaves opposite, reduced to scales; flowers perfect, together and sunken in each hollow of the upper joints of the stem forming a spike, with scalelike bracts; calyx usually early deciduous; petals 0; stamens 2; ovary superior, 1-locular; styles 2–3; fruit 1-seeded, the fruit wall free from the seed; seed vertical.

About thirteen species found throughout the world comprise this genus.

Only the following species has been found in Illinois.

1. **Salicornia europaea** L. Sp. Pl. 3. 1753. Fig. 17.
Salicornia herbacea L. Sp. Pl. ed. 2, 5. 1762.

Annual herb; stems succulent, to 30 cm tall, branched, the internodes 10–18 mm long, 1–3 mm wide, green, turning red in autumn; scale leaves obtuse, inconspicuous; spikes narrowly cylindrical, to 10 cm long, with obtuse bracts; central flower in the cluster of 3 slightly elevated above the lateral 2; seeds 1–2 mm long, puberulent.

Common Name: Samphire; Glasswort.
Habitat: Along a river (in Illinois).
Range: Nova Scotia south along the coast to Georgia; Illinois; Michigan; Wisconsin.
Illinois Distribution: Known from a single collection in Cook County.

This remarkably distinct species has jointed, succulent stems with flowers sunken in each hollow of the upper part of the stem.

This is essentially a species of the Atlantic coast. It was found by Glen Winterringer in 1948 along the Little Calumet River north of Harvey in Cook County where it was growing in soil contaminated by industrial discharge.

This species flowers from August to October.

2. **Polycnemum** A. Br.—Polycnemum

Annual herbs; stems low, branching; leaves opposite, subulate to filiform; flowers perfect, solitary in the axils of the upper leaves, each subtended by a pair of scarious bracts; calyx scarious, deeply 5-parted; petals 0; stamens 3; ovary superior, 1-locular; styles 3; fruit 1-seeded, completely surrounded by the calyx; seed vertical, pitted.

This small genus consists of five species native to the Mediterranean region.

Only the following species has been found in Illinois.

1. **Polycnemum majus** A. Br. in Koch, Syn. Fl. Germ. ed. 2, 695. 1843–45. Fig. 18.

Annual herb from a slender taproot; stems spreading to erect, to 20 cm tall, glabrous or nearly so; leaves opposite, simple, subulate to filiform, very crowded, 1-nerved, up to 10 mm long; flowers solitary, axillary, subtended by an ovate-lanceolate,

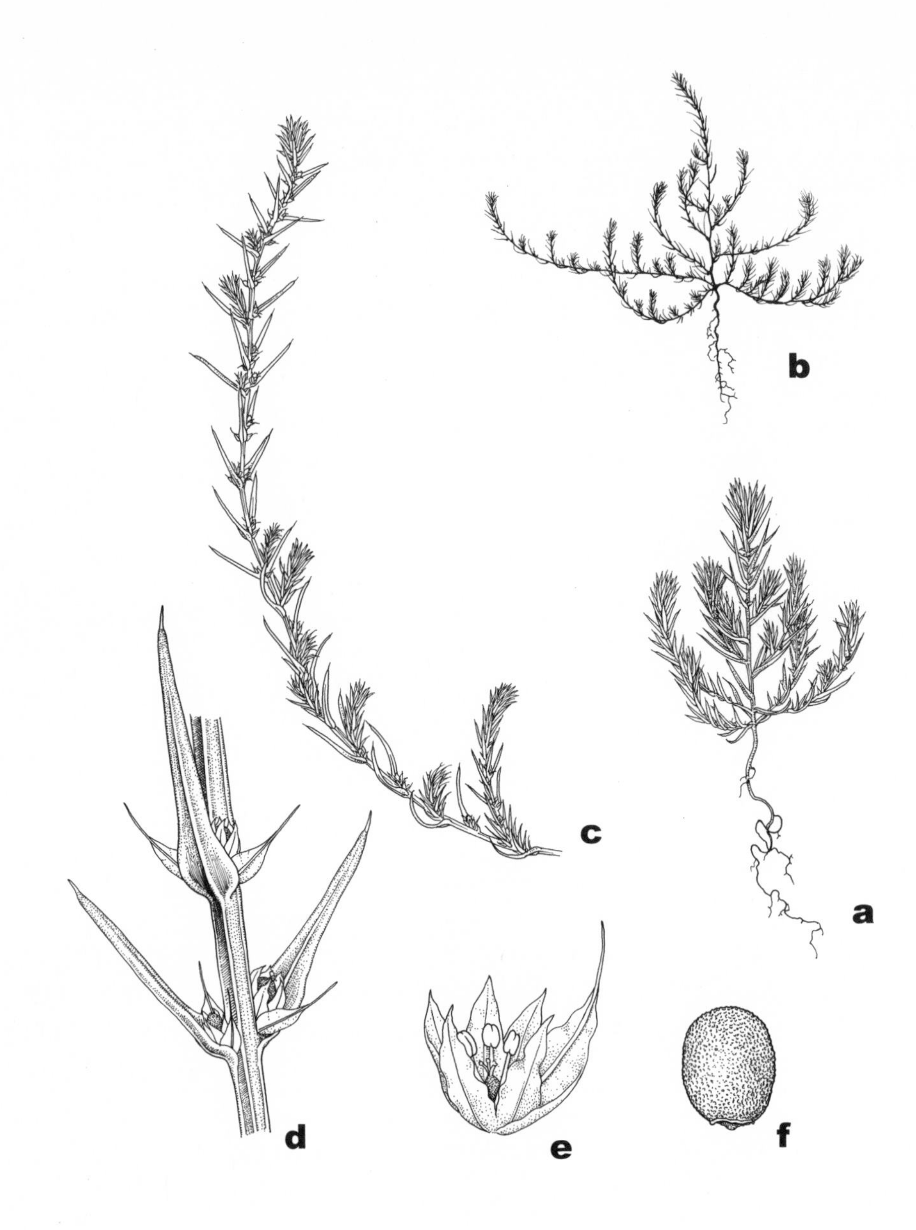

18. **Polycnemum majus**
a. Habit
b. Habit
c. Habit
d. Flowering nodes
e. Flower
f. Seed

awned bract longer than the flower; calyx deeply 5-lobed, scarious, 1–2 mm long; seed lenticular, black, pitted, 1.3–1.5 mm long.

Common Name: Polycnemum.
Habitat: Along a railroad.
Range: Native to Europe; adventive in North America in Ontario and Illinois.
Illinois Distribution: Known from a single collection along a railroad in Monroe County.

This slender little annual with crowded, needle-shaped leaves has its solitary, axillary fruits completely enclosed by the calyx. Records from Ontario and Monroe County, Illinois, are apparently the only ones known from North America.

Polycnemum majus flowers from July to September.

3. **Salsola** L.—Russian Thistle

Annual herbs; stems much branched; leaves alternate, simple, rather succulent, linear, spine-tipped; flowers perfect, solitary or few in the upper leaf axils, each subtended by a pair of small bracts; calyx deeply 5-parted, incurved over the fruit at maturity; petals 0; stamens 5; ovary superior, 1-locular; styles 2; fruit not completely enclosed by the sepals, 1-seeded; seed horizontal.

About forty species comprise this genus that is native primarily to desert and saline areas of Europe, Asia, and northern Africa.

The following two species in Illinois may be separated by the following key:

1. Inflorescence loosely spicate, with spreading bracts . . . . 1. *S. tragus*
1. Inflorescence densely spicate, with appressed bracts . . . . 2. *S. collina*

1. Salsola tragus L. Sp. Pl. ed. 2, 322. 1762. Fig. 19.
Salsola kali L. var. *tenuifolia* Tausch, Fl. 11:326. 1828.
Salsola iberica Sennen & Pau, Bull. Acad. Geogr. Bot. 18:476. 1908.
Salsola pestifer A. Nels. Rocky Mt. Bot. 169. 1909.

Annual herbs with slender to stout taproots; stems ascending to erect, bushy, glabrous or puberulent, to 75 cm tall, sometimes purple-striate; leaves linear to filiform, spine-tipped, to 3 (–7) mm long, rather thickish, sessile or nearly clasping at the base; flower solitary in the leaf axils, forming loose spikes, subtended by 2 small, spreading bracts; calyx lobes 5, oblong to lanceolate, concave, in fruit becoming winged and inflexed over the fruit; fruit a 1-seeded utricle up to 8 mm in diameter, white to pinkish; seed 1.5–3.0 mm long, orbicular, shiny, black.

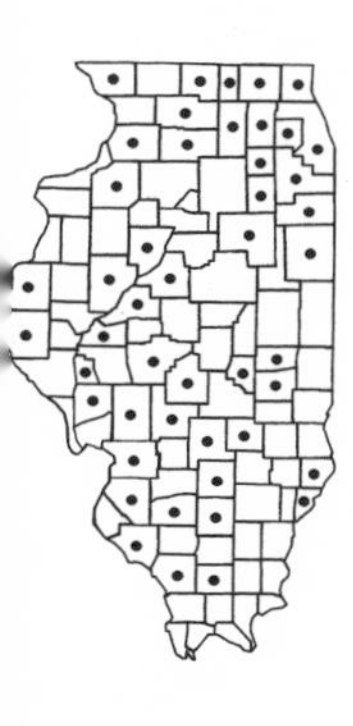

Common Name: Russian Thistle; Tumbleweed.
Habitat: Along railroads; sandy fields; on sand beaches.
Range: Native to Asia; adventive from Quebec to British Columbia, south to California, Arizona, Texas, Missouri, Illinois, and New York.
Illinois Distribution: Scattered throughout the state.

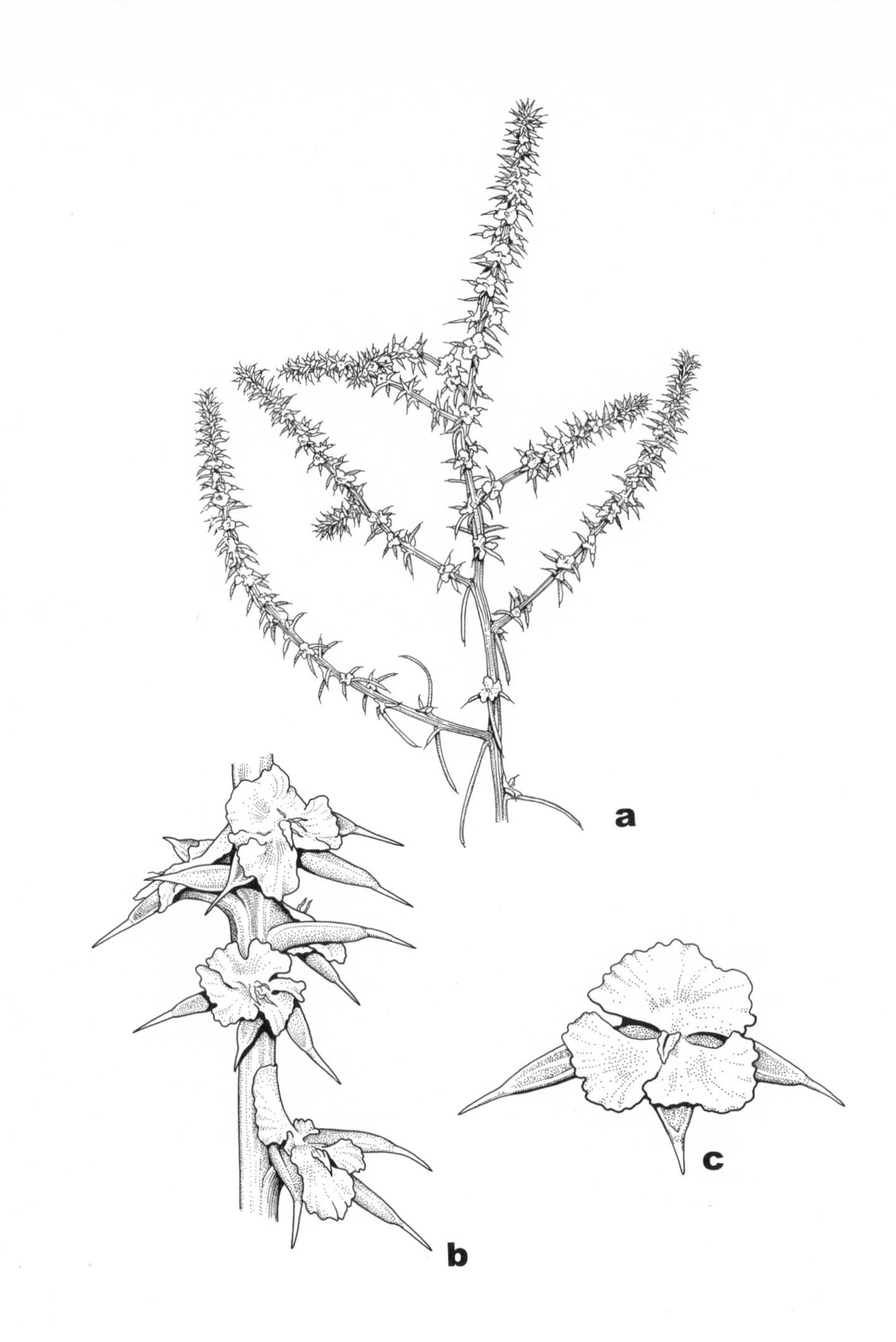

19. **Salsola tragus**
a. Flowering branch
b. Flowering branch
c. Flower

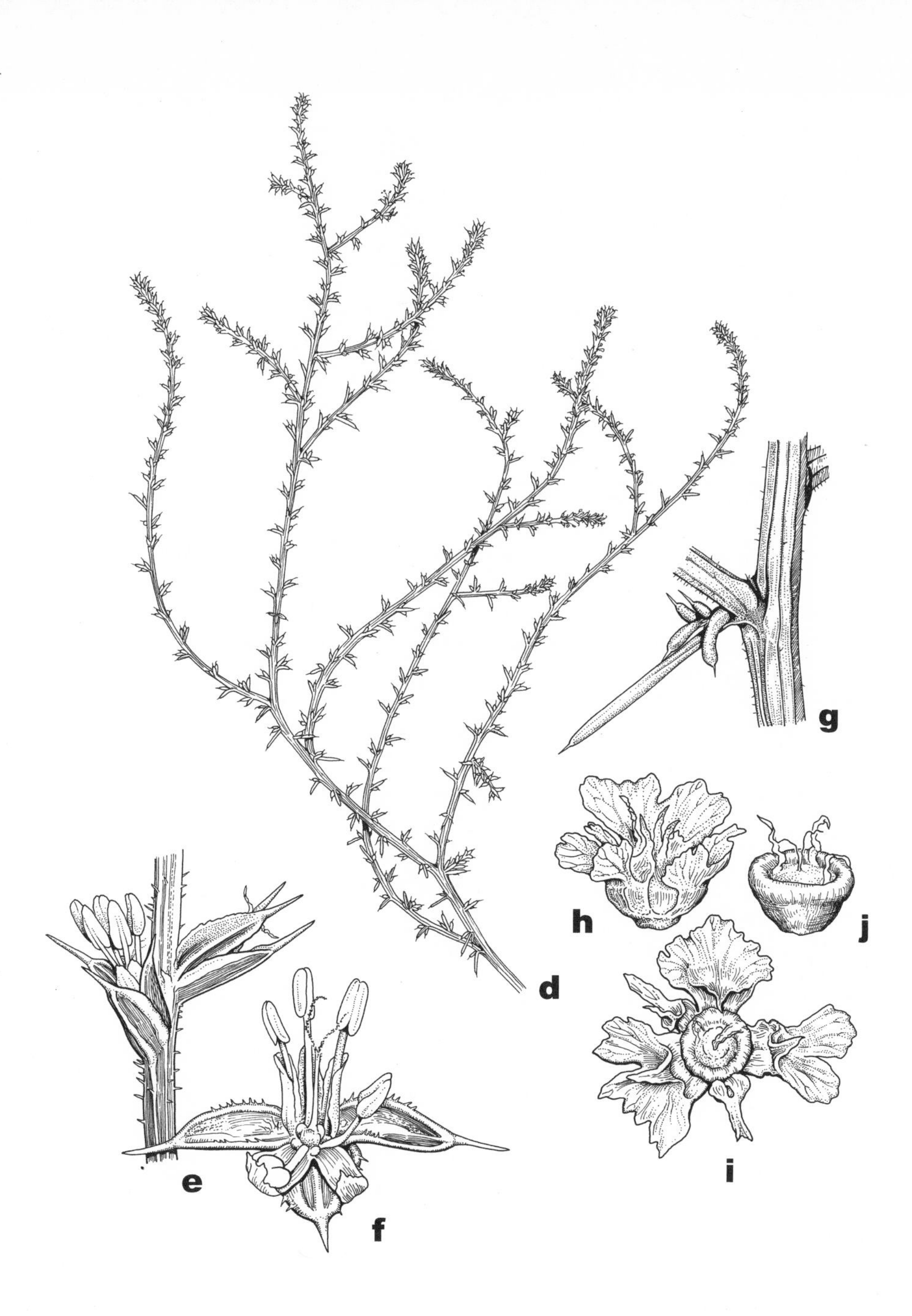

d. Habit
e. Flowering node
f. Flowers with sepals and bracts pulled back to show stamen and pistil
g. Fruiting node
h. Flower
i. Sepals pulled down to expose capsule
j. Capsule

Salsola tragus differs from the very similar *S. collina* by its interrupted spikes and spreading bracts.

The nomenclature for this species has been frequently altered. It has been called the same as *S. kali* or a variety of *S. kali. Salsola kali,* however, has more awl-shaped, spine-tipped leaves. It has also been called *S. iberica* and *S. pestifer,* but these two binomials seem to refer to the same plant as *S. tragus.*

Although this prickly-leaved adventive is frequently found in railroad ballast and sandy disturbed fields, it also occurs in sand adjacent to Lake Michigan with such natives as *Cakile edentula, Chamaesyce polygonifolia,* and *Potentilla anserina.*

This species flowers from July to September.

2. **Salsola collina** Pallas, Ill. Pl. 34, t. 26. 1803. Fig. 20.

Annual herbs from taproots; stems ascending to erect, somewhat bushy, glabrous or puberulent, to 60 cm tall; leaves linear, spine-tipped, to 2 (–3) cm long, more or less thickish, sessile or nearly clasping at the base; flower solitary in the leaf axils, forming dense spikes, subtended by 2 usually appressed bracts; calyx lobes 5, oblong to lanceolate, concave and becoming inflexed in fruit, not winged; fruit a 1-seeded utricle, up to 6 mm in diameter, pale; seed 1.5–2.5 mm long, orbicular, shiny, black.

Common Name: Saltwort.
Habitat: Disturbed sandy soil (in Illinois).
Range: Native to Europe and Asia; sparingly adventive in the United States.
Illinois Distribution: Known from Bureau, Cass, Henry, Kankakee, Lee, Madison, and Whiteside counties.

Although similar to *S. tragus,* this species may be distinguished by its wingless calyx lobes and its dense spikes with appressed bracts.

Salsola collina flowers from July to September.

4. **Kochia** Roth—Kochia

Annual (in Illinois) or perennial herbs; stems usually branched, erect; leaves alternate, simple, usually not succulent, not spine-tipped; flowers perfect or pistillate, clustered in the leaf axils, bracteate; calyx shallowly 5-lobed, inflexed and enclosing the fruit; stamens 3–5; ovary superior, 1-locular; styles 2; fruit a 1-seeded utricle; seed horizontal, free from the pericarp.

There are about twenty species in the genus, native to the western United States, Europe, and Asia. Blackwell, Baechle, and Williamson (1978) have studied the North American species.

Only the following species occurs in Illinois.

1. **Kochia scoparia** (L.) Roth, Neues Journ. Bot. Schrad. 3:85. 1809. Fig. 21.
Chenopodium scoparia L. Sp. Pl. 221. 1753.

Annual herbs from slender to stout taproots; stems to 1 m tall, branched or occasionally unbranched, usually short-pubescent; leaves alternate, simple, narrowly

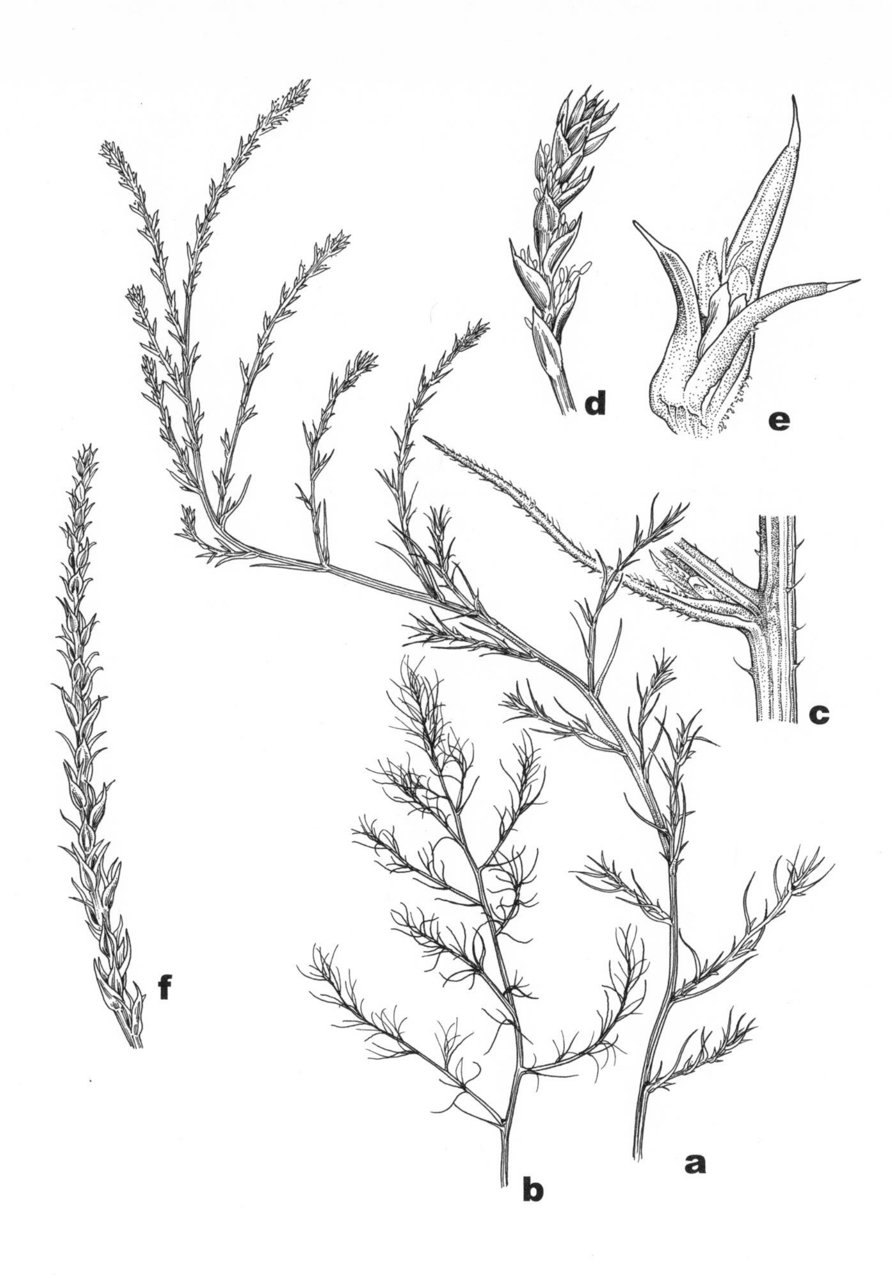

20. **Salsola collina**
a. Habit in flower
b. Leaf variations
c. Node
d. Flowering branch
e. Young flower
f. Fruiting branch

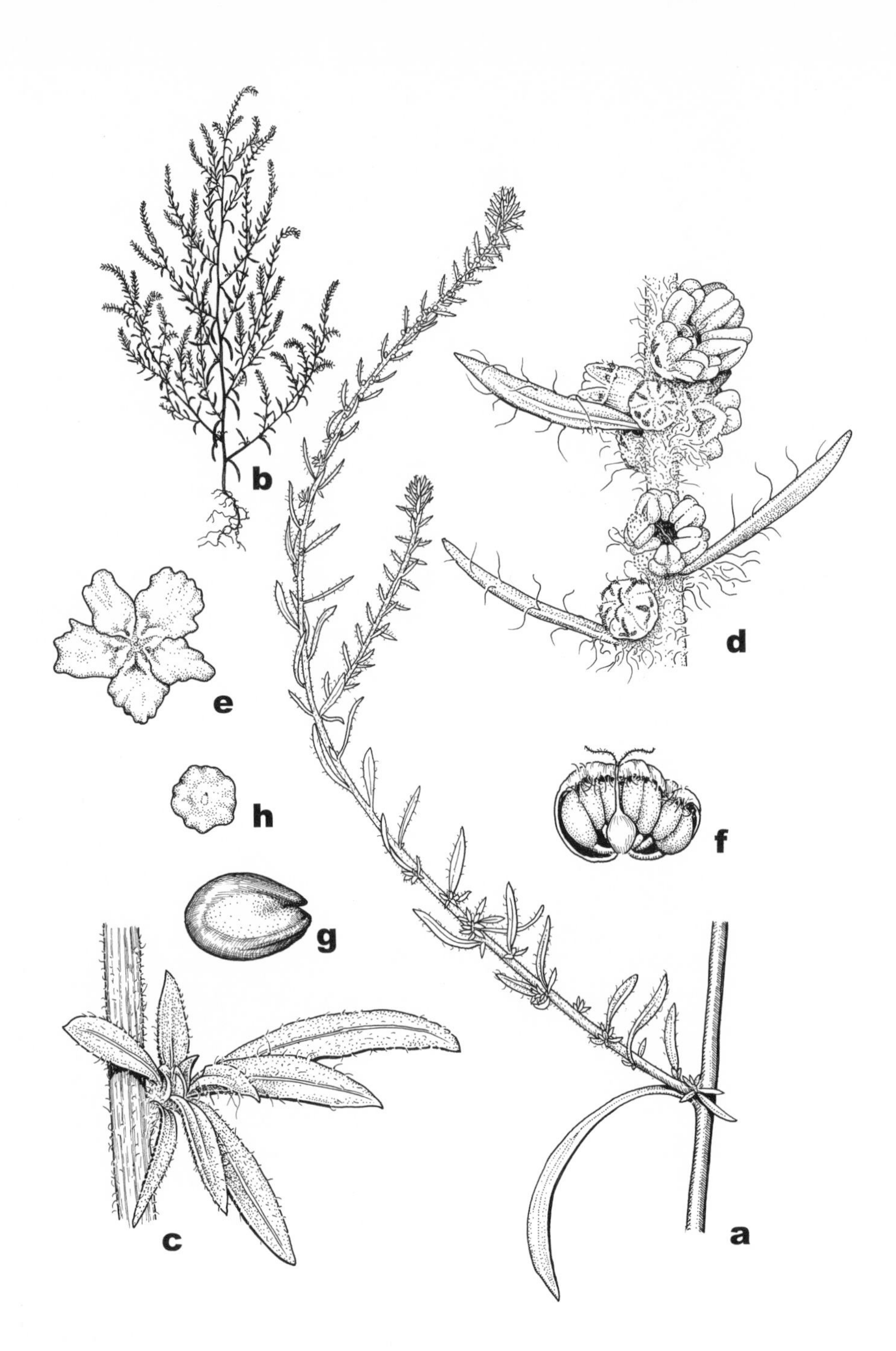

21. **Kochia scoparia**
a. Flowering branch
b. Habit
c. Axillary leaves
d. Flowering branch
e. Flower
f. Longitudinal section with ovary hidden among stamens
g. Seed
h. Aborted seed

lanceolate to lanceolate, acute at the apex, cuneate at the base, to 7 cm long, 3–8 mm broad, glabrous to short-pubescent, often ciliate at the base, the lowest on short petioles and 3- to 5-nerved, often becoming reddish; flowers up to 7 per leaf axil; calyx shallowly 5-lobed, puberulent, becoming winged in fruit; seed black, shiny.

Common Name: Kochia.
Habitat: Disturbed soil; old fields.
Range: Native to Asia; adventive throughout North America.
Illinois Distribution: Scattered throughout the state.

This is a rather common but often overlooked species because of its inconspicuous flowers. It occurs primarily in old fields and other disturbed areas. The calyx lobes are usually winged in fruit, and the stems and leaves are usually short-hairy.

Cultivated garden plants, called var. *culta* Farw., have a spherical growth form and numerous soft leaves that turn scarlet in autumn. These cultigens are known as burning bush or summer cypress.

Kochia scoparia flowers from July to September.

Chenopodium L.—Goosefoot; Lamb's Quarters

Annual or perennial herbs, often white-mealy, sometimes glandular-pubescent; leaves alternate, simple, entire or toothed or even pinnatifid; flowers perfect, borne in glomerules, spikes, or panicles, usually without bracts; calyx 5-cleft, sometimes keeled; petals 0; stamens 5; ovary superior, 1-locular; styles 2–3; fruit a 1-seeded utricle; seed vertical or horizontal, sometimes completely enclosed by the free or firmly adherent pericarp.

Approximately 150 species comprise this genus. Almost all are native to temperate regions of the World.

The genus consists of several clearly distinct entities as well as many species that have a strong resemblance to each other. In this book I am essentially following the philosophy of the four botanists who have worked diligently on this genus during the twentieth century—Standley (1915), Aellen (1929), Aellen and Just (1943), and Wahl (1954). Diagnostic characters utilized by these botanists included degree of white-mealiness (or lack of it) on the plant, especially on the calyx, whether the seed is positioned vertically or horizontally, whether the pericarp is firmly attached to or readily falls from the seed, and whether the seed is fully or partially enclosed by the pericarp. If the seed is present, the plant will fairly readily be identified. I have used these same characters to delineate the species in Illinois.

In order to break the genus down into workable groups, the following summary of species in Illinois is provided.

Plants with some or all seeds vertical: *ambrosioides, botrys, pumilio, glaucum, bonus-henricus, capitatum, rubrum.*

Pericarp free or only weakly attached to the seed: *ambrosioides, desiccatum, pratericola, standleyanum, simplex, glaucum, bonus-henricus, capitatum, rubrum.*

Calyx not white-mealy: *glaucum, bonus-henricus, capitatum, polyspermum, rubrum, urbicum, murale, ambrosioides, botrys, pumilio, simplex, lanceolatum.*

Leaves never white-mealy or sometimes not white-mealy: *ambrosioides, botrys, pumilio, standleyanum, lanceolatum, simplex, berlandieri, strictum, missouriense, bonus-henricus, capitatum, polyspermum.*

Major leaves entire (upper leaves often entire in some species are not included here): *polyspermum, botrys, pallescens, desiccatum, standleyanum, lanceolatum, foggii.*

Calyx glandular: *botrys, pumilio.*

Calyx lobes sometimes or always acute to acuminate: *botrys, album, berlandieri, strictum, bushianum, polyspermum, missouriense, foggii, macrocalycium.*

Calyx completely enclosing the fruit: *bushianum, capitatum, ambrosioides, desiccatum, pratericola, pallescens, standleyanum* (sometimes), *rubrum* (sometimes), *album, lanceolatum, opulifolium, berlandieri, macrocalycium.*

Seeds dull: *botrys, bushianum, murale, macrocalycium.*

Seeds reticulate or rugulate: *berlandieri, bushianum, foggii, macrocalycium.*

Key to the Species of **Chenopodium** in Illinois

1. Stems and leaves glandular-pubescent, or at least glandular; plants aromatic.
    2. All seeds vertical; glands of calyx bright yellow, sessile .... 1. *C. pumilio*
    2. At least some of the seeds horizontal as well as vertical; glands of calyx not bright yellow, stipitate, or calyx merely puberulent.
        3. Leaves deeply sinuate-dentate to pinnatifid; calyx lobes obtuse, pubescent; calyx completely enclosing fruit; pericarp free or nearly so from the seed; seeds shiny, black .... 2. *C. ambrosioides*
        3. Leaves shallowly sinuate-dentate to entire; calyx lobes acute, with stipitate glands; calyx only partially enclosing fruit; pericarp firmly adherent to the seed; seeds dull, dark brown .... 3. *C. botrys*
1. Stems and leaves not glandular; plants not aromatic (except for the fetid *C. berlandieri*).
    4. Calyx completely covering fruit at maturity, or barely reaching the tip in *C. standleyanum.*
        5. All seeds vertical; calyx strawberry-red .... 4. *C. capitatum*
        5. Seeds horizontal (some seeds vertical in *C. rubrum*); calyx not strawberry-red.
            6. Pericarp free from seed or only very weakly adherent.
                7. Leaves coarsely and regularly toothed or even hastate; some of the seeds vertical; leaves rhombic to ovate; calyx fleshy; seeds 0.8–1.0 mm broad, with sharp margins .... 13. *C. rubrum*
                7. Leaves entire, sparsely low-serrate, or with a single rounded tooth on each side; all seeds horizontal; leaves lanceolate to oblong to ovate; calyx not fleshy; seeds 1.0–1.5 mm broad, with rounded margins.
                    8. Leaves more or less thick, lanceolate to elliptic, with one low, rounded tooth on each side; calyx sparsely white-mealy ...... 5. *C. pratericola*
                    8. Leaves thin, lanceolate to ovate, entire or sparsely low-serrate; calyx usually densely white-mealy.
                        9. Calyx completely covering fruit; leaves obtuse to subacute at the tip .... 6. *C. desiccatum*
                        9. Calyx barely covering top of fruit; leaves acute .... 15. *C. standleyanum*
            6. Pericarp firmly adherent to the seed.
                10. Leaves linear to lanceolate, entire or with a few low teeth; calyx glabrous or sparsely white-mealy.
                    11. Leaves linear; calyx sparsely white-mealy; seeds 1.2–1.5 mm broad ... . 7. *C. pallescens*

11. Leaves lanceolate; calyx not white-mealy; seeds 1.0–1.2 mm broad . . . . 8. *C. lanceolatum*

10. Leaves rhombic to broadly ovate, regularly sinuate-dentate to 3-lobed; calyx densely white-mealy.

12. Leaves never 3-lobed, merely sinuate-dentate; plants fetid . . . . 9. *C. berlandieri*

12. At least some of the leaves 3-lobed and regularly toothed; plants not fetid.

13. Calyx cleft nearly to the middle; leaves usually as broad as long; stems often glabrate and not white-mealy; seeds smooth, not puncticulate . . . . 10. *C. opulifolium*

13. Calyx cleft nearly to the base; leaves usually up to 1 ½ times longer than broad; stems usually sparsely to densely white-mealy; seeds smooth or puncticulate.

14. Seeds (1.3–)1.5–2.3 mm broad, dull; leaves sparsely white-mealy on the lower surface; seeds reticulate.

15. Seeds 1.5–2.3 mm broad; inflorescence pendulous at maturity; leaves ovate to rhombic . . . . 11. *C. bushianum*

15. Seeds 1.3–1.7 mm broad; inflorescence erect at maturity; leaves lanceolate to lance-elliptic . . . . 12. *C. macrocalycium*

14. Seeds 1.3–1.5 mm broad, shiny; leaves densely white-mealy on the lower surface; seeds smooth or pitted . . . . 13. *C. album*

4. Calyx partially covering fruit at maturity.

16. Some or all of the seeds vertical.

17. Leaves triangular, thick; seeds 1.3–1.5 mm broad, the margins rounded . . . . 16. *C. bonus-henricus*

17. Leaves rhombic to oblong to ovate, thin; seeds 0.5–1.0 mm broad, the margins acute.

18. Seeds 0.5–0.6 mm broad; calyx glabrous, not fleshy; leaves densely white-mealy . . . . 17. *C. glaucum*

18. Seeds 0.8–1.0 mm broad; calyx sometimes white-mealy, fleshy; leaves sparsely white-mealy on the lower surface . . . . 14. *C. rubrum*

16. All seeds horizontal.

19. Seeds free or easily separated from the pericarp.

20. Leaves up to 8 cm long, entire or shallowly toothed; calyx white-mealy; seeds 1.0–1.5 mm broad . . . . 15. *C. standleyanum*

20. Leaves up to 20 cm long, usually with 4 large teeth on each side; calyx glabrous; seeds 1.8–2.0 mm broad . . . . 18. *C. simplex*

19. Seeds firmly adherent to the pericarp.

21. Leaves thick, entire; inflorescence often from base of the plant to the top . . . . 19. *C. polyspermum*

21. Leaves thin, toothed, rarely entire or with 1–2 pairs of teeth; inflorescence not at base of the plant.

22. Calyx sparsely white-mealy; leaves usually glabrous or only sparsely white-mealy.

23. Leaves with irregular small teeth; seeds smooth . . . . . . 20. *C. strictum*

23. Leaves with regular coarse teeth, or entire or with 1–2 pairs of teeth; seeds puncticulate or rugulate.

24. All except the upppermost leaves coarsely toothed; seeds puncticulate . . . . 21. *C. missouriense*

24. All leaves entire or with 1–2 pairs of teeth; seeds rugulate .... 22. *C. foggii*

22. Calyx glabrous; leaves densely white-mealy on the lower surface.

25. Seeds 0.9–1.0 mm broad, smooth, shiny, the margin rounded .... 23. *C. urbicum*

25. Seeds 1.2–1.5 mm broad, puncticulate, dull, the margin sharp .... 24. *C. murale*

1. Chenopodium pumilio R. Br. Prodr. 1:407. 1810. Fig. 22.

Annual herb from slender taproots, aromatic; stems rather slender, prostrate to spreading to ascending, up to 40 cm long, branched or unbranched, glandular-pubescent, not white-mealy; leaves lanceolate to ovate, obtuse to acute at the apex, cuneate to rounded at the base, to 30 mm long, to 25 mm broad, with 2–4 coarse teeth or lobes on each margin, rather thin, 3-veined, densely glandular-pubescent, not white-mealy; flowers in dense glomerules forming short spikes up to 5 mm long; calyx densely yellow-glandular, not white-mealy, the lobes obtuse; seed vertical, firmly attached to the pericarp.

Common Name: Aromatic Goosefoot.
Habitat: Disturbed soil.
Range: Native of Eurasia; occasionally found as an adventive in North America.
Illinois Distribution: Known from a single collection from McDonough County.

The distinguishing features of *C. pumilio* are the glandular stems and leaves and the bright yellow sessile glands on the calyx. The similar *C. botrys* has stipitate glands on the calyx. *Chenopodium ambrosioides* is also aromatic but lacks glands on the calyx.

Chenopodium pumilio flowers during July and August.

2. Chenopodium ambrosioides L. Sp. Pl. 219. 1753. Fig. 23.
Chenopodium anthelminticum L. Sp. Pl. 220. 1753.
Ambrina ambrosioides (L.) Spach. Hist. Veg. Phan. 5:295. 1836.
Ambrina anthelmintica (L.) Spach. Hist. Veg. Phan. 5:295. 1836.
Chenopodium ambrosioides L. var. *anthelminitcum* (L.) A. Gray, Man. Bot. ed. 5, 408. 1867.

Annual herb with stout taproots, less commonly persisting as a perennial, strongly scented; stems stout, ascending to erect, to 1 m tall, branched, sparsely glandular or eglandular, tomentulose or glabrous, not white-mealy; leaves lanceolate to ovate, obtuse to acute at the apex, rounded to cuneate at the base, to 10 cm long, to 5 cm broad, the margins sinuate-dentate or even somewhat pinnatifid, more or less thin, 3-veined, often glandular-dotted, not white-mealy; flowers in glomerules, rarely solitary, in interrupted leafy or leafless spikes; calyx glabrous or short-pubescent, the lobes obtuse, with or without a low keel, completely enclosing the fruit; fruit

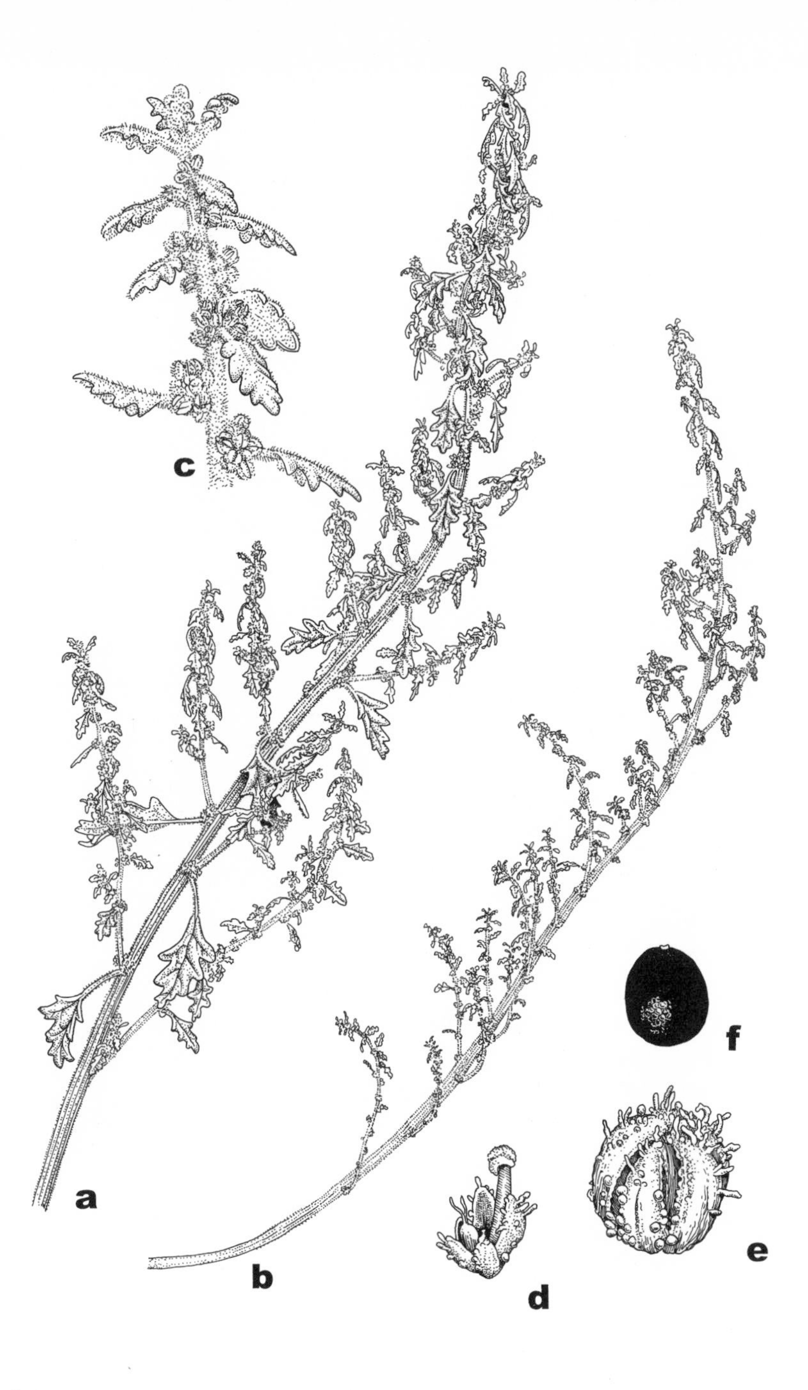

22. **Chenopodium pumilio**
a. Flowering branch
b. Flowering branch
c. Flowering branch
d. Flower
e. Fruit
f. Seed

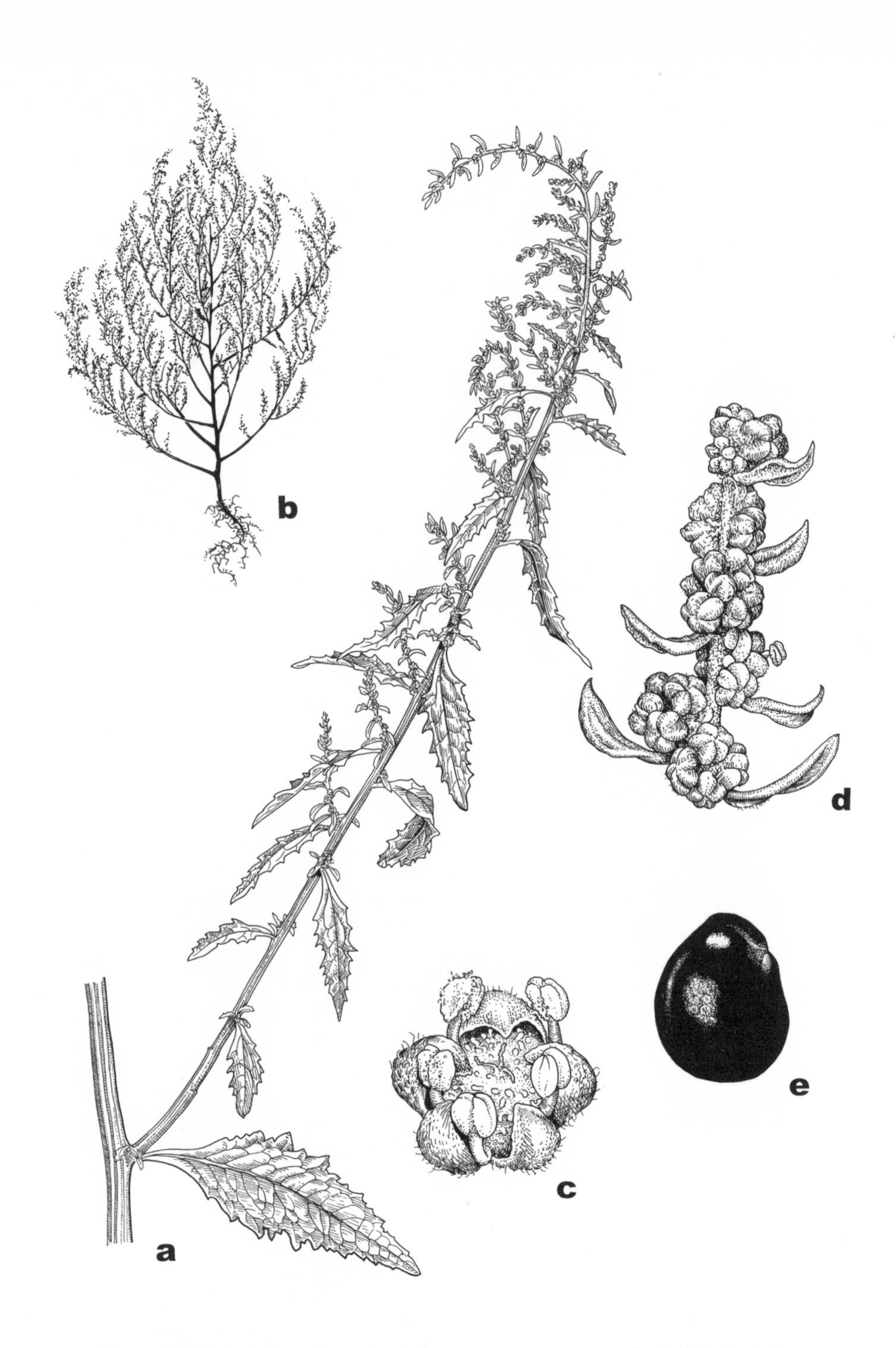

23. **Chenopodium ambrosioides**
a. Flowering branch
b. Habit
c. Flower
d. Fruiting branch
e. Seed

with nearly free pericarp; seeds horizontal or vertical, 0.6–0.8 mm broad, smooth, shiny, black, the margins rounded.

Common Name: Mexican Tea.
Habitat: Waste ground.
Range: Native to tropical America; adventive throughout North America, Europe, and Asia.
Illinois Distribution: Occasional to common in most of Illinois, although rare or absent in most of the northernmost counties.

The features that distinguish this species are the lack of white-mealiness on the stems, leaves, and calyx; the pubescent calyx; and the strong medicinal aroma. *Chenopodium botrys* and *C. pumilio* are somewhat similar, but both of these species have glandular calyces. In *C. ambrosioides,* the calyx completely covers the fruit, and the seed is nearly free from the pericarp. In *C. botrys,* the calyx only partially covers the fruit, and the pericarp is firmly adherent to the seed.

Specimens without leafy bracts in the inflorescence and with no keel on the calyx lobes have been called var. *anthelminticum.* This variety, which has been collected a few times, particularly in the southern counties of Illinois, is not recognized in this work, even though I have recognized it in previous publications on the Illinois flora. Linnaeus actually thought that this variety was worthy of species status.

Chenopodium ambrosioides and *C. botrys,* because of their aromatic nature, were at one time segregated into the genus *Ambrina.*

This species has been used medicinally for a number of purposes, including a tonic for worms in humans.

Chenopodium ambrosioides flowers from May to November.

3. **Chenopodium botrys** L. Sp. Pl. 219. 1753. Fig. 24.
Ambrina botrys (L.) Moq. Chen. Mon. Enum. 373. 1840.

Annual herb with taproots, strongly aromatic; stems rather slender, erect, up to 60 cm tall, branched, densely glandular-pubescent, viscid, not white-mealy; leaves oval to oblong, up to 5 cm long, up to 4 cm broad, obtuse to subacute at the apex, cuneate to truncate at the base, the margins entire to sinuately lobed, more or less thin, 3-veined, often glandular, not white-mealy; flowers in dense cymes forming an elongated panicle; calyx densely glandular-pubescent, the glands stipitate, not white-mealy, the lobes acute to acuminate, scarcely keeled, partially enclosing the fruit; fruit with a white pericarp firmly attached to the seed; seeds horizontal or vertical, 0.5–0.7 mm in diameter, smooth, dull, dark brown.

Common Name: Jerusalem Oak.
Habitat: Disturbed soil.
Range: Native to Europe and Asia; adventive in much of North America.
Illinois Distribution: Occasional throughout the state.

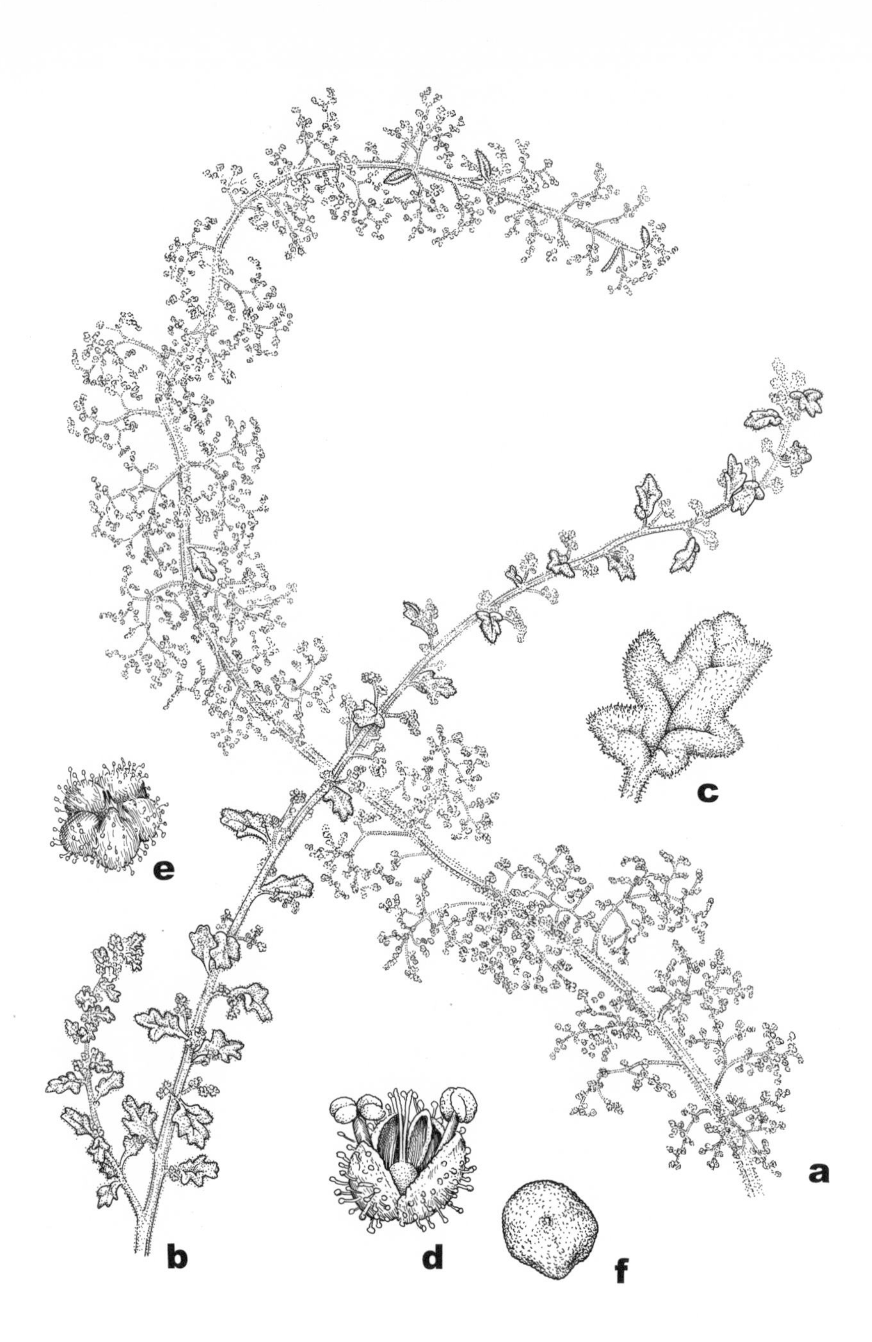

24. **Chenopodium botrys**
a. Flowering branch
b. Flowering branch
c. Leaf
d. Flower
e. Fruit
f. Seed

Chenopodium botrys differs from *C. ambrosioides* by its glandular calyx and its usually entire or merely sinuate leaves, while *C. ambrosioides* has a pubescent calyx and dentate to pinnatifid leaves. *Chenopodium botrys* has stipitate glands on its calyx, while *C. pumilio* has sessile glands.

Mead (1846) was the first to report this species from Illinois.

Chenopodium botrys flowers from July to October.

4. Chenopodium capitatum (L.) Aschers. Fl. Brand. 572. 1864. Fig. 25.
Blitum capitatum L. Sp. Pl. 2. 1753.

Annual herb from slender taproots; stems prostrate to ascending to erect, more or less fleshy, to 60 cm tall, branched or unbranched, glabrous, not white-mealy; leaves triangular, acute at the apex, truncate at the base, to 6 cm long, to 5 cm broad, sharply toothed, sometimes hastate, bright green, not white-mealy, more or less fleshy, 3-veined, the petioles often as long as the blades; flowers in globose clusters in the leaf axils, forming interrupted spikes; calyx bright red, fleshy, obtuse at the tip, not keeled, somewhat resembling bright strawberries at maturity, completely enclosing the fruit; fruit with pericarp free from the seed; seeds vertical, compressed, ovate.

Common Name: Strawberry Blite.
Habitat: Disturbed soil.
Range: Native to Europe and Asia; sparingly adventive in North America.
Illinois Distribution: Known only from McHenry and Peoria counties.

This species is readily distinguished by its bright red fleshy calyx and its triangular leaves that are coarsely and regularly dentate. When the fruits mature, forming spherical clusters, they resemble strawberries. Since this is the only species of *Chenopodium* with fleshy red calyces, Linnaeus created a new genus for it that he called *Blitum.*

The only known Illinois collections were made during the nineteenth century by Frederic Brendel in Peoria County and George Vasey in McHenry County, although Mead (1846) listed it from Illinois earlier than these collections.

Chenopodium capitatum flowers from May to August.

5. Chenopodium pratericola Rydb. Bull. Torrey Club 39 : 310. 1912. Fig. 26.
Chenopodium leptophyllum Nutt. ex Moq. in DC. Prodr. 13 : 71. 1849, in synon.
Chenopodium petiolare HBK. var. *leptophylloides* Murr. Bull. Herb. Boiss. II (4):306. 1916.
Chenopodium desiccatum A. Nels. var. *leptophylloides* (Murr.) Wahl, Field Lab. 23 : 22. 1955.

Annual herb with slender taproots; stems erect, branched, to 75 cm tall, white-mealy to glabrate, rarely reddish; leaves lanceolate to elliptic, obtuse to acute at the apex, cuneate at the base, to 6 cm long, to 18 mm broad, usually with a rounded tooth on each side, rarely entire, thick, 3-nerved, white-mealy on the lower surface,

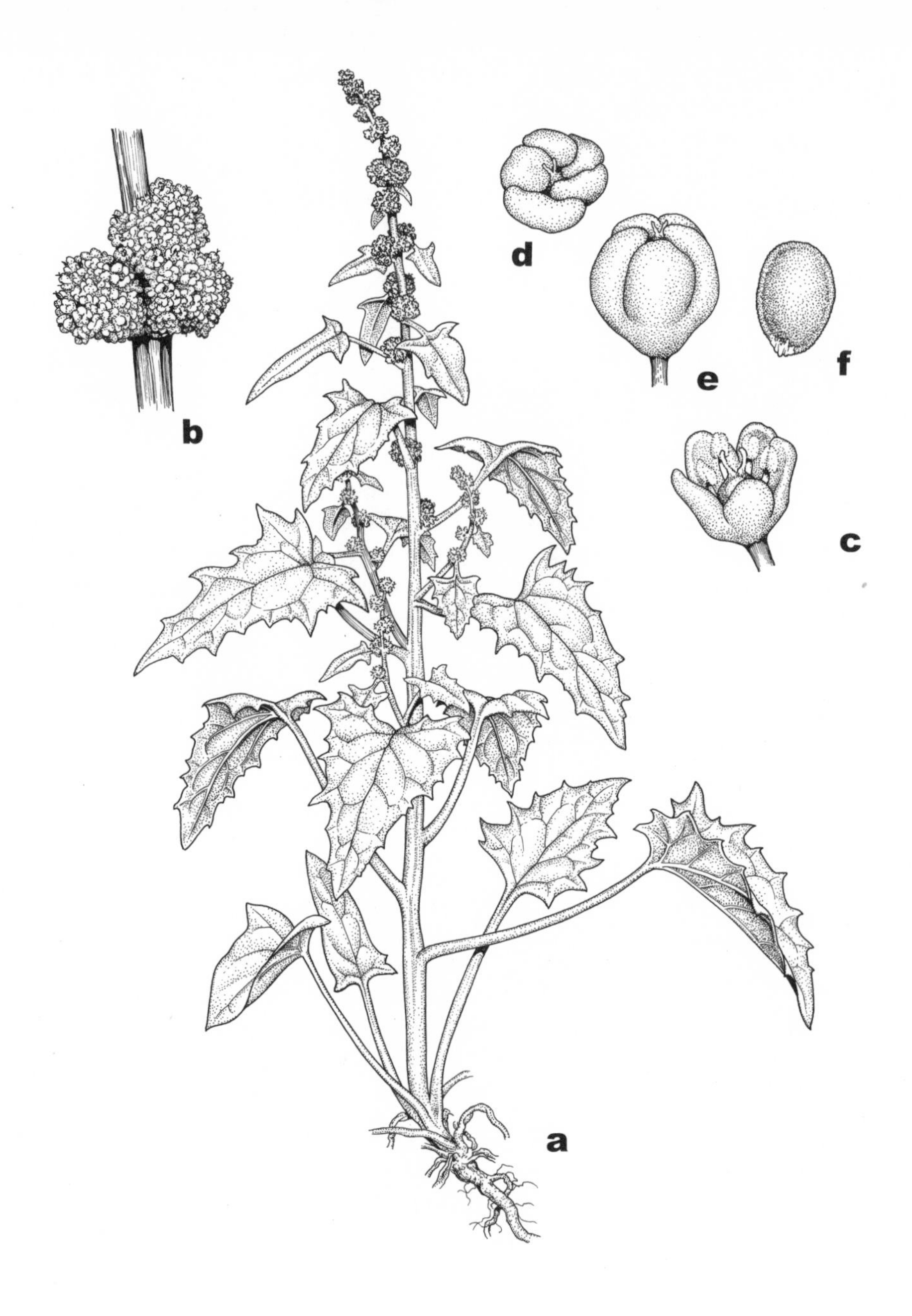

25. **Chenopodium capitatum**
a. Habit
b. Flowering node
c. Flower
d. Fruit
e. Fruit
f. Seed

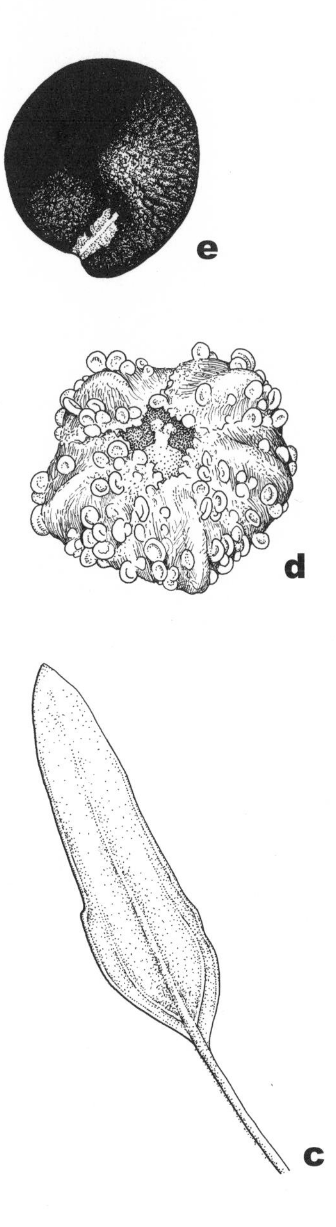

26. **Chenopodium pratericola**
a. Habit
b. Habit
c. Leaf
d. Flower
e. Seed

green and glabrous on the upper surface, with petioles about half as long as the blade; flowers in dense glomerules forming paniculate spikes; calyx somewhat mealy although usually green, the lobes obtuse, more or less keeled, completely enclosing the fruit; fruit with a free pericarp; seeds horizontal, 1.0–1.5 mm broad, smooth, shiny, black, the margins obtuse.

Common Name: Narrow-leaved Goosefoot.
Habitat: Disturbed, often sandy soil.
Range: Manitoba to Washington, south to Arizona and Missouri; adventive in the eastern United States, including Illinois.
Illinois Distribution: Occasional throughout the state, where some of the populations may be native.

This species is readily recognized by its rather thick, lanceolate to elliptic leaves that usually have a single tooth on each margin. The calyx is sparsely white-mealy. The similar *C. pallescens* has thin, entire, oblong to ovate leaves and a calyx that is densely white-mealy.

This species has had an unstable taxonomic history. Nuttall first described it in synonymy as *C. leptophyllum,* thus invalidating that binomial. Murray then called it *C. petiolare* var. *leptophylloides.* At the species level, however, the correct binomial appears to be Rydberg's *C. pratericola.* In 1955, Wahl treated this plant as a variety of the western *C. desiccatum,* a decision I followed in 1986. I now believe that it is best to consider *C. pratericola* and *C. desiccatum* as two distinct species.

Chenopodium pratericola occurs in Illinois primarily in sandy soil where there is considerable disturbance. Some collections from along the Mississippi River may be from native populations.

This species flowers from June to October.

6. Chenopodium desiccatum A. Nels. Bot. Gaz. 34:362. 1902. Fig. 27.

Annual herb with slender taproots; stems erect, branched, to 50 cm tall, white-mealy to glabrate, often reddish; leaves oblong to ovate, obtuse to subacute at the apex, cuneate at the base, to 3 (–4) cm long, to 10 mm broad, entire, thin, 3-veined, white-mealy on both surfaces, with petioles about half as long as the blades; flowers in dense glomerules forming paniculate spikes; calyx densely white-mealy, the lobes obtuse, keeled, completely enclosing the fruit; fruit with a free pericarp; seeds horizontal, 1.0–1.5 mm broad, nearly smooth, shiny, reddish brown to black, the margins obtuse.

Common Name: Entire-leaved Goosefoot.
Habitat: Sandy soil along rivers; black oak savannas.
Range: Manitoba to Washington, south to California, Arizona, and Illinois.
Illinois Distribution: Known only from Jackson County and from the northeastern corner of the state.

Chenopodium desiccatum is readily distinguished by its thin, entire, oblong to ovate leaves that are white-mealy, its densely white-mealy calyx,

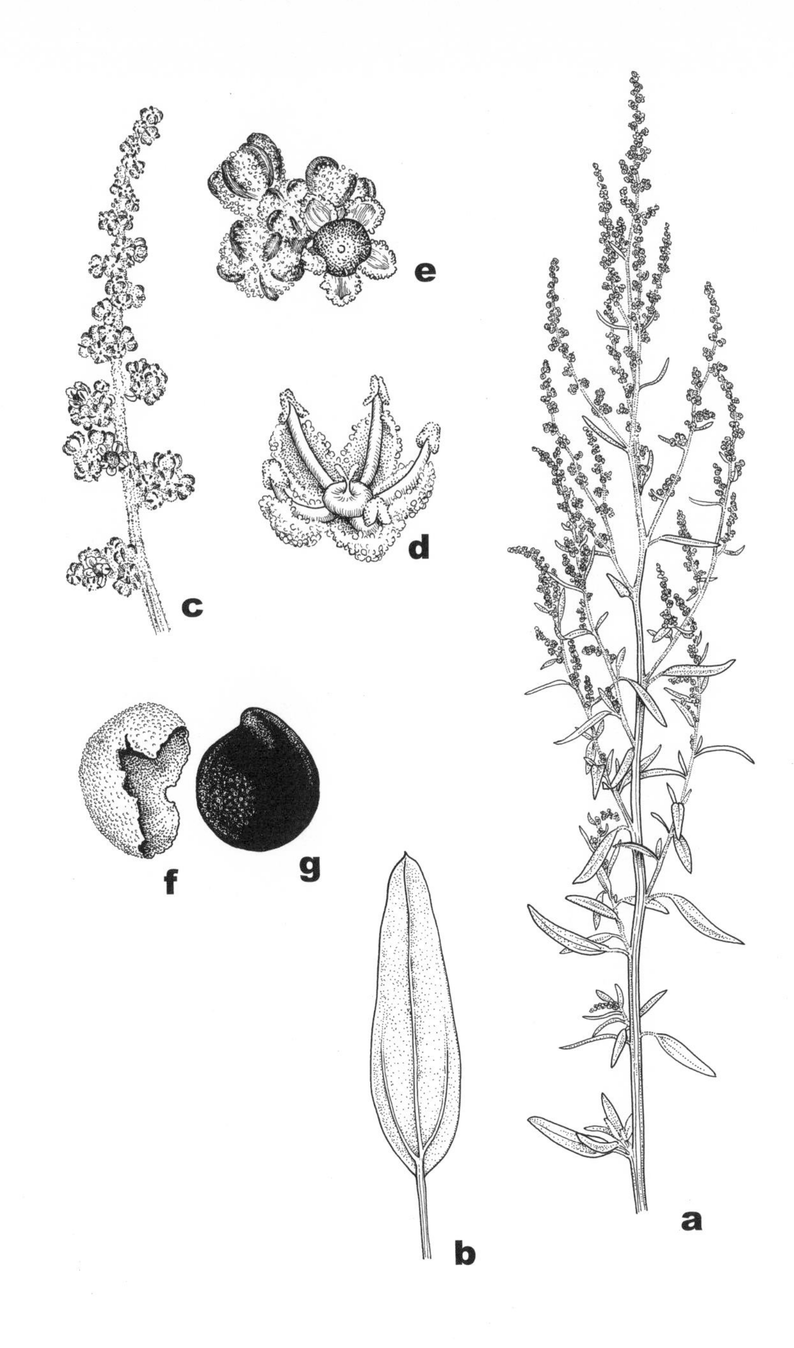

27. **Chenopodium desiccatum**
a. Habit
b. Leaf
c. Fruiting branch
d. Flower
e. Fruit
f. Pericarp
g. Seed

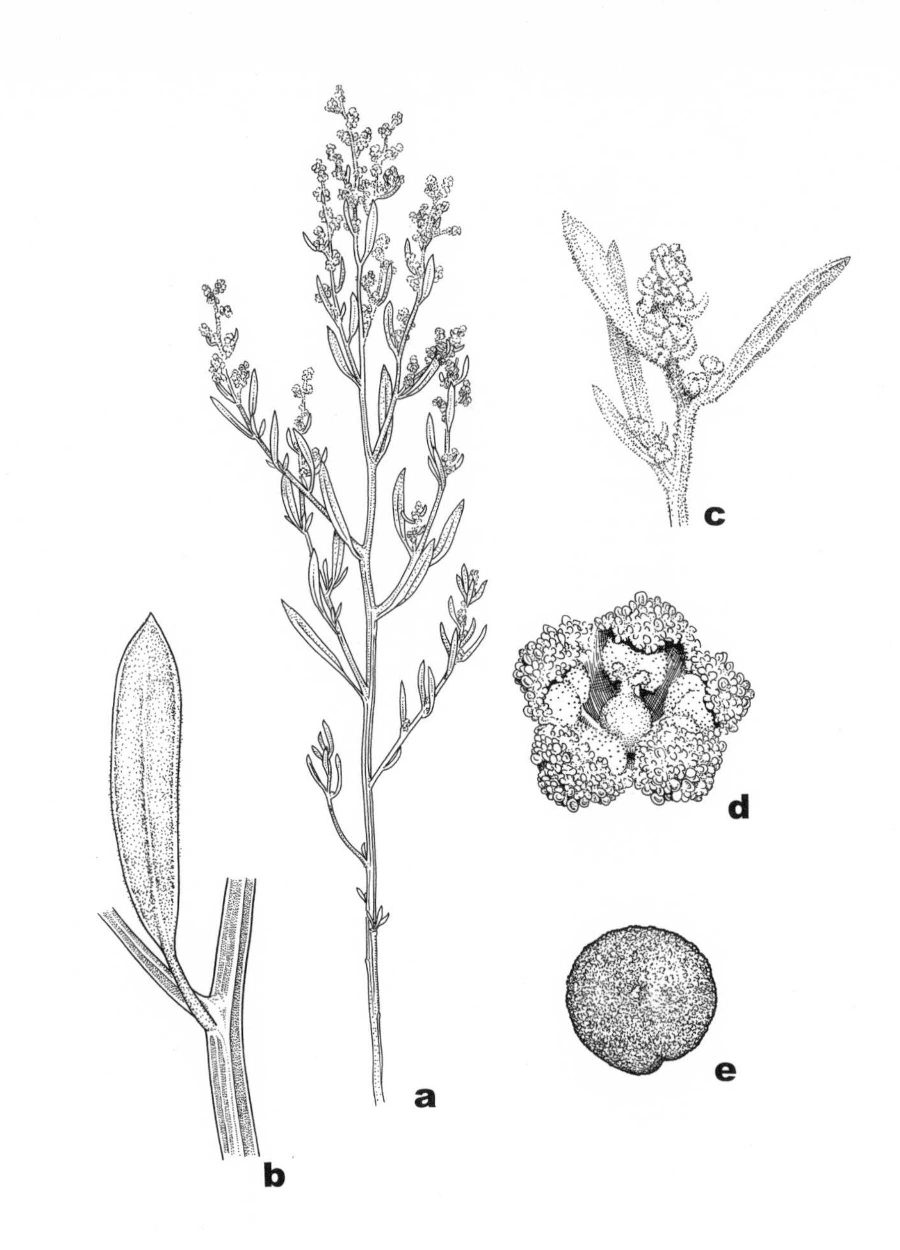

28. **Chenopodium pallescens**
a. Habit
b. Leaf and node
c. Flowering branch
d. Flower
e. Fruit

and its reddish brown seeds. The similar *C. pratericola* differs in its thick, lanceolate to elliptic leaves with one tooth on each side, its sparsely white-mealy calyx, and its black seeds.

The range of this species is primarily west of the Mississippi River. In Illinois, it was collected along the sandy banks of the Mississippi River in the Grand Tower Park in Jackson County. This same species apparently occurs in black oak savannas in the Chicago area, where Swink and Wilhelm (1994) call it *C. leptophyllum.*

Chenopodium desiccatum flowers in June and July.

7. Chenopodium pallescens Standl. N. Am. Fl. 21 (1):15. 1916. Fig. 28.

Annual herb with slender taproots; stems erect, branched, to 60 cm tall, sparsely white-mealy to glabrate, pale green; leaves linear, obtuse to subacute at the apex, cuneate at the base, to 3.5 cm long, to 3 mm broad, entire, thick, 1-nerved, sparsely white-mealy, pale green, the petioles not more than one- third as long as the blades; flowers in dense glomerules forming paniculate spikes; calyx sparsely white-mealy, the lobes obtuse to subacute, keeled, completely enclosing the fruit; fruit with pericarp firmly attached to the seed; seeds horizontal, 1.2–1.5 mm broad, smooth, shiny, black, the margins obtuse.

Common Name: Pale-leaved Goosefoot.
Habitat: Rocky ground; disturbed soil.
Range: Indiana to Colorado, south to New Mexico, Oklahoma, and Missouri.
Illinois Distribution: Confined to the northern half of Illinois.

This species is recognized by its thick, entire, linear leaves; its sparsely white-mealy calyx; and its pericarp that is firmly attached to the seed. *Chenopodium desiccatum* is very similar in appearance, but it has a densely white-mealy calyx and its pericarp is free from the seed.

This plant grows in open, sandy or rocky habitats in the northern half of the state.

My report of this species in 1986 from Jackson County is based on a plant I have now redetermined as *C. desiccatum.*

Chenopodium pallescens flowers from June to September.

8. Chenopodium lanceolatum Muhl. ex Willd. Enum. Hort. Berol. 291. 1809. Fig. 29.

Annual herb from slender or stout taproots; stems erect, branched, to 1.5 m tall, glabrate or rarely very sparsely white-mealy; leaves green, lanceolate to narrowly ovate, obtuse to subacute at the apex, tapering or rounded at the base, to 7.5 cm long, to 5 cm broad, entire or with a few low, rounded teeth, rather thick, usually 3-nerved, rarely sparsely white-mealy on the lower surface, with petioles up to half or more as long as the blade; flowers in small glomerules in interrupted spikes and cymes; calyx glabrate, rarely sparsely white-mealy, the lobes obtuse, low-keeled, completely enclosing the fruit; fruit with the pericarp adherent to the seed; seeds horizontal, 1.0–1.2 mm broad, smooth or pitted, shiny, black.

29. **Chenopodium lanceolatum**
a. Habit
b. Flower
c. Fruiting branch
d. Fruit
e. Seed with pericarp

Common Name: Lance-leaved Lamb's Quarters.
Habitat: Disturbed soil.
Range: Newfoundland to British Columbia, south to California, Texas, and Florida.
Illinois Distribution: Scattered throughout the state but apparently not common.

Although most botanists combine *C. lanceolatum* with *C. album*, I am recognizing them as two different species. The following chart summarizes the differences as I see them:

Characteristic	*Chenopodium lanceolatum*	*Chenopodium album*
Stems	glabrate to sparsely mealy	sparsely to densely mealy
Leaf indument	glabrate to rarely mealy	densely mealy on lower surface
Leaf shape	lanceolate to narrow-ovate	rhombic to ovate
Leaf margins	entire or with a few low, rounded teeth	irregularly dentate to 3-lobed
Calyx	glabrate or rarely mealy	densely white-mealy
Seed length	1.0–1.2 mm	1.3–1.5 mm
Flower clusters	interrupted	continuous
Calyx margins	white	yellowish

Chenopodium lanceolatum flowers from May to October.

9. **Chenopodium berlandieri** Moq. Chen. Mon. Enum. 23. 1840. Fig. 30.
Chenopodium zschackei Murr. Deutsch. Bot. Monatsschr. 19:39. 1901.
Chenopodium berlandieri Moq. ssp. *zschackei* (Murr.) Zobzschackel, Verz. Anh. Phan. 3:70. 1909.

Annual herb from slender taproots, often somewhat fetid; stems erect, branched, to 1 m tall, pale to deep green, glabrate or sparsely white-mealy; leaves rhombic to ovate to oblong, obtuse to acute and mucronulate at the apex, cuneate to rounded at the base, to 3.5 cm long, to 15 mm broad, irregularly dentate but not 3-lobed, more or less thin, not shiny, densely white-mealy when young, soon becoming glabrate; flowers in small glomerules in slender interrupted paniculate spikes; calyx densely white-mealy, the lobes obtuse to acute, sharply keeled, completely enclosing the fruit; fruit with pericarp adhering to the seed; seeds horizontal, 0.8–1.2 mm broad, reticulate, shiny, black, the margins rounded.

Common Name: Stinking Lamb's Quarters.
Habitat: Disturbed soil.
Range: New York to North Dakota, south to Texas and Florida; Mexico.
Illinois Distribution: Scattered in Illinois.

This is another species in the *C. album* complex, and it is sometimes not considered distinct from *C. album*. Despite some intergradation, there

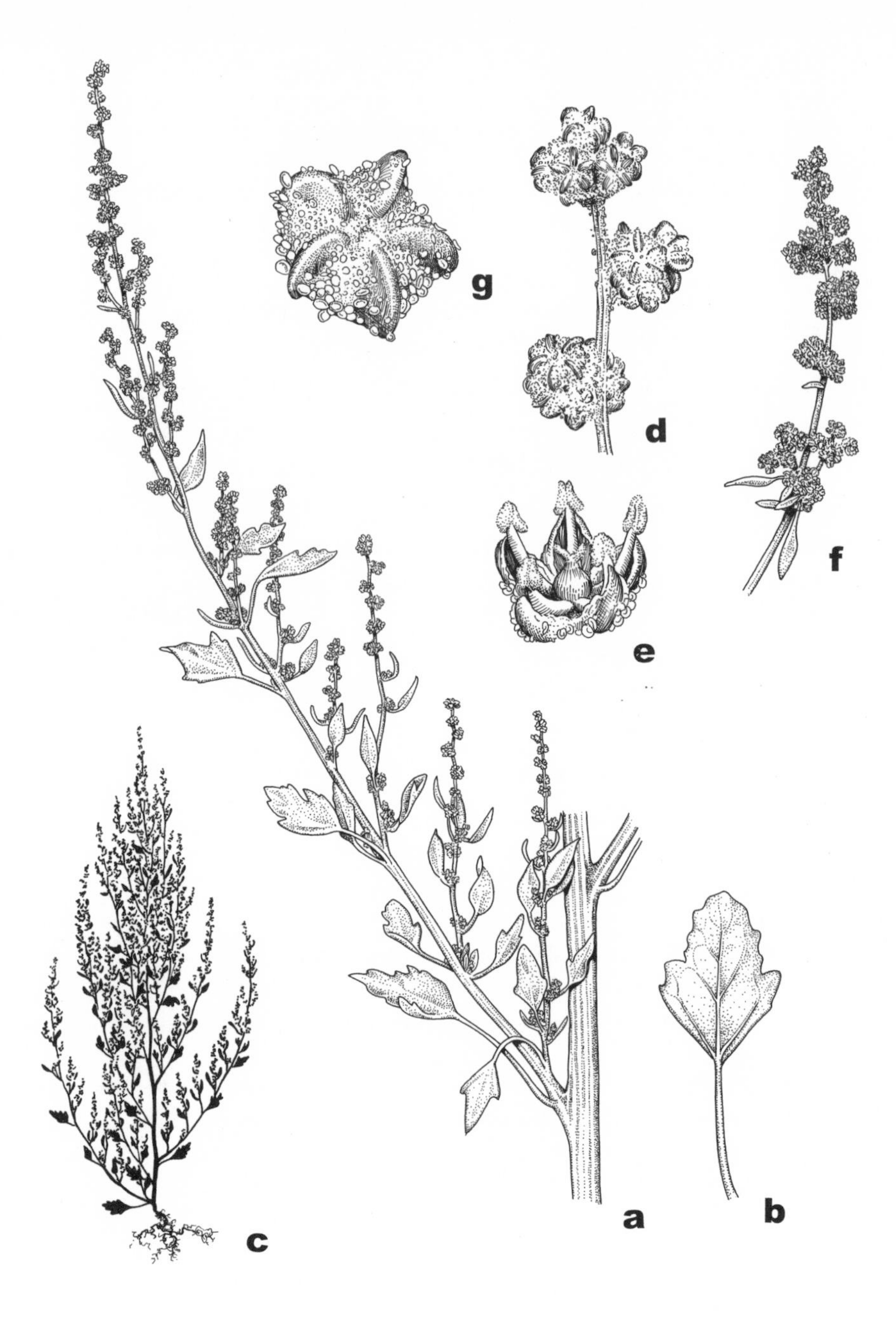

30. **Chenopodium berlandieri**
a. Flowering branch
b. Leaf
c. Habit
d. Flowering branch
e. Flower
f. Stout fruiting branch
g. Mature fruit

are enough differences between various entities in the complex to be able to recognize several species.

Chenopodium berlandieri is distinguished by its fetid odor, its glabrate to sparsely white-mealy stems and leaves, the usual absence of any lobes on the leaves, the sharp keel on the lobes of the calyx, and its rather small seeds. I am, however, unable to recognize the various varieties Wahl (1954) attributes to *C. berlandieri*.

This species flowers from June to October.

10. **Chenopodium opulifolium** Schrad. in DC. Fl. Fr. 5:372. 1805. Fig. 31.

Annual herb from slender taproots; stems ascending to erect, branched, to 1 m tall, sparsely white-mealy or glabrate, dark green; leaves rhombic to broadly ovate, obtuse at the apex, rounded to truncate at the base, to 4 cm long, nearly as broad, sparsely white-mealy on the lower surface, otherwise glabrate, shallowly 3-lobed to sinuate-dentate, thin, 3-veined; flowers in small glomerules in dense paniculate spikes; calyx densely white-mealy, cleft only halfway to the base, the lobes obtuse, keeled, completely enclosing the fruit; fruit with the pericarp adherent to the seeds; seeds horizontal, 1.0–1.5 mm broad, smooth, shiny, black, the margins rounded.

Common Name: Goosefoot.
Habitat: Disturbed soil.
Range: Native to Europe and Asia; adventive along the eastern seaboard; also Illinois, Missouri, Texas.
Illinois Distribution: Known from Wabash and Winnebago counties.

This species is often included in *C. album*, but it differs in its leaves, which are as broad as long; its calyx, which is split only halfway to the base; and its seeds, which are smooth and not puncticulate.

The first Illinois collection was made at Rockford by M. S. Bebb in 1880.

Chenopodium opulifolium flowers during July and August.

11. **Chenopodium bushianum** Aellen in Fedde, Rep. Sp. Nov. 26:63. 1929. Fig. 32.
Chenopodium paganum Reichenb. Fl. Germ. 579. 1832, misapplied.

Annual herb from stout taproots; stems stout, erect, branched, to 2 m tall, sparsely to densely white-mealy, often becoming purplish; leaves rhombic to ovate, obtuse or acute at the apex, tapering or rounded at the base, to 7.5 cm long, to 5 cm broad, irregularly sinuate-dentate to shallowly 3-lobed, rarely entire, rather thin, usually 3-nerved, sparsely white-mealy, otherwise dark green to yellow-green, with petioles up to half or more as long as the blade; flowers in glomerules forming axillary or terminal paniculate spikes, the spikes pendulous at maturity; calyx densely white-mealy, the lobes obtuse to acute, sharply keeled, completely enclosing the fruit; fruit with the pericarp adherent to the seed; seeds horizontal, 1.5–2.0 mm broad, reticulate, rather dull, black.

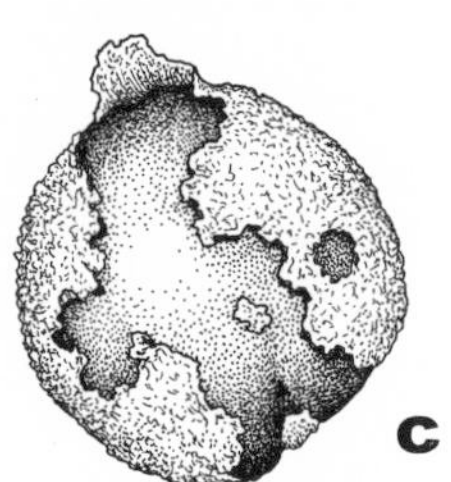

31. **Chenopodium opulifolium**
a. Habit
b. Flower
c. Seed with pericarp

32. **Chenopodium bushianum**
a. Habit
b. Fruiting branch
c. Mature fruit cluster
d. Flower
e. Seed

Common Name: Goosefoot.
Habitat: Disturbed soil.
Range: Quebec to Alaska, south to California, Florida, and Texas.
Illinois Distribution: Scattered in Illinois but particularly common in counties that border the Mississippi and Illinois rivers.

Chenopodium bushianum is similar to *C. berlandieri* in that both have reticulate seeds and a sharp keel on each lobe of the calyx. It differs in its larger, dull seeds (1.5–2.0 mm broad). *Chenopodium berlandieri* has smaller, shiny seeds (0.8–1.2 mm broad).

Chenopodium macrocalycium is also similar to *C. bushianum* but is smaller in all respects, has narrower leaves, and has erect inflorescences.

Although *C. paganum* predates *C. bushianum,* it does not have priority because *C. paganum* already applies to a different species.

Chenopodium bushianum flowers from August to October.

12. Chenopodium macrocalycium Aellen, Rep. Spec. Nov. Regni. Veg. 26(1–6):119. 1929. Fig. 33.
Chenopodium berlandieri Moq. var. *macrocalycium* (Aellen) Cronq. Man. Vasc. Pl. NE. US. & Adj. Can., ed. 2, 863. 1991.

Annual herb from stout taproots; stems erect, branched, to 1.2 m tall, sparsely white-mealy; leaves narrowly ovate to lanceolate, obtuse to acute at the apex, tapering or rounded at the base, to 4.5 cm long, to 3 cm broad, irregularly sinuate-dentate to shallowly 3-lobed, rather thin, usually 3-nerved, sparsely white-mealy, otherwise dark green to yellow-green, with petioles up to half or more as long as the blade; flowers in terminal paniculate spikes, the spikes erect at maturity; calyx white-mealy, the lobes obtuse to acute, sharply keeled, completely enclosing the fruit; fruit with the pericarp adherent to the seed; seeds horizontal, 1.3–1.7 mm broad, reticulate, rather dull, black.

Common Name: Goosefoot.
Habitat: Along a road (in Illinois).
Range: Coastal plain of the eastern United States and around the southern end of Lake Michigan; central Illinois.
Illinois Distribution: Known only from Champaign Co.: along Route 45, eight miles northeast of Rantoul, July 30, 1949, H. M. Franklin.

This species is in the *C. berlandieri*—*C. bushianum* complex. It lacks the fetid odor of *C. berlandieri.* The seeds of *C. macrocalycium* are broader than those of *C. berlandieri,* and the leaves of *C. macrocalycium* often are 3-lobed. It differs from *C. bushianum* by its smaller stature, smaller leaves, erect inflorescences, and its usually narrower seeds.

The Illinois specimen was originally identified as *C. berlandieri.*

Chenopodium macrocalycium flowers from July to September.

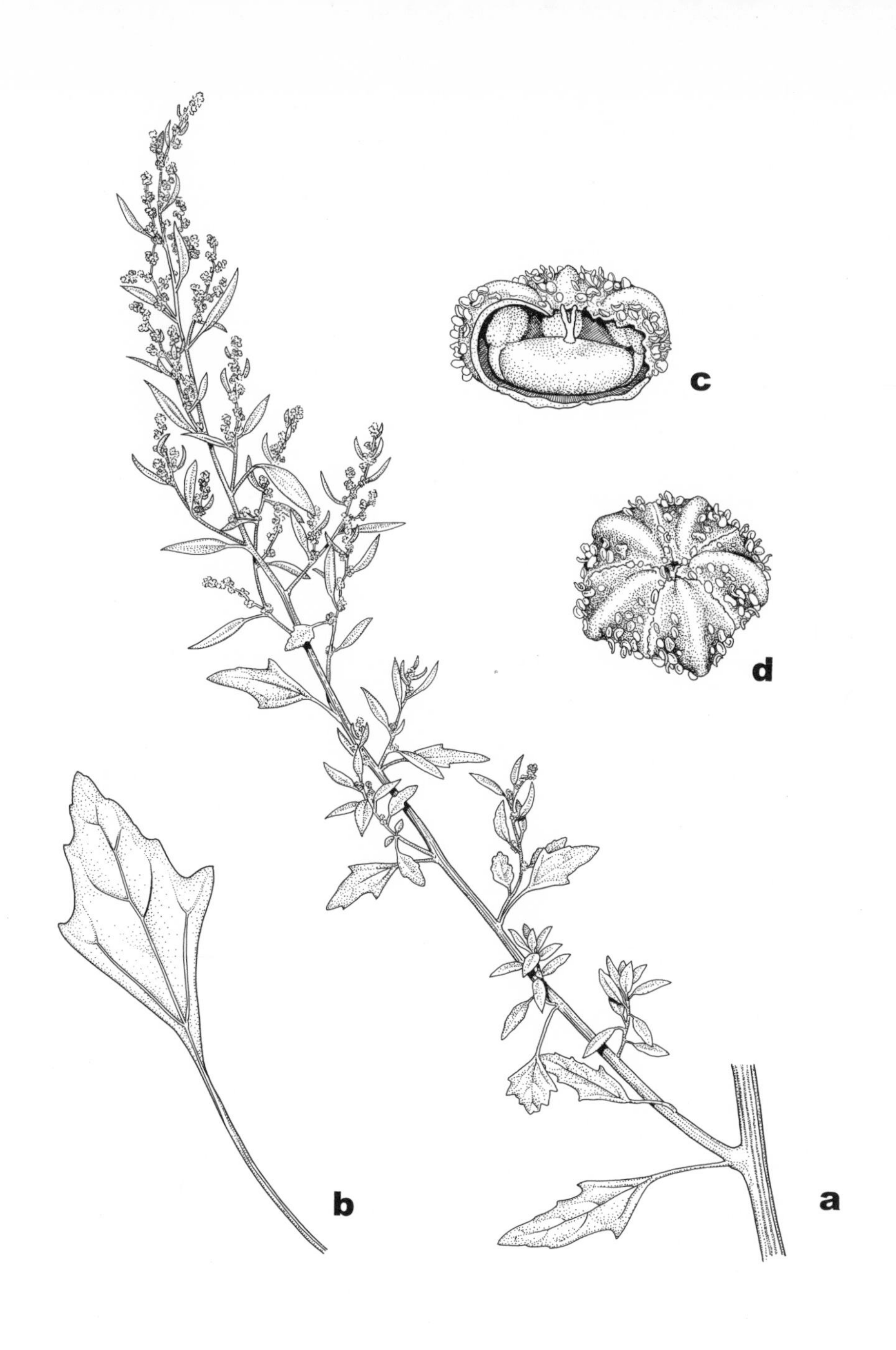

33. **Chenopodium macrocalycium**

a. Habit
b. Leaf
c. Flower with two calyx lobes removed
d. Fruit

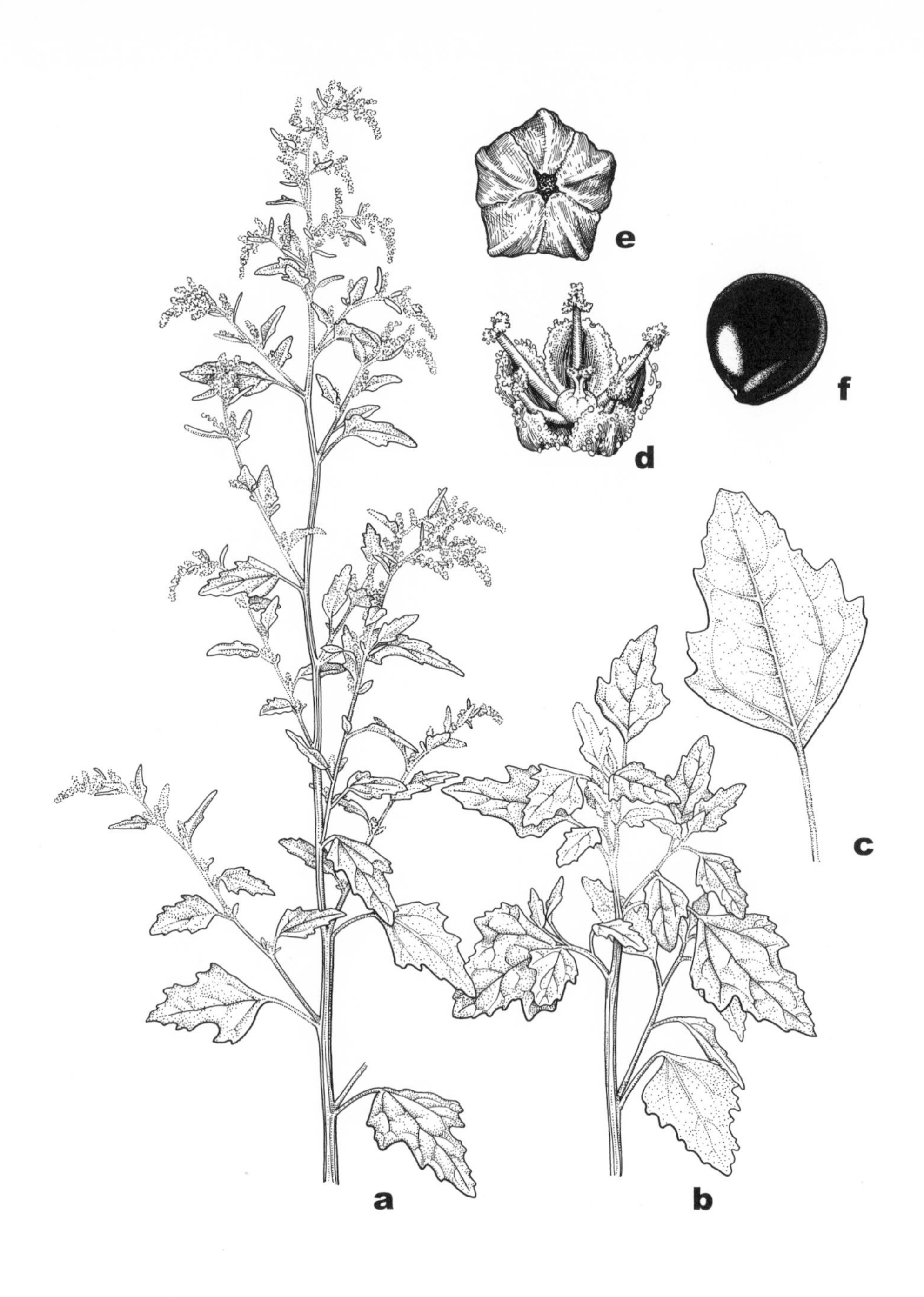

34. **Chenopodium album**
a. Habit
b. Branch of young plant
c. Leaf
d. Flower
e. Fruit
f. Seed

13. Chenopodium album L. Sp. Pl. 219. 1753. Fig. 34.

Annual herb from slender or stout taproots; stems erect, branched, to 2 m tall, sparsely to densely white-mealy, sometimes glaucous or pale green, often becoming purplish; leaves glaucous, rhombic to ovate, obtuse or apiculate at the apex, tapering or rounded at the base, to 7.5 cm long, to 5 cm broad, irregularly sinuate-dentate to 3-lobed, rather thick, usually 3-nerved, usually densely white-mealy, at least on the lower surface, with petioles up to half or more as long as the blade; flowers in large glomerules forming axillary or terminal paniculate spikes; calyx densely white-mealy, the lobes obtuse to acute, keeled, completely enclosing the fruit; fruit with the pericarp adherent to the seed; seeds horizontal, 1.3–1.5 mm broad, smooth or pitted, shiny, black.

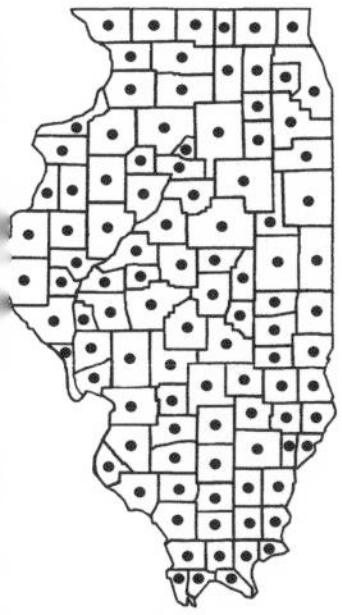

Common Name: Lamb's Quarters.
Habitat: Disturbed soil.
Range: Native of Europe and Asia; adventive throughout North America.
Illinois Distribution: Common; in every county.

Many botanists consider *C. album* to be a highly variable species that has been split unjustifiably into a number of segregate species, some or all of which include *C. lanceolatum, C. opulifolium, C. berlandieri, C. bushianum,* and *C. macrocalycium.* All of these species are alike in that each has the calyx completely covering the fruit and each has the pericarp firmly adherent to the seed. The only other Illinois species that has these two characteristics and is not in the *C. album* complex is *C. pallescens. Chenopodium pallescens* is readily distinguished by its thick, linear, entire leaves.

I recognize *C. album, C. lanceolatum, C. opulifolium, C. berlandieri, C. bushianum,* and *C. macrocalycium* as distinct species. The last five differ from *C. album* as follows: *C. lanceolatum* has lanceolate leaves, less densely white-mealy stems and leaves, fewer and lower teeth on the leaf margins, smaller seeds with a whitish rather than yellowish margin, and interrupted spikes; *C. opulifolium* has leaves as broad as long, a calyx that is split halfway to the base, and completely smooth seeds; *C. berlandieri* has smaller seeds that are reticulate, calyx lobes with a sharp keel, leaves that are never 3-lobed, and a fetid odor; *C. bushianum* has seeds that are reticulate, larger, and dull, leaves that are more sparsely white-mealy, and calyx lobes with a sharp keel.

Chenopodium macrocalycium differs from *C. bushianum* by its narrower leaves and erect inflorescences.

Instead of lumping together several obviously diverse species, I choose to try and sort them out as best as possible. Although there are some questionable specimens in Illinois, most of them may be determined using the characters noted above.

When young, the leaves and stems of *C. album* may be cooked and eaten as a vegetable. Its edibility is probably why early immigrants brought the plant to this country.

Chenopodium album flowers from May to October.

14. Chenopodium rubrum L. Sp. Pl. 218. 1753. Fig. 35.
Blitum rubrum (L.) Reichb. Fl. Germ. Exc. 582. 1830–32.

Annual herb from slender taproots; stems erect, branched or unbranched, to 75 cm tall, glabrous, green to reddish brown, not white-mealy; leaves rhombic to ovate, acuminate at the tip, cuneate at the base, to 9 cm long, to 6 cm broad, coarsely toothed to even hastate, rather thin, 3-veined, finely white-mealy on the lower surface, otherwise glabrate, the uppermost narrower and entire, the petioles at least half as long as the blade; flowers in glomerules in axillary spikes forming a panicle; calyx more or less fleshy, glabrous or sparsely white-mealy, barely enclosing the fruit; fruit with the pericarp weakly attached to the seed; seeds vertical and horizontal, 0.8–1.0 mm broad, smooth, shiny, with sharp margins.

Common Name: Coast Blite.
Habitat: Disturbed soil.
Range: Newfoundland to Washington, south to California, Arizona, New Mexico, Nebraska, Missouri, and New Jersey; adventive in Illinois.
Illinois Distribution: Known only from Cook and Peoria counties.

This species has leaves that are coarsely toothed or sometimes hastate, a fleshy calyx that is sometimes sparsely white-mealy, and both vertical and horizontal seeds.

Chenopodium rubrum flowers from July to October.

15. Chenopodium standleyanum Aellen in Fedde, Rep. Sp. Nov. 26:153. 1929. Fig. 36.
Chenopodium boscianum Moq. Enum. Chen. 21. 1840, misapplied.

Annual herb with slender taproots; stems ascending to erect, slender, usually branched, to 1.5 m tall, usually glabrous, rarely sparsely white-mealy; leaves lanceolate to narrowly ovate, acute at the apex, cuneate at the base, to 8 cm long, to 1 cm broad, entire or sparsely serrulate, thin, 3-veined, glabrous or sparsely white-mealy, the petioles usually about half as long as the blade; flowers single or in small glomerules forming short, interrupted spikes; calyx white-mealy, the lobes obtuse, barely covering the margins of the fruit; fruit with the pericarp easily separated from the seed; seeds horizontal, 1.0–1.5 mm broad, mostly smooth, shiny, black.

Common Name: Goosefoot.
Habitat: Woodlands; roadsides.
Range: Quebec to North Dakota, south to Texas and Georgia.
Illinois Distribution: Scattered throughout the state.

Chenopodium standleyanum has a calyx that barely covers the fruit and seeds that are easily separated from the pericarp. Its thin, lanceolate to narrowly ovate leaves are entire or sparsely serrulate.

Chenopodium boscianum, which is an earlier binomial, was misapplied to this species.

35. **Chenopodium rubrum**
a. Flowering branch
b. Flowering branch
c. Flower with 5-lobed calyx
d. Flower with 3-lobed calyx
e. Fruit
f. Seed

36. **Chenopodium standleyanum**
a. Flowering branch
b. Habit
c. Flowering branch
d. Flower
e. Seed

This is one of the few native species of *Chenopodium* in Illinois, often occurring in woodlands.

This species flowers from June to October.

16. Chenopodium bonus-henricus L. Sp. Pl. 218. 1753. Fig. 37.

Perennial herb from thickened rhizomes; stems stout, erect, to 75 cm tall, dark green, glabrous, rarely sparsely white-mealy, unbranched or sparsely branched; leaves broadly triangular to hastate, acute to acuminate at the apex, with acute basal lobes, to 15 cm long, nearly as broad, entire or shallowly sinuate, fleshy, glabrous, not white-mealy, some of the petioles as much as twice as long as the blade; flowers in dense glomerules in short spikes forming a panicle; calyx green, not white-mealy, the lobes obtuse, not keeled, not completely enclosing the fruit; fruit with the seed free from the pericarp; seeds vertical, or the uppermost horizontal, 1.3–1.5 mm broad, smooth, shiny, black, with rounded margins.

Common Name: Good King Henry.
Habitat: Disturbed soil.
Range: Native to Europe; sparingly adventive in the United States.
Illinois Distribution: Known only from Cook and DuPage counties.

Chenopodium bonus-henricus is distinguished by its lack of white-mealiness, its traingular to hastate leaves, and the predominance of vertical seeds.

This species flowers from July to October.

17. Chenopodium glaucum L. Sp. Pl. 220. 1753. Fig. 38.

Annual herb with slender taproots; stems prostrate to erect, to 50 cm long, branched, white-mealy, becoming reddish or purplish; leaves oblong to narrowly ovate, obtuse at the apex, cuneate at the base, to 5 cm long, to 1.5 cm broad, the margins sinuate-dentate or at least sinuate-angled, thin, pale green, 3-veined, densely white-mealy on the lower surface, otherwise glabrate; flowers in dense glomerules in interrupted axillary and terminal spikes; calyx glabrous, not white-mealy, with obtuse lobes, not keeled, partially enclosing the fruit; fruit with pericarp free from the seed; seeds vertical or horizontal, 0.5–0.6 mm broad, smooth, dark red-brown, the margins acute.

Common Name: Oak-leaved Goosefoot.
Habitat: Disturbed soil.
Range: Native to Europe and Asia; scattered as an adventive in the United States.
Illinois Distribution: Scattered throughout the state but not common.

The distinguishing features of this species are the calyx only partially enclosing the fruit, the vertical seeds that are the smallest in the genus, the completely glabrous calyx, and the densely white-mealy undersurface of the leaves.

Chenopodium glaucum flowers from July to September.

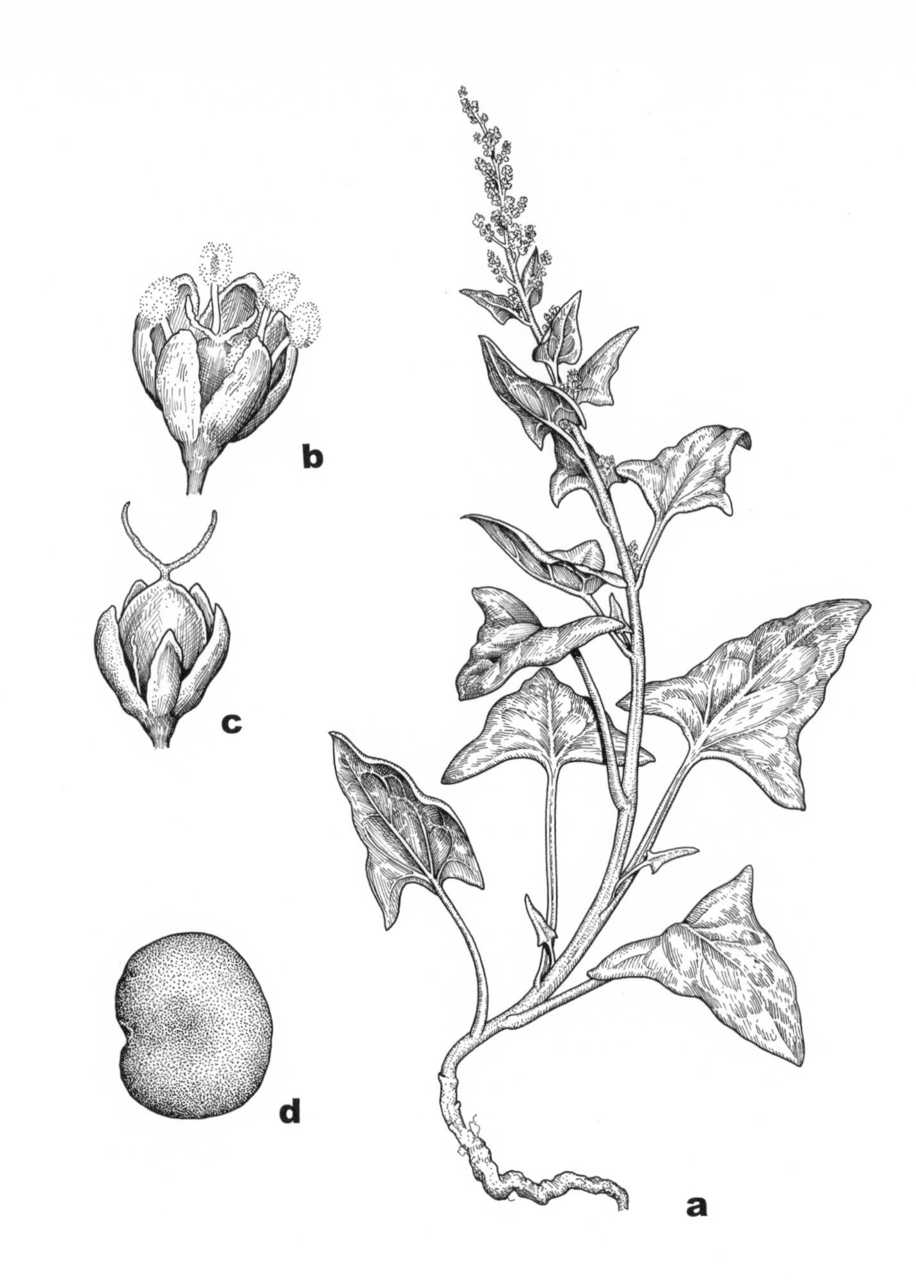

37. **Chenopodium bonus-henricus**

a. Habit
b. Flower
c. Young fruit
d. Seed

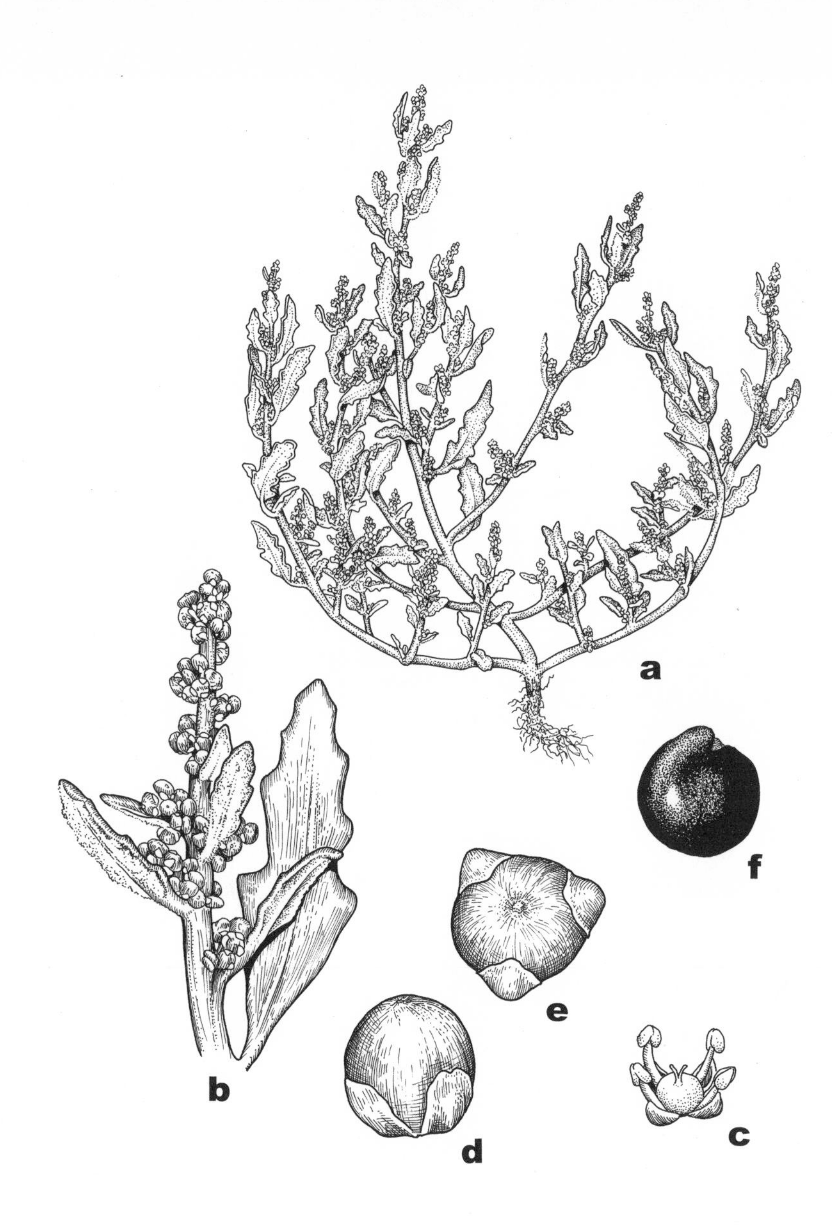

38. **Chenopodium glaucum**
a. Habit
b. Flowering branch
c. Flower
d. Vertical fruit
e. Horizontal fruit
f. Seed

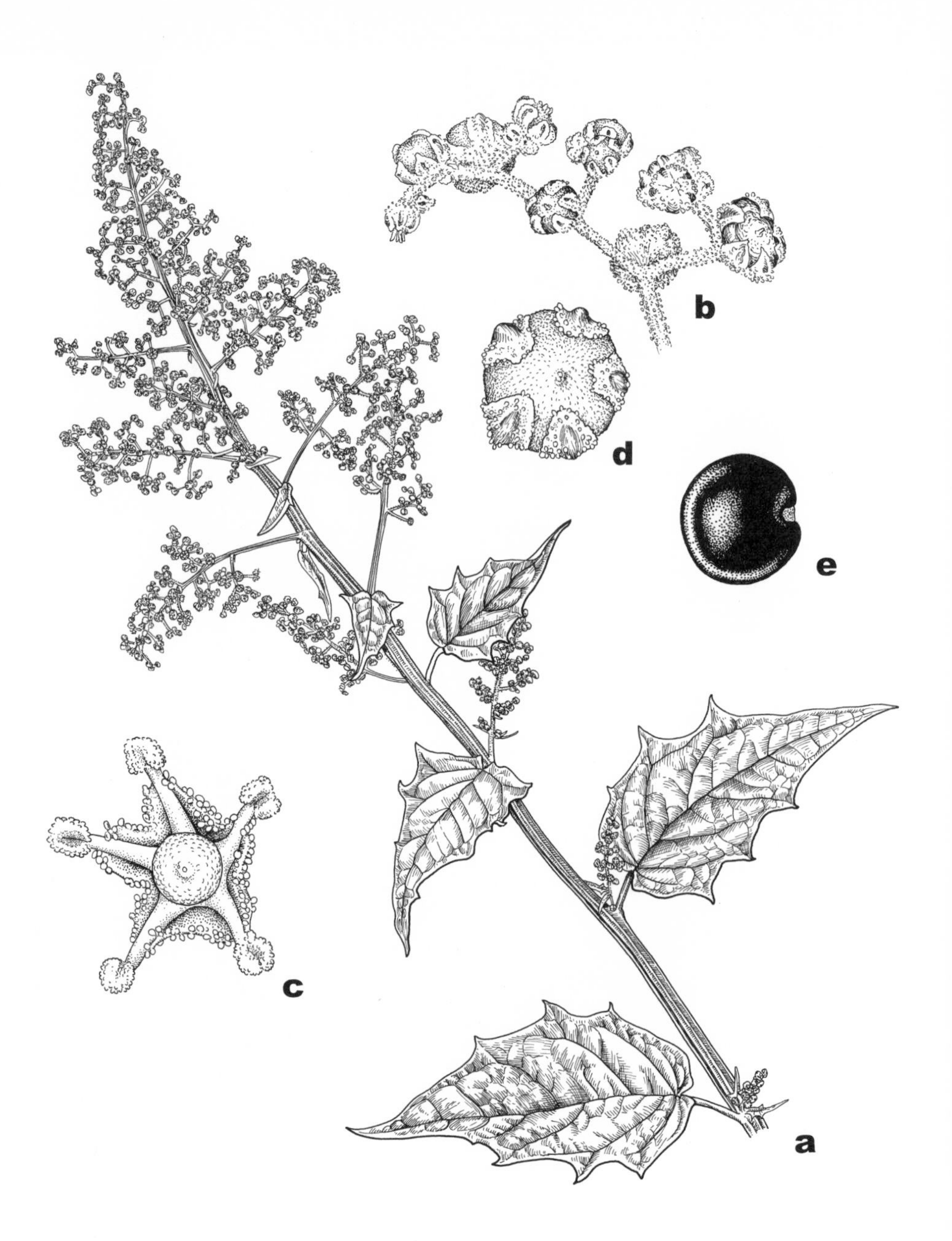

39. **Chenopodium simplex**
a. Flowering branch
b. Flowering branch
c. Flower
d. Fruit with pericarp
e. Seed

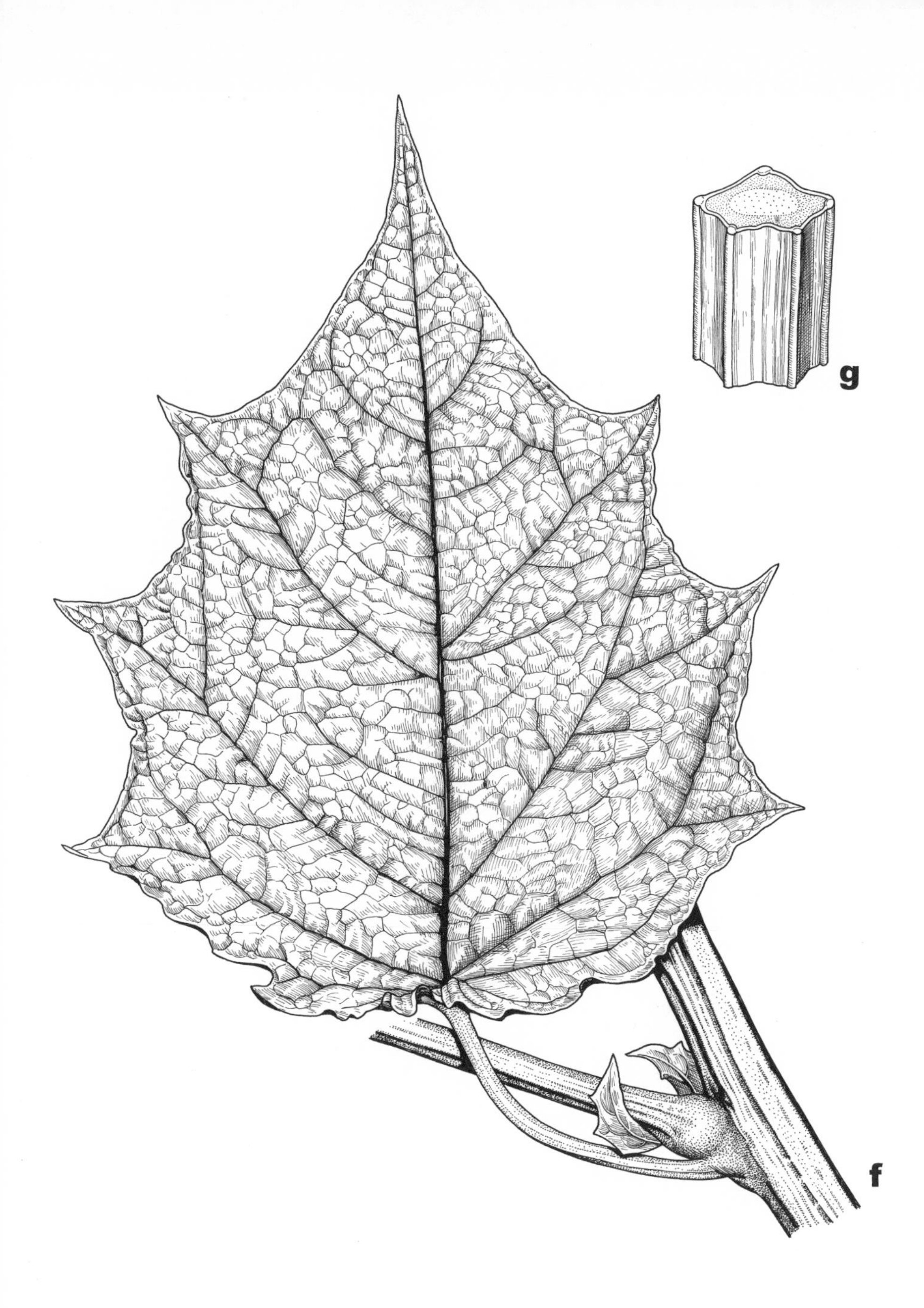

f. Leaf

g. Section of stem

18. Chenopodium simplex (Torr.) Raf. Atl. Journ. 1:146. 1836. Fig. 39.
Chenopodium hybridum L. Sp. Pl. 219. 1753, misapplied.
Chenopodium hybridum L. var. *simplex* Torr. Ann. Lyc. N.Y. 2:239. 1827.
Chenopodium gigantospermum Aellen in Fedde, Rep. Sp. Nov. 26:144. 1929.
Chenopodium hybridum L. var. *gigantospermum* (Aellen) Rouleau, Nat. Can. 71:268. 1944.

Annual herb from taproots; stems erect, branched, to 1.2 m tall, glabrous, green, not white-mealy; leaves ovate to deltate-ovate, long-acuminate at the apex, rounded to subcordate at the base, to 20 cm long, often nearly as broad, with up to 4 large teeth on each margin, thin, 3- to 5-veined, glabrous, green, never white-mealy; flowers in large panicles; calyx green, not white-mealy, the lobes obtuse, slightly keeled, only partially enclosing the fruit; fruit with a thin, reticulate pericarp that is attached to or easily separated from the seed; seeds horizontal, 1.8–2.0 mm broad, punctate, shiny, black.

Common Name: Maple-leaved Goosefoot.
Habitat: Shaded ledges; rocky woods.
Range: Quebec to British Columbia, south to California, New Mexico, and Virginia.
Illinois Distribution: Occasional throughout the state.

This species is easily recognized because of its very large leaves that usually have four large teeth on each margin, its large seeds (1.8–2.0 mm broad), and the complete absence of white-mealiness.

The nomenclature for this species is complicated, but *C. simplex* appears to be the correct binomial.

Chenopodium simplex is a native species of rocky woodlands. It also frequents shaded ledges.

This species flowers from June to October.

19. Chenopodium polyspermum L. Sp. Pl. 220. 1753. Fig. 40.

Annual herb from slender taproots; stems prostrate to spreading to erect, branched, to 80 cm tall, green, not white-mealy; leaves oblong to oval, obtuse at the apex, rounded or tapering at the base, to 6.5 cm long, to 4.5 cm broad, entire, rather thick, 3-veined, dull, green, never white-mealy, the petioles about half as long as the blade; flowers in dense glomerules forming dense cymes or spikes nearly from the base of the plant to the top; calyx green, never white-mealy, subacute or obtuse, not keeled, not completely covering the fruit; fruit with pericarp firmly attached to the seed; seeds horizontal, 0.8–1.2 mm broad, shiny, smooth, black, the margins rounded.

Common Name: Many-seeded Goosefoot.
Habitat: Disturbed soil along a railroad (in Illinois).
Range: Native to Europe; sparingly adventive in the United States.
Illinois Distribution: Known only from Jackson County.

40. **Chenopodium polyspermum**
a. Habit
b. Habit
c. Flower
d. Fruits
e. Seed

The distinguishing features of this species are the absence of white-mealiness, the rather thickish oblong to oval, entire leaves, and the inflorescence that usually begins near the base of the plant and extends to the top.

Although Kibbe (1952) reported this species from Hancock County, I have seen no specimens to verify this. The only extant collection was made along a railroad in Murphysboro, Jackson County, in 1955.

Chenopodium polyspermum flowers from June to October.

20. Chenopodium strictum Roth var. **glaucophyllum** (Aellen) Wahl, Bartonia 27: 38. 1954. Fig. 41.
Chenopodium glaucophyllum Aellen in Fedde, Rep. Sp. Nov. 26:155. 1929.

Annual herb from slender taproots; stems erect, to 1 m tall, branched or unbranched, glabrous or sparsely white-mealy; leaves ovate, acute at the apex, more or less rounded at the base, to 20 mm long, to 15 mm broad, glabrous or sparsely white-mealy, thin, 3-veined, irregularly serrate or the uppermost entire; flowers in glomerules forming spikes arranged in panicles; calyx sparsely white-mealy, the lobes obtuse to subacute, weakly keeled, not enclosing the fruit; fruit with pericarp adherent to the seed; seeds horizontal, 0.9–1.1 mm broad, smooth, shiny, black.

Common Name: Erect Goosefoot.
Habitat: Disturbed soil.
Range: Quebec to British Columbia, south into the United States.
Illinois Distribution: Known only from Peoria and St. Clair counties.

This plant is sparsely white-mealy with thin leaves that are irregularly toothed.

The calyx only partially covers the fruit, and the pericarp is firmly attached to the seed.

The only Illinois collections were made during the nineteenth century.

Chenopodium strictum var. *glaucophyllum* flowers from July to September.

21. Chenopodium missouriense Aellen, Bot. Notiser 1928:206. 1928. Fig. 42.
Chenopodium album L. var. *missouriense* (Aellen) Bassett & Crompton, Can. Journ. Bot. 60:603. 1982.

Annual herb from taproots; stems erect, branched, to 1.5 m tall, white-mealy to glabrate; leaves rhombic to ovate, obtuse to acute at the apex, more or less rounded at the base, to 6 cm long, to 4 cm broad, all but the uppermost coarsely toothed, usually 3-veined, white-mealy to glabrate; flowers in dense glomerules; calyx sparsely white-mealy, the lobes acute, strongly keeled, covering most of the fruit; fruit with pericarp closely adhering to the seed; seeds horizontal, 0.9–1.4 mm broad, puncticulate.

Common Name: Missouri Goosefoot.
Habitat: Fields, disturbed soil.

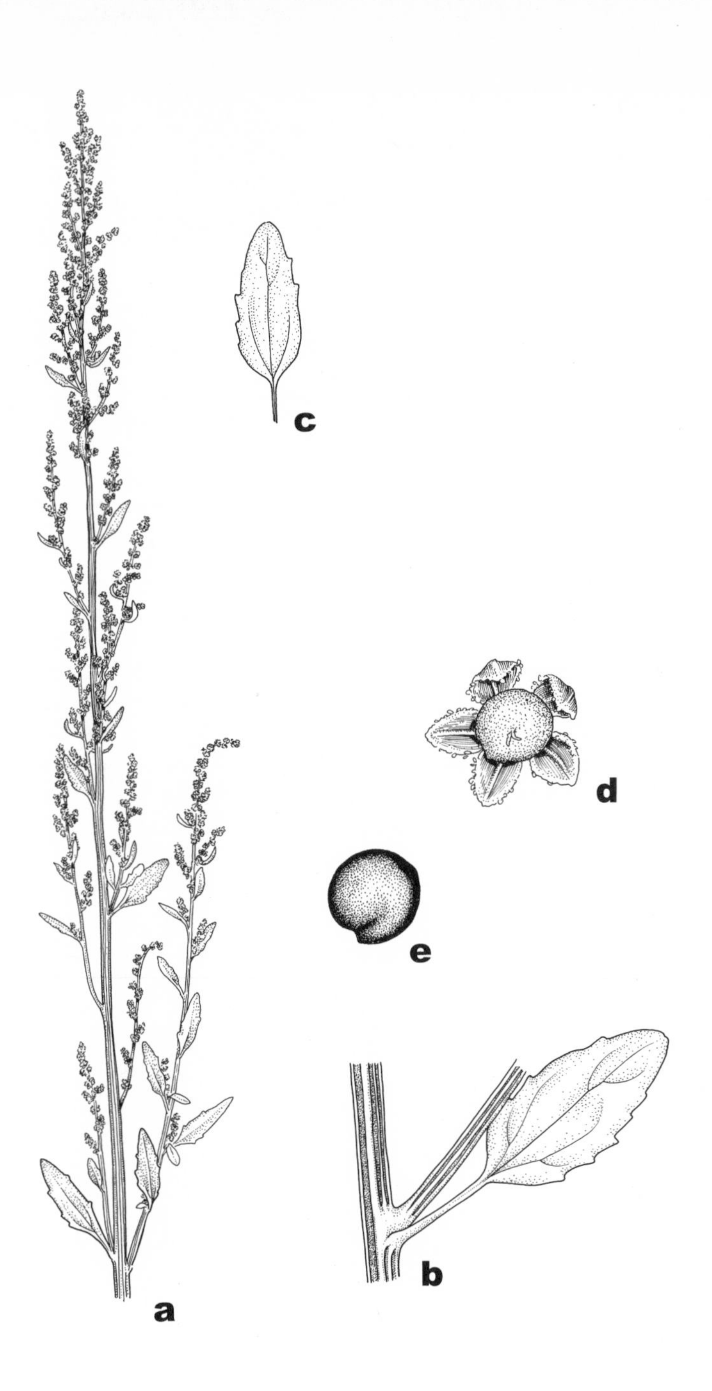

41. **Chenopodium strictum** var. **glaucophyllum**
a. Habit
b. Leaf and node
c. Leaf
d. Flower
e. Seed

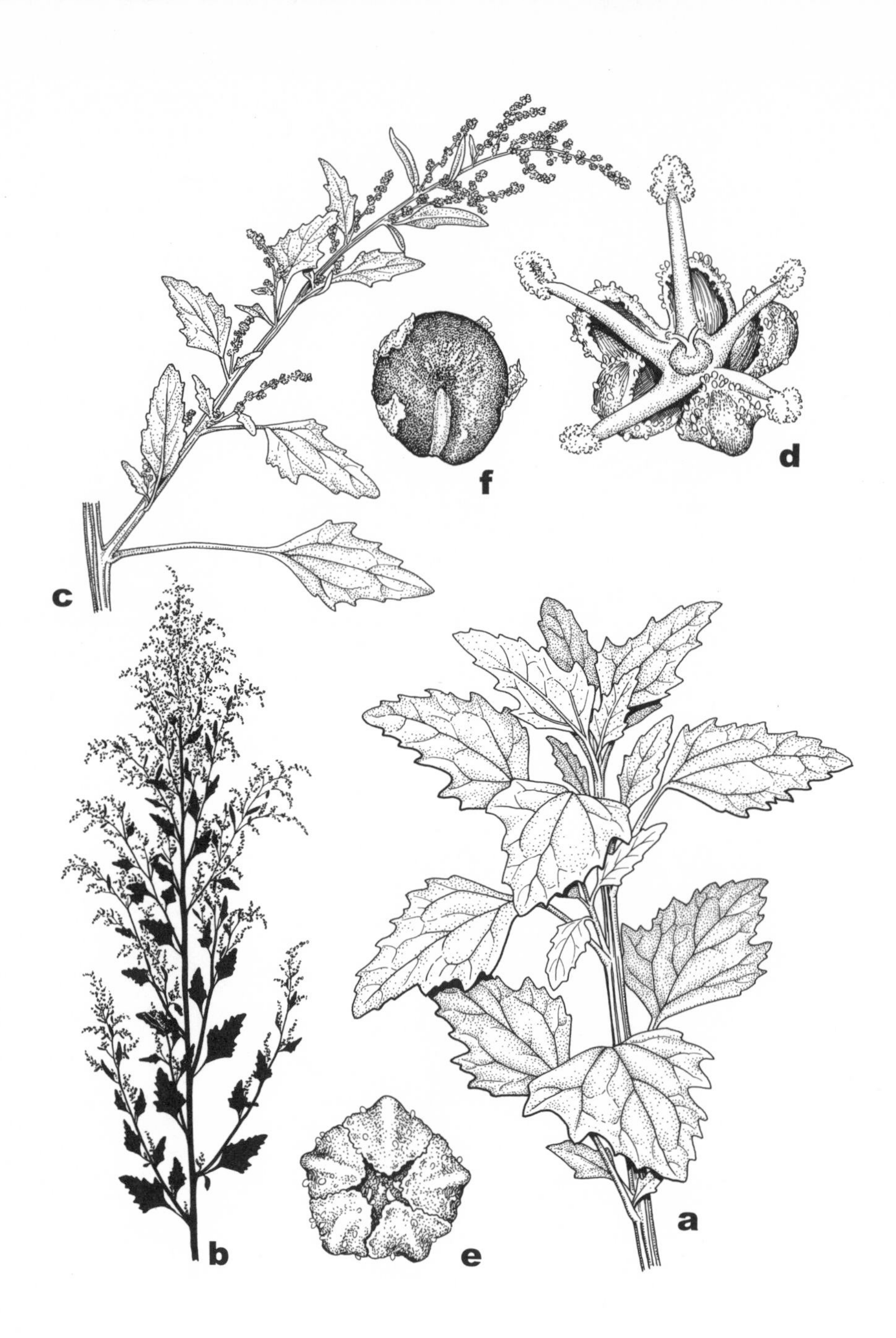

42. **Chenopodium missouriense**
a. Young branch
b. Habit
c. Flowering branch
d. Flower
e. Seed with pericarp
f. Seed with pericarp

Range: New England to Michigan, south to Texas and Illinois; Mexico.
Illinois Distribution: Scattered throughout the state.

Although *C. missouriense* is sometimes placed in either *C. album* or *C. berlandieri* or is considered to be a variety of *C. album*, it differs from all other members of the *C. album* complex except *C. foggii* in its calyx that only partially covers the fruit. It is similar to *C. berlandieri* by virtue of its sharply keeled calyx lobes. However, *C. missouriense* does not have the fetid odor that *C. berlandieri* has. *Chenopodium foggii* differs in its entire or nearly entire leaves and its rugulate seeds.

Chenopodium missouriense flowers during September and October.

22. Chenopodium foggii Wahl, Bartonia 27:19. 1954. Fig. 43.

Annual herb from taproots; stems erect, branched, to 1.2 m tall, sparsely white-mealy to glabrate; leaves rhombic to ovate, obtuse to acute at the apex, more or less rounded at the base, to 5.5 cm long, to 3.5 cm broad, entire or with 1–2 pairs of teeth on each margin, thin, usually 3-veined, sparsely white-mealy to glabrate; flowers in dense glomerules; calyx sparsely white-mealy, the lobes acute, strongly keeled, covering most of the fruit; fruit with pericarp closely adhering to the seed; seeds horizontal, 0.9–1.2 mm broad, rugulate.

Common Name: Fogg's Goosefoot.
Habitat: Fields.
Range: Quebec to Minnesota, south to Oklahoma and Arkansas.
Illinois Distribution: Known from Mason and Will counties.

This species is usually included within *C. album*, but it differs from *C. album* in that the calyx does not completely cover the fruit and its seeds are rugulate. Its leaves are also smaller and without teeth. It appears to be most closely related to *C. missouriense* but differs by its usually entire leaves and its rugulate seeds.

Chenopodium foggii flowers from August to October.

23. Chenopodium urbicum L. Sp. Pl. 218. 1753. Fig. 44.

Annual herb from slender taproots, more or less fleshy; stems ascending to erect, branched or unbranched, to 75 cm tall, glabrate or very sparsely white-mealy; leaves rhombic to ovate, obtuse or acute at the apex, cuneate to truncate at the base, to 7.5 cm long, to 5 cm broad, coarsely sinuate-dentate, thin, 3-veined, shiny, usually densely white-mealy on the lower surface, otherwise glabrate, the petioles sometimes nearly as long as the blade; flowers in small glomerules in axillary or terminal interrupted spikes; calyx green, rarely white-mealy, the lobes obtuse, scarcely keeled, only partly enclosing the fruit; fruit with the pericarp adhering to the seed; seeds horizontal, 0.9–1.0 mm broad, bluntly angular, shiny, smooth, black.

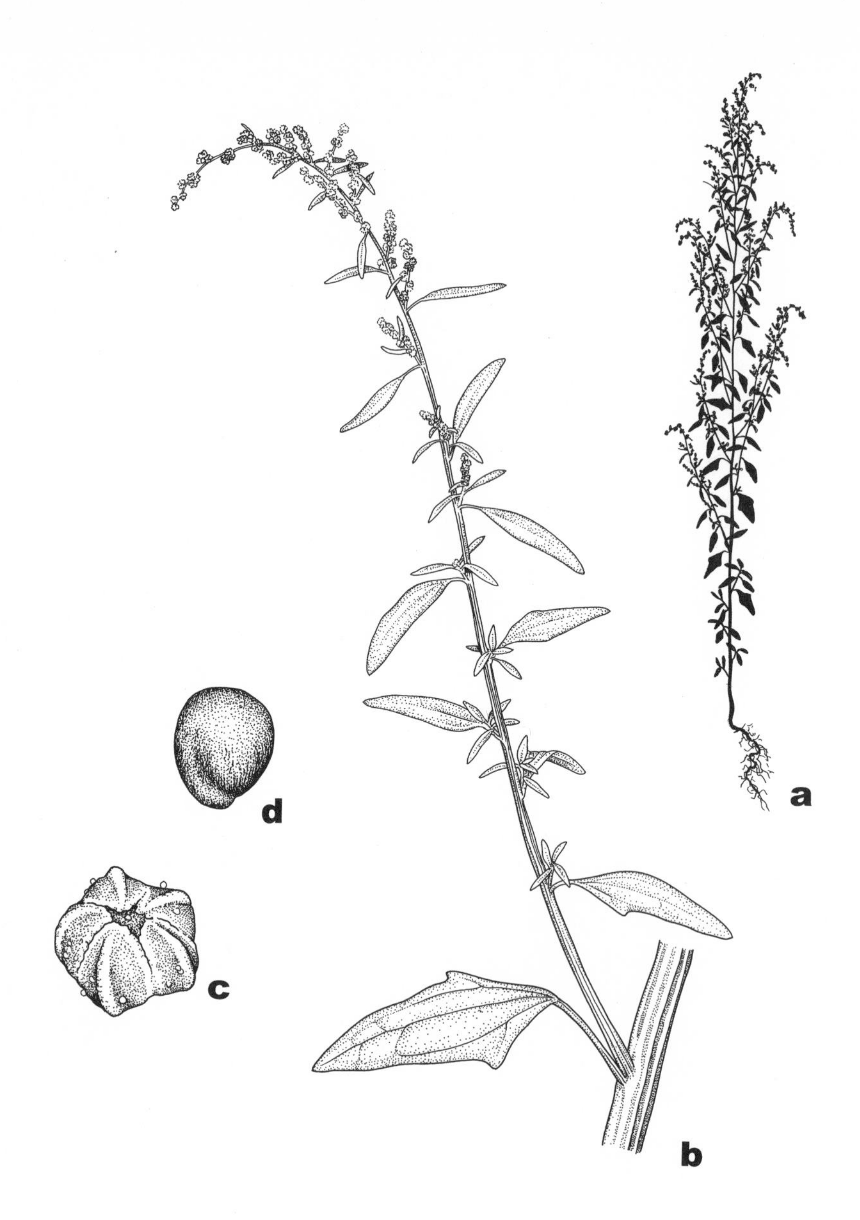

43. **Chenopodium foggii**
a. Habit
b. Habit
c. Fruit
d. Seed

44. **Chenopodium urbicum** b. Flower c. Seed
a. Flowering branch

Common Name: City Goosefoot.
Habitat: Disturbed soil.
Range: Native to Europe; widely scattered as an adventive in the United States, particularly around metropolitan areas.
Illinois Distribution: Scattered but not common in all but the southernmost counties.

Chenopodium urbicum is distinguished by its shiny, rhombic to ovate leaves that are sharply sinuate-dentate; by its nearly glabrous calyx lobes that are only weakly keeled; and by its smooth shiny seeds that are about 0.9–1.0 mm broad. The very similar *C. murale* has puncticulate dull seeds that are 1.2–1.5 mm broad.

This adventive species was reported from Illinois as early as 1860 by George Vasey.

Chenopodium urbicum flowers from July to October.

24. Chenopodium murale L. Sp. Pl. 219. 1753. Fig. 45.

Annual herb from slender taproots, more or less fleshy; stems spreading to ascending to erect, branched or unbranched, to 50 cm tall, glabrate or very sparsely white-mealy; leaves ovate, obtuse or acute at the apex, cuneate to truncate at the base, to 7.5 cm long, to 5 cm broad, coarsely sharply sinuate-dentate, thin, 3-veined, shiny, usually densely white-mealy on the lower surface, otherwise glabrate, the petiole sometimes nearly as long as the blade; flowers in small glomerules in axillary and terminal spreading cymes or panicles; calyx green, rarely white-mealy, the lobes obtuse, scarcely keeled, only partially enclosing the fruit; fruit with the pericarp adherent to the seed; seeds horizontal, 1.2–1.5 mm broad, sharply angular, dull, puncticulate, black.

Common Name: Nettle-leaved Goosefoot.
Habitat: Disturbed soil.
Range: Native to Europe; sparingly adventive in the United States.
Illinois Distribution: Scattered but not common throughout the state.

This species is very similar to *C. urbicum* but differs in its dull, puncticulate seeds that are 1.2–1.5 mm broad.

Chenopodium murale flowers from June to November.

6. **Monolepis** Schrad.—Poverty-weed

Annual herbs; stems decumbent to erect, sometimes fleshy; leaves alternate, simple, much reduced on the upper part of the stem; flowers borne in clusters of 1–15 in the axils of the leaves, with leaflike bracts; flowers perfect or pistillate; calyx lobes 1–3, persistent, bractlike; stamen 1 (absent in pistillate flowers); ovary superior, 1-locular; styles 2; fruit a 1-seeded utricle; pericarp free or adherent to the seed; seeds vertical.

Three species comprise the genus. All are native to the western United States. Only the following species is known from Illinois.

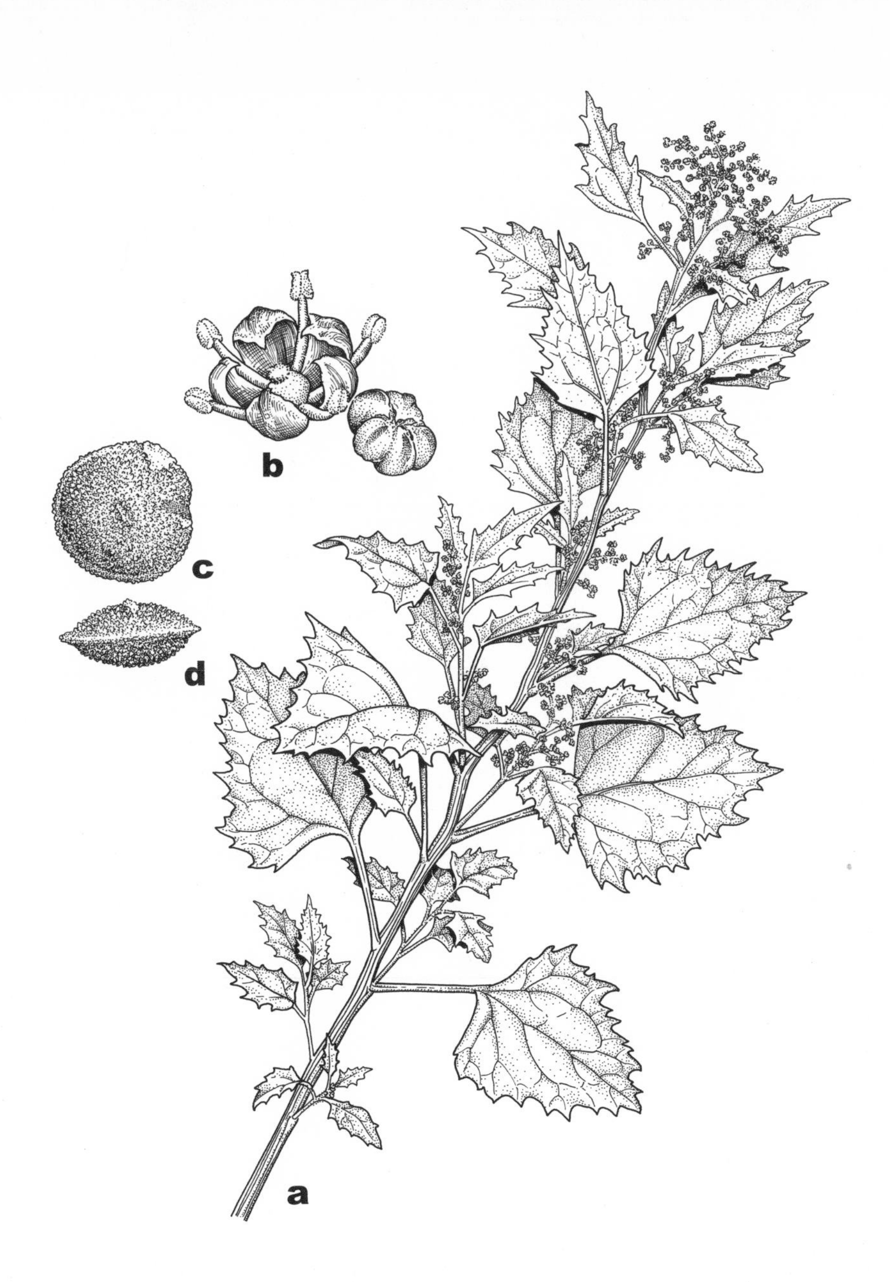

45. **Chenopodium murale**
a. Flowering branch
b. Flower
c. Seed
d. Seed

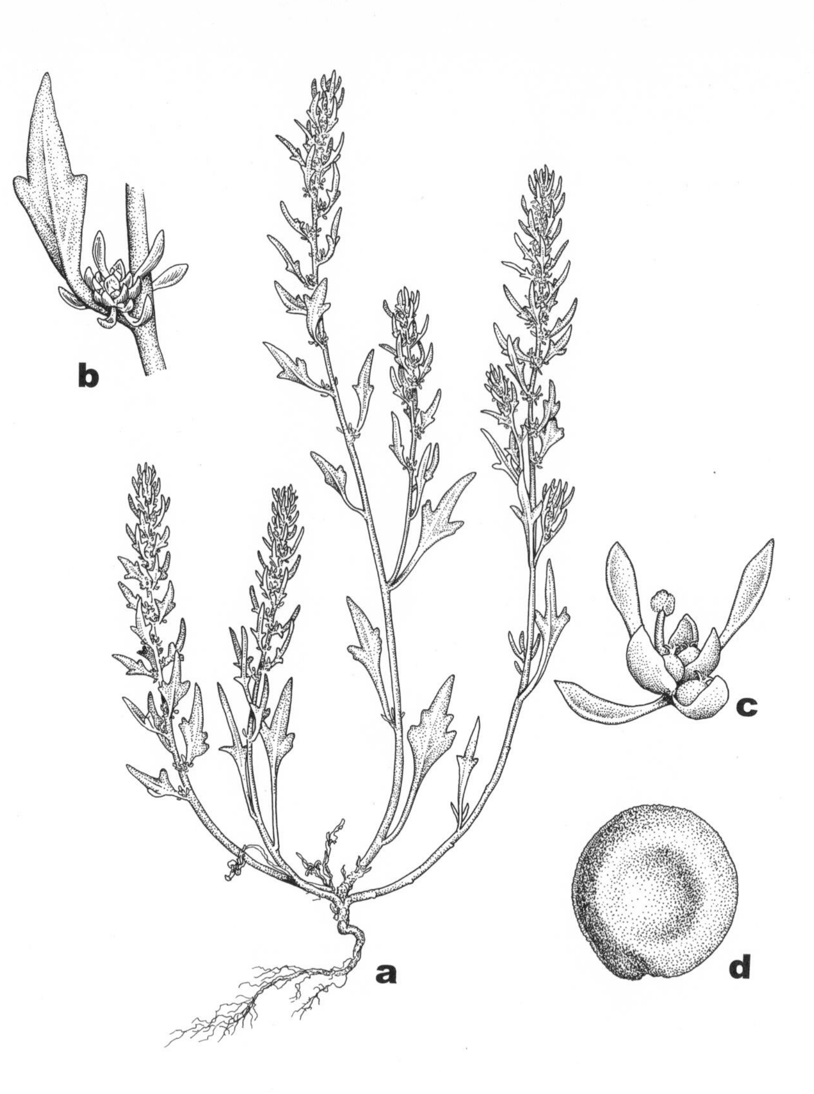

46. **Monolepis nuttalliana**
a. Habit
b. Flowering node
c. Flower cluster with single sepal and bracts
d. Seed

1. Monolepis nuttalliana (Roemer & Schultes) E. Greene, Fl. Fran. 168. 1891. Fig. 46.
Blitum nuttallianum Roemer & Schultes, Mant. 1:65. 1822.

Annual herb from a slender taproot; stems decumbent to ascending, succulent, to 30 cm tall, much branched, white-mealy when young, becoming glabrate; leaves fleshy, triangular to lanceolate, to 6 cm long, acute at the apex, tapering to nearly truncate at the base, white-mealy when young, becoming glabrate, entire or sparsely toothed or hastate, sessile or on petioles to 2 mm long; flowers in dense, sessile, axillary clusters, subtended by usually red bracts at maturity; sepal 1, persistent, bractlike, more or less fleshy, oblanceolate to spatulate, not enclosing the fruit; fruit 1–2 mm broad, the pericarp pitted, firmly attached to the seed; seeds vertical, about 1 mm in diameter, compressed, dark brown to black.

Common Name: Poverty-weed.
Habitat: Disturbed dry, alkaline soil.
Range: Native to the western United States; scattered as an adventive in the eastern United States.
Illinois Distribution: Known only from Grundy, McHenry, and Sangamon counties.

This species is distinguished by its fleshy leaves and stems that are white-mealy at first, its single sepal, and its often hastate leaves. It is reminiscent of a *Chenopodium* by virtue of its white-mealiness.

The collections from Grundy and McHenry counties were made at the county fairgrounds.

The original Illinois collection was made in Sangamon County.

Monolepis nuttalliana flowers from June to October.

7. **Cycloloma** Moq.—Winged Pigweed

Annual herb; stems erect; leaves alternate, simple; flowers perfect or pistillate, sessile and borne singly in paniculate spikes, without bracts; calyx composed of 5 broadly winged sepals, completely covering the fruit; petals 0; stamens 5; ovary superior, 1-locular; styles 2–3; fruit a 1-seeded utricle; pericarp free from the seed; seeds horizontal.

Only the following species comprises the genus.

1. Cycloloma atriplicifolium (Spreng.) Coult. Mem. Torrey Club 5:143. 1894. Fig. 47.
Salsola atriplicifolia Spreng. Nachtr. Bot. Gart. Halle 1:35. 1803.
Salsola platyphylla Michx. Fl. Bor. Am. 1:174. 1803.
Cycloloma platyphyllum (Michx.) Moq. Enum. Chen. 18. 1842.

Annual herb from a taproot; stems up to 75 cm tall, spreading to erect, much branched, usually cobwebby-pubescent, becoming glabrate; leaves lanceolate to oblong to ovate, acute at the apex, cuneate at the base, to 7.5 cm long, 6–15 mm broad, sinuate-dentate, cobwebby-pubescent to glabrate, sessile or on petioles up to 15 mm long; flowers in paniculate spikes, without bracts; calyx up to 5 mm in diameter, 5-parted, each lobe broadly hyaline-winged and serrulate, completely

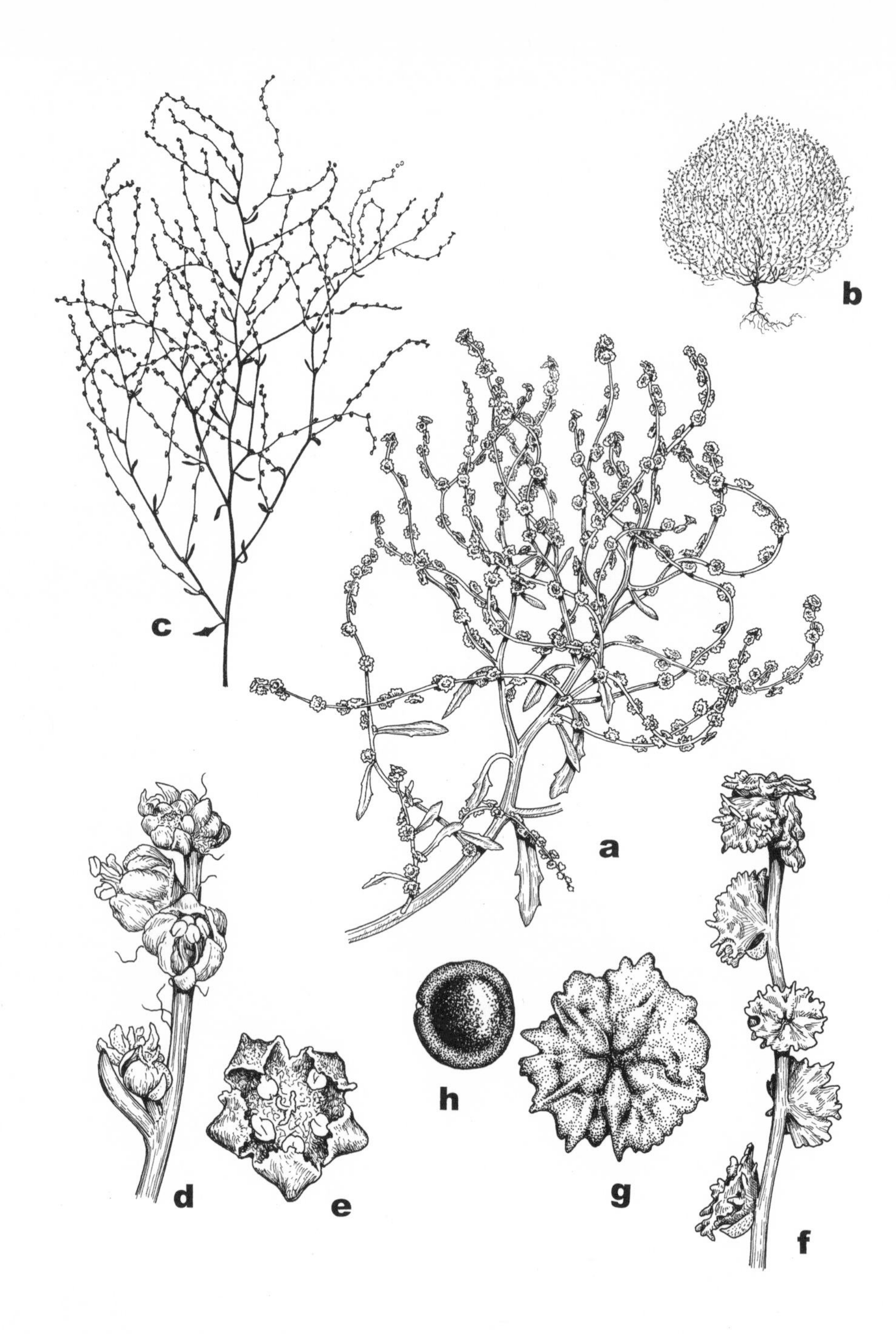

47. **Cycloloma atriplicifolium**
a. Flowering branch
b. Habit
c. Flowering branch
d. Flowering branch
e. Flower
f. Fruiting branch
g. Fruit
h. Seed

covering the fruit; fruit a depressed-globose, 1-seeded utricle, 1.8–2.0 mm in diameter; pericarp free from the seed; seeds horizontal, 1.3–1.6 mm broad, flat, smooth, black, shiny.

Common Name: Winged Pigweed.
Habitat: Dry, sandy soil, often along rivers.
Range: Manitoba south to Arizona, Texas, and Arkansas; adventive in the eastern United States; also in Mexico.
Illinois Distribution: Scattered throughout the state.

This species occurs in extremely sandy soil. The entire plant may be uprooted during the summer or autumn and tumble along, being blown by the wind.

This species is unique in the family because the fruit is surrounded by the horizontal, broadly white-hyaline calyx.

Cycloloma atriplicifolium flowers from July to October.

8. **Atriplex** L.—Orache; Saltbush

Annual or perennial herbs (in Illinois), or shrubs; stems usually scurfy-scaly; leaves alternate or opposite, simple, variously toothed or entire; flowers unisexual, the plants monoecious or dioecious; staminate flowers in spikes or spherical glomerules, without bracts; pistillate flowers in spikes or spikelike panicles or spherical glomerules, with 2 bracts per flower; staminate flower consisting of a 3- to 5-lobed calyx and 3–5 stamens; pistillate flower consisting of usually no calyx, a 1-locular, superior ovary, and 2 styles; petals 0; fruit a 1-seeded utricle; pericarp membranaceous, free from the seed; seeds usually vertical.

This is a genus of primarily saline or alkaline soils found in most dry regions of the World. There are about 250 species worldwide.

The species in Illinois may be distinguished by the following key:

1. All leaves alternate.
    2. Plants silvery-pubescent; leaves entire or nearly so . . . . 1. *A. argentea*
    2. Plants hoary-mealy; leaves sinuate or dentate . . . . 2. *A. rosea*
1. At least the lower leaves opposite.
    3. Bracts subtending the fruits orbicular, thin, reticulate; pistillate flower with a calyx . . . . 3. *A. hortensis*
    3. Bracts subtending the fruits deltate to rhombic, thick, smooth or tuberculate but scarcely reticulate; pistillate flower with no calyx.
        4. Bracteoles subtending the fruit 5–12 mm long, smooth or sparsely tuberculate . . . . 4. *A. glabriuscula*
        4. Bracteoles subtending the fruit up to 5 mm long, distinctly tuberculate (except in *A. patula*).
            5. Lowest leaves linear, usually entire . . . . 5. *A. littoralis*
            5. Lowest leaves lanceolate to ovate, usually hastate, entire or dentate.
                6. Leaves lance-hastate, entire or very sparsely toothed . . . . 6. *A. patula*
                6. Leaves deltate-hastate to oval-hastate, dentate . . . . 7. *A. prostrata*

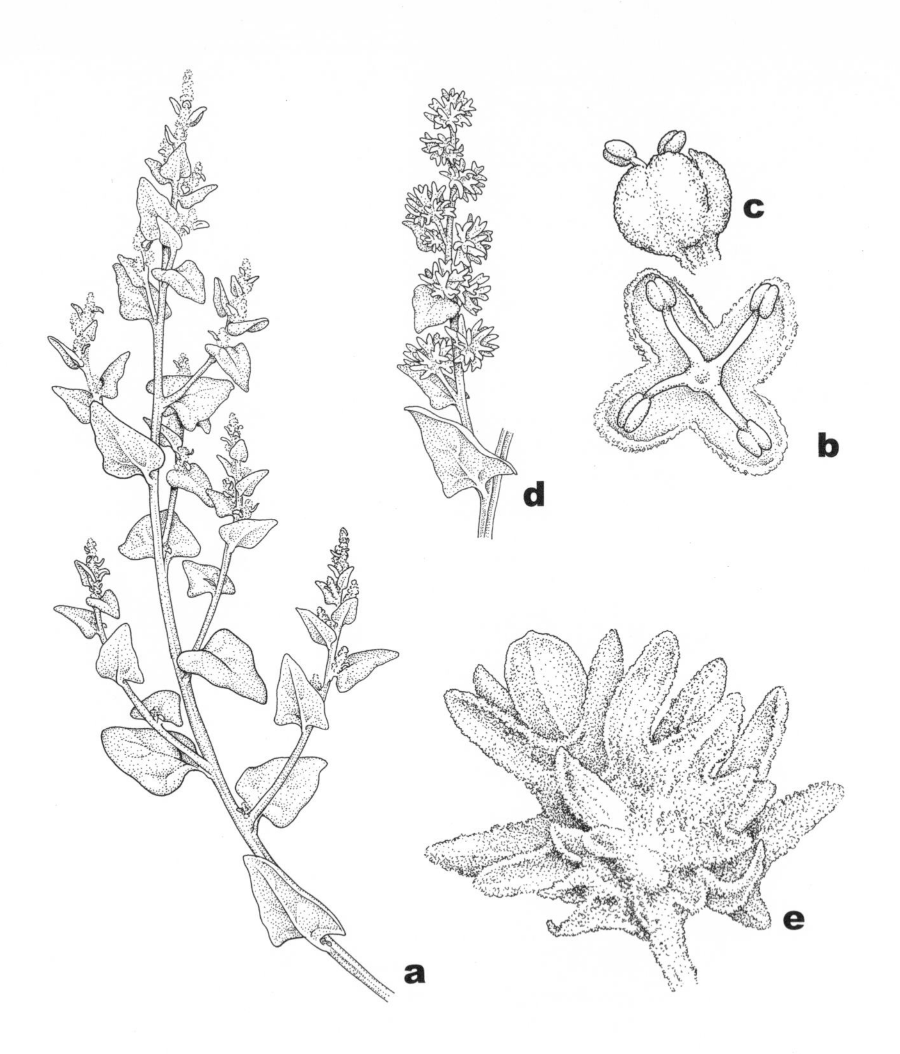

48. **Atriplex argentea**
a. Habit
b. Staminate flower, face view
c. Staminate flower, side view
d. Fruiting branch
e. Fruiting bracts

1. Atriplex argentea Nutt. Gen. 1:198. 1818. Fig. 48.

Annual monoecious herb from a taproot; stems erect, to 60 cm tall, branched, silvery-pubescent, at least when young; leaves alternate, deltate-ovate to ovate, acute at the apex, rounded at the base, entire or sparingly dentate, silvery-pubescent, at least when young, to 5 cm long, to 4 cm broad, sessile or petiolate; staminate flowers usually in short, dense spikes; pistillate flowers usually in capitate, axillary clusters; bracts of fruits 4–8 mm long, nearly as broad, obovate, entire to laciniate; seed about 1.5 mm long, brown.

Common Name: Silver Orache.
Habitat: Disturbed soil.
Range: Native west of the Mississippi River; adventive in the eastern United States.
Illinois Distribution: Known only from Cook, Hancock, and Menard counties.

This species is recognized by its silvery-pubescent stems and leaves.

The Cook County collection by Umbach and the Menard County collection by Hall were made during the nineteenth century.

Kibbe (1952) reported this species from Hancock County, but I have not seen a specimen.

Atriplex argentea flowers from July to October.

2. Atriplex rosea L. Sp. Pl. ed. 2, 1493. 1763. Fig. 49.

Annual monoecious herb from a taproot; stems erect, to 1 m tall, usually rather stout, branched, hoary-mealy, at least when young; leaves alternate, often turning red, oval to ovate to rhombic, obtuse to acute at the apex, mucronulate, rounded or cuneate at the base, sinuate to dentate, hoary-mealy, at least when young, to 8 cm long, to 5 cm broad, petiolate; staminate and pistillate flowers in axillary glomerules; bracts of fruits 4–5 mm long, rhombic to suborbicular, short-tuberculate along the sides; seeds 1.5–2.0 mm long, dull, dark brown.

Common Name: Red Orache.
Habitat: Disturbed soil.
Range: Native to Europe and Asia; sparingly adventive in other parts of the world.
Illinois Distribution: Known from Cook, Kane, and Will counties.

The common name red orache refers to the leaves that usually turn red during the season. The hoary-mealy leaves serve to distinguish this species.

Although the Cook and Will counties collections were made during the nineteenth century, the Kane County collection was made at the south edge of Aurora during the twentieth century.

Atriplex rosea flowers from August to October.

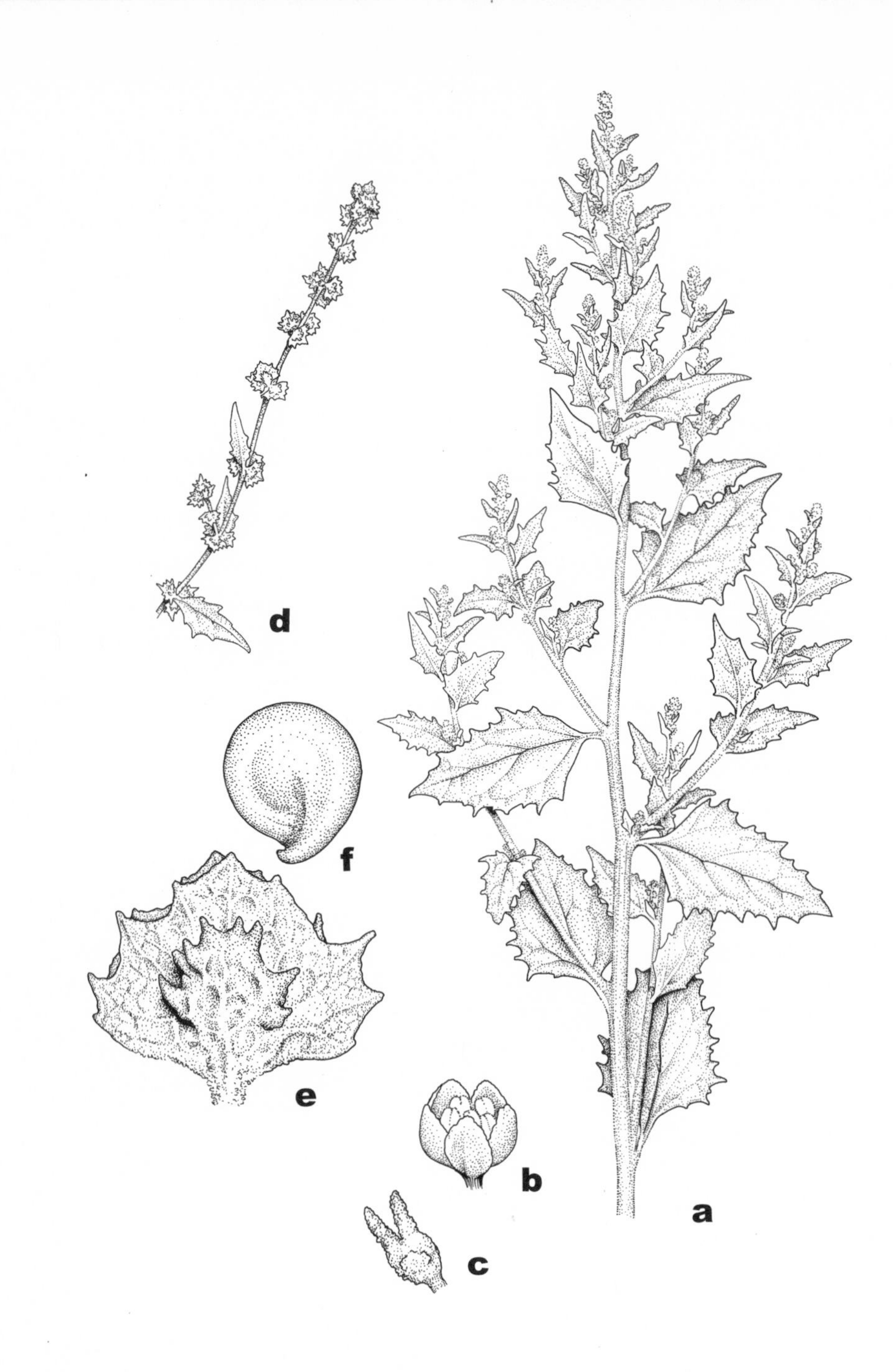

49. **Atriplex rosea**
a. Habit
b. Staminate flower
c. Portion of pistillate flower
d. Fruiting branch
e. Fruiting bract
f. Seed

3. Atriplex hortensis L. Sp. Pl. 1053. 1753. Fig. 50.

Annual monoecious herb from a taproot; stems ascending to erect, to 1.5 m tall, branched, sparsely scaly; lower leaves opposite, upper leaves alternate, deltate to ovate, acute at the apex, rounded at the base, entire to serrulate, glabrous or sparsely scaly, to 1.2 cm long, often nearly as broad, subsessile to petiolate; pistillate flowers with a 3- to 5-lobed calyx, the flowers borne in terminal spikes; staminate flowers in axillary spikes; bracts of fruits 8–18 mm long, rotund, entire, thin, reticulate; seeds 1–4 mm long, black or brown.

Common Name: Garden Orache.
Habitat: Disturbed soil.
Range: Native to Europe and Asia; escaped from cultivation in North America.
Illinois Distribution: Known from DuPage, Fayette, Lake, St. Clair, and Vermilion counties.

This garden ornamental rarely escapes from gardens into disturbed soil. There are two DuPage County collections, both made during the last decade of the nineteenth century. Glen Winterringer collected the Lake County specimen at Deerfield during the middle of the twentieth century. Harry Ahles collected the Vermilion County specimen at Danville in 1951.

This species differs from all others in the genus in Illinois by its reticulate-veined bracts that subtend the fruit.

Atriplex hortensis flowers during August and September.

4. Atriplex glabriuscula Edmonston, Fl. Shetl. 39. 1845. Fig. 51.

Annual monoecious herb from a taproot; stems erect, to 1.5 m tall, rather stout, simple or branched, glabrous at maturity; lower leaves opposite, upper leaves alternate, fleshy, deltate to ovate, acute at the apex, rounded or truncate at the base, entire or low-sinuate, glabrous at maturity, to 15 cm long, often nearly as broad, petiolate; staminate and pistillate flowers in interrupted spikes; pistillate flowers without a calyx; bracts of fruit 5–12 mm long, smooth or sparsely tuberculate; seeds 2–4 mm broad, black.

Common Name: Smooth Orache.
Habitat: In roadside gravel.
Range: Greenland to southern New England; adventive in Illinois.
Illinois Distribution: Known only from Kane County.

This fleshy species, a native of the upper Atlantic coast, was found by Dick Young at Gilberts, Kane County, in 1976.

Atriplex glabriuscula differs by its smooth or sparsely tuberculate fruiting bracts 5–12 mm long.

This species flowers during August.

50. **Atriplex hortensis**
a. Habit in flower
b. Habit in fruit
c. Staminate flower
d. Pistillate flower
e. Bracts surrounding seed
f. Seed

51. **Atriplex glabriuscula** a. Habit b. Habit

5. Atriplex littoralis L. SP. Pl. 1054. 1753. Fig. 52.
Atriplex patula L. var. *littoralis* (L.) A. Gray, Man. Bot. ed. 5, 409. 1867.

Annual monoecious herb from a taproot; stems ascending to erect, to 75 cm tall, branched, scurfy-scaly, at least when young; lower leaves opposite, upper leaves alternate, linear, acute at the apex, cuneate at the base, entire, rarely sparsely toothed or hastate, glabrous or scurfy-scaly, to 5 cm long, to 5 mm broad; staminate and pistillate flowers in interrupted spikes; pistillate flowers without a calyx; bracts of fruit 4–5 mm long, narrowly ovate, tuberculate; seeds 1.4–1.7 mm long, suborbicular, dull dark brown.

Common Name: Seaside Orache.
Habitat: Disturbed sandy soil.
Range: New Brunswick to New Jersey and Pennsylvania; Indiana; Wisconsin; Illinois.
Illinois Distribution: Sparingly found in a few northeastern counties.

Atriplex littoralis is readily recognized by its linear, usually entire leaves and its fruiting bracts 4–5 mm long. It is sometimes considered a variety of *A. patula.*

Atriplex littoralis flowers from July to October.

6. Atriplex patula L. Sp. Pl. 1053. 1753. Fig. 53.

Annual monoecious herb from a taproot; stems ascending to erect, to nearly 1 m tall, branched, scurfy-scaly when young, becoming glabrate; lower leaves opposite, upper leaves alternate, lance-hastate, acute at the apex, truncate at the base, entire or very sparsely toothed, glabrous or scurfy-scaly, to 6 cm long, to 4 cm broad, petiolate; staminate and pistillate flowers in interrupted spikes; pistillate flowers without a calyx; bracts of fruit 4–5 mm long, ovate, scarcely tuberculate; seeds 1.4–1.8 mm long, orbicular, dull, dark brown.

Common Name: Spear Scale.
Habitat: Disturbed soil, often in saline areas.
Range: Native of Europe, Asia, and northern Africa; frequently adventive in North America.
Illinois Distribution: Scattered throughout the state.

This species differs from *A. prostrata* by its lance-hastate leaves that are entire or very sparsely toothed. Some botanists combine these two species.

Atriplex patula flowers from July to September.

7. Atriplex prostrata Moq. in DC. Prodr. 13:99. 1849. Fig. 54.
Atriplex hastata L. Sp. Pl. 1053. 1753, misapplied.
Atriplex patula L. var. *hastata* (L.) A. Gray, Man. Bot. ed. 5, 409. 1867.

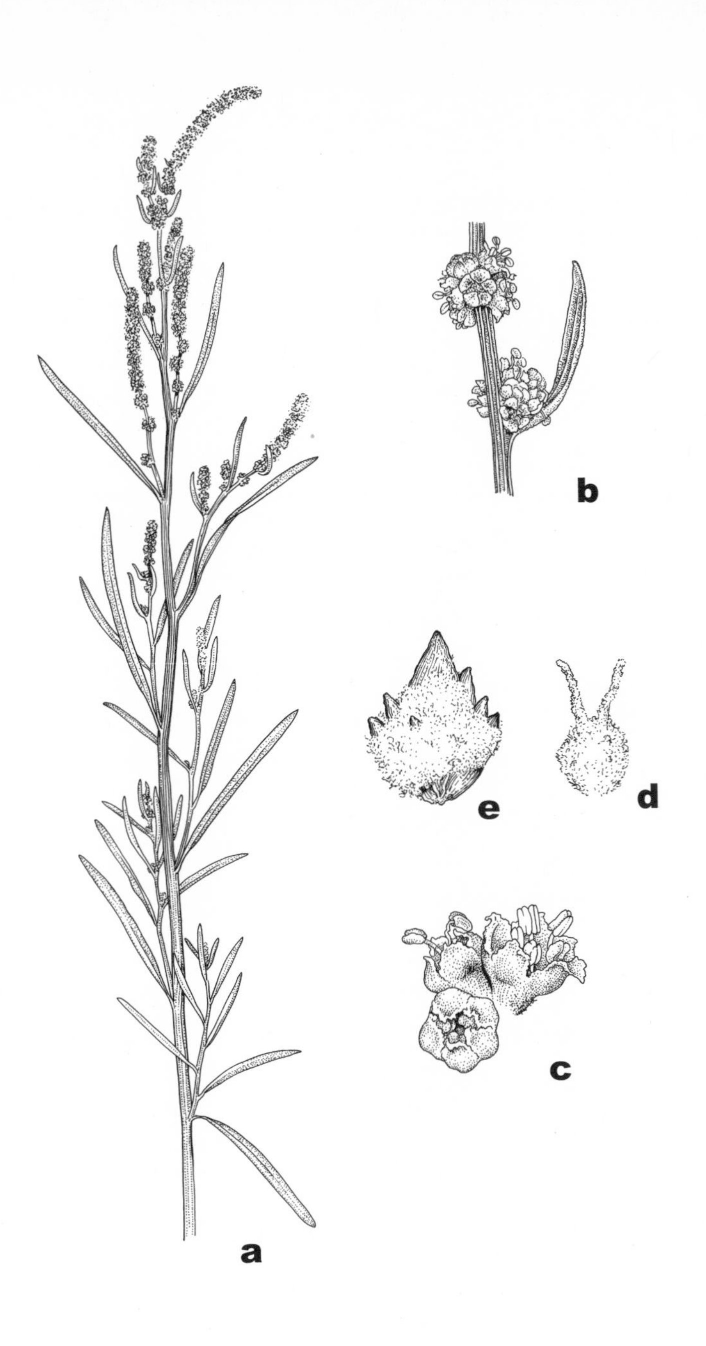

52. **Atriplex littoralis**
a. Habit
b. Flowering nodes
c. Cluster of staminate flowers
d. Pistillate flower
e. Fruit

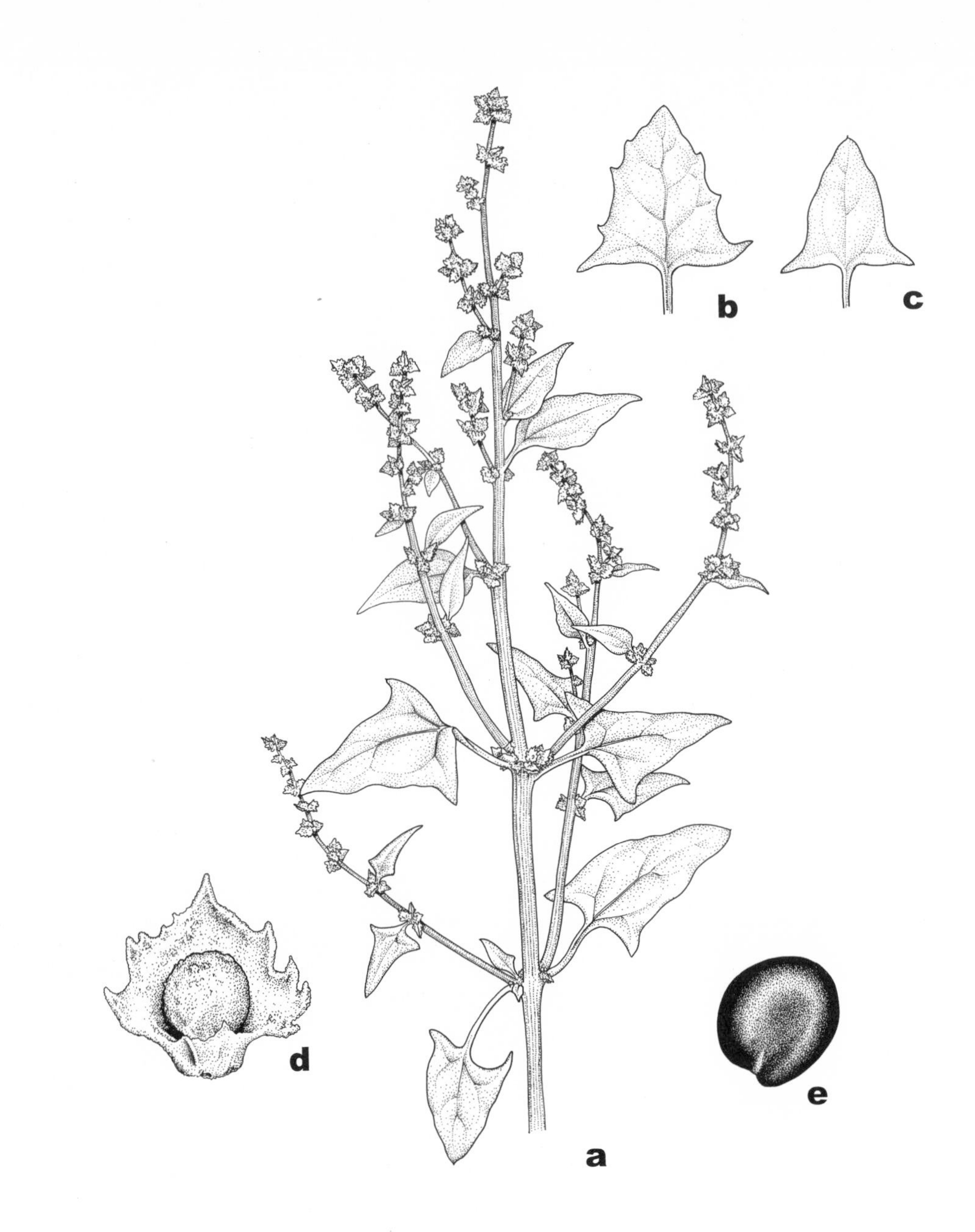

53. **Atriplex patula**
a. Habit
b. Leaves
c. Leaves
d. Half of bract removed to show seed
e. Seed

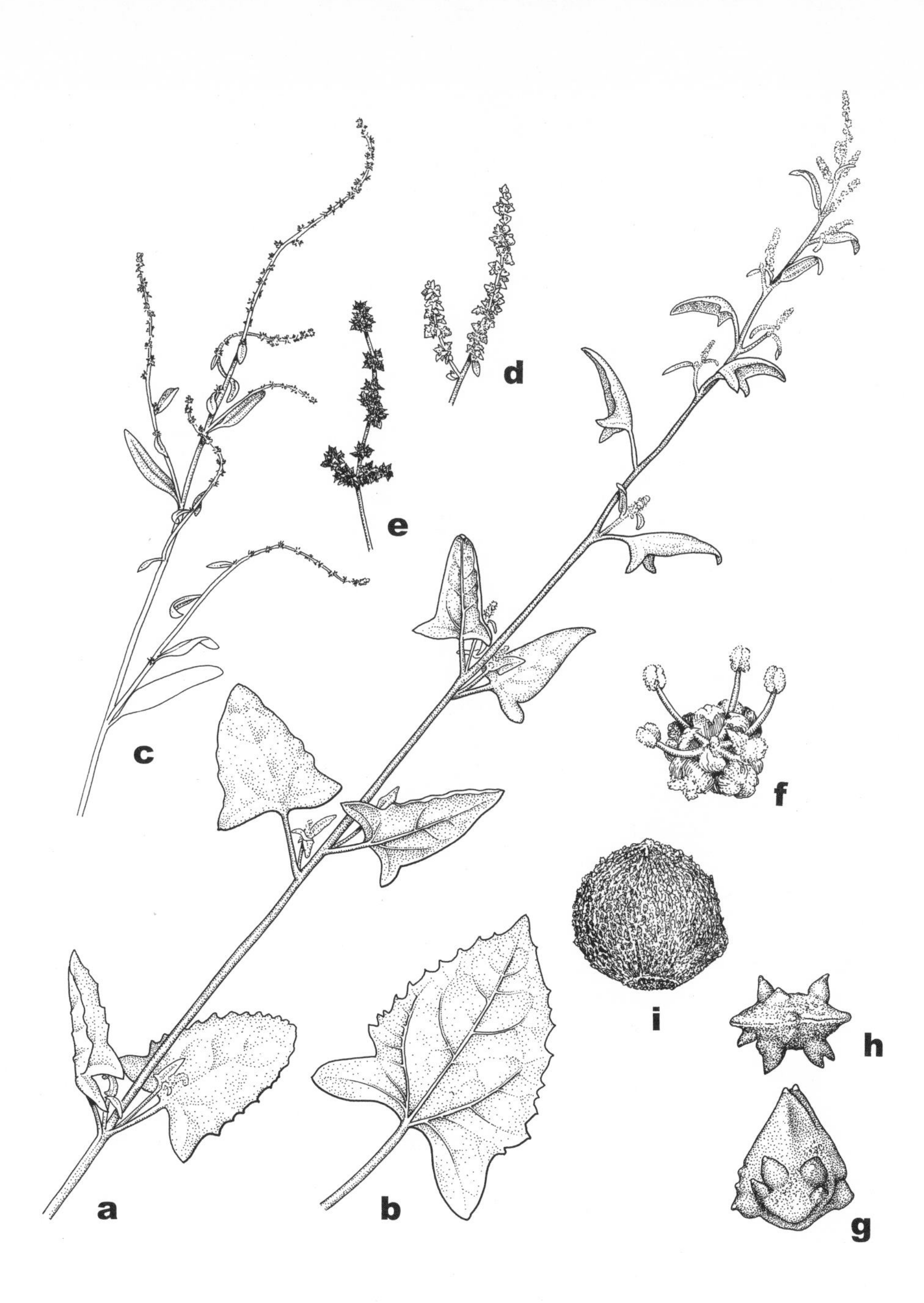

54. **Atriplex prostrata**
a. Habit
b. Leaf
c. Fruiting branch
d. Fresh fruiting branch (purple)
e. Dry fruiting branch (black)
f. Flowering cluster
g. Fruit, side view
h. Fruit, top view
i. Seed

Annual monoecious herb from a taproot; stems prostrate to ascending, to nearly 1 m long, branched, scurfy-scaly when young, becoming glabrate; lower leaves opposite, upper leaves alternate, deltate-hastate to oval-hastate, acute at the apex, truncate at the base, dentate, glabrous or scurfy-scaly at maturity, to 7 cm long, to 5 cm broad, petiolate; staminate and pistillate flowers in interrupted spikes; pistillate flowers without a calyx; bracts of fruit 4–5 mm long, rhombic, tuberculate; seeds 1.5–2.0 mm long, orbicular, dull, dark brown.

Common Name: Common Orache; Spear Scale.
Habitat: Disturbed soil, often in saline areas.
Range: Native of Europe; adventive throughout most of the United States.
Illinois Distribution: Scattered throughout the state.

While some botanists combine this species with *A. patula* and others make it a variety of *A. patula,* Voss (1985) gives reasons why this diploid plant could be given species status.

Atriplex prostrata differs from *A. patula* in the broader, dentate leaves.

This species flowers from July to September.

9. **Suaeda** Forsk.—Sea Blite

Annual or perennial herbs; leaves alternate, thick and more or less fleshy, nearly terete; flowers perfect or polygamous, solitary or clustered in the upper axils, bracteate; calyx 5-parted, the lobes unequal, keeled or winged, at least in fruit, completely enclosing the fruit; petals 0; stamens 5; ovary superior, 1-locular; styles 2; fruit a 1-seeded utricle; pericarp separating from the seed; seeds vertical or horizontal.

About 115 species comprise this genus. They occur worldwide in saline and alkaline soils.

Only the following species occurs in Illinois.

1. Suaeda calceoliformis (Hook.) Moq. Chen. Mon. Enum. 128. 1840. Fig. 55.
Salsola depressa Pursh, Fl. Am. Sept. 197. 1814, misapplied.
Chenopodium calceoliforme Hook. Fl. Bor. Am. 2:126. 1838.
Suaeda depressa (Pursh) S. Wats. Bot. King's Exp. 294. 1871.

Annual herb from slender taproots; stems prostrate to erect, to 75 cm tall, glabrous, glaucous, branched; leaves ascending, linear, subacute at the apex, cuneate at the base, to 3 (–4) cm long, entire, glabrous, glaucous, flat on the upper surface, sessile; flowers clustered in dense spikes, each cluster with 3–7 flowers, bracteate; calyx cleft to about the middle, each lobe 1.5–2.0 mm broad, of different sizes, at least some of the lobes hooded, keeled, narrowly winged; seeds lenticular, 1.0–1.7 mm long, black, shiny.

Common Name: Sea Blite.
Habitat: Along roadsides where salt has been applied during the winter.
Range: Native to the western United States; adventive in Illinois and other areas in eastern North America.
Illinois Distribution: Confined to the northeastern counties of Illinois.

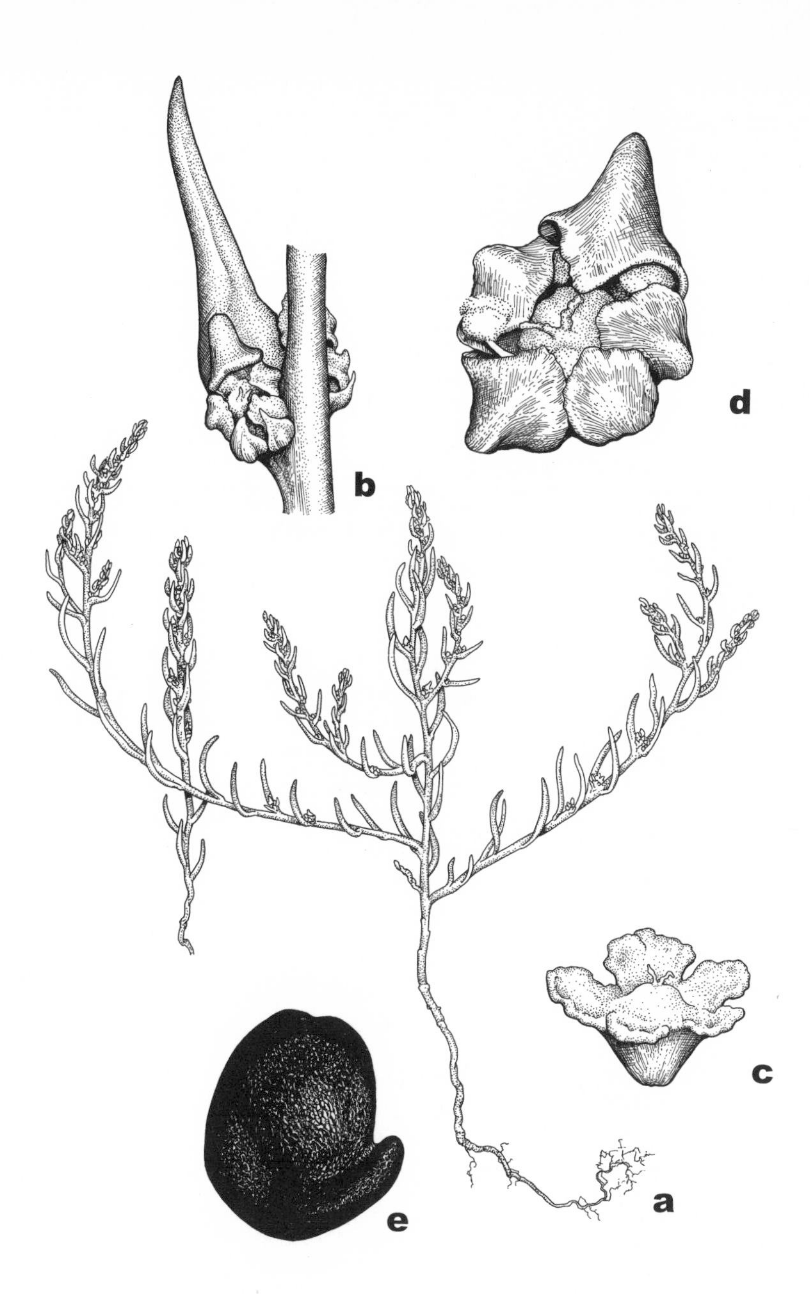

55. **Suaeda calceoliformis**
a. Habit
b. Flowering node
c. Flower
d. Flower
e. Seed

This species is distinguished by its bracts shorter than the calyx and its unequally sized calyx lobes.

This species has adapted to the highway shoulders in the Chicago area where great amounts of salt are applied during the winter.

For years this plant was known as *S. depressa*, but McNeill, Bassett, and Crompton (1977) have shown that the correct binomial should be *C. calceoliformis*.

This species flowers from August to October.

10. **Corispermum** L.—Bugseed

Annual herbs; leaves alternate, simple, linear to linear-lanceolate, 1-nerved; flowers perfect, borne in spikes; each flower subtended by leaflike bracts; sepals 1–5, scarious; petals 0; stamens 1–5; ovary superior, 1-locular; styles 2; fruit a 1-seeded utricle; seeds vertical, flat, sometimes winged.

There are about sixty species in this genus, all native to the northern hemisphere. Maihle and Blackwell (1978) have studied the North American species.

The three species in Illinois may be separated by the following key:

1. Spikes dense, 4–8 mm thick; fruit 3.5–4.5 mm long.
   2. Seeds winged . . . . 1. *C. hyssopifolium*
   2. Seeds unwinged . . . . 2. *C. orientale*
1. Spikes more slender and open, 3–4 mm thick; fruit 2.0–3.5 mm long . . . . 3. *C.nitidum*

1. Corispermum hyssopifolium L. Sp. Pl. 4. 1753. Fig. 56.

Annual herb from a slender taproot; stems slender, spreading, to 60 cm tall, much branched, glabrous or stellate-pubescent, often reddish-tinged; leaves linear to linear-lanceolate, acute and cuspidate at the apex, cuneate to rounded at the sessile base, up to 7 cm long, 1–3 mm broad, entire, glabrous or stellate-pubescent; spikes densely flowered, up to 8 cm long, 4–8 mm thick, the bracts ovate to lanceolate, up to 10 mm long, concealing the fruit; fruits 3.5–4.5 mm long, narrowly winged, ovoid to ellipsoid to nearly orbicular, the styles persistent; seeds vertical, about 1 mm long, black.

Common Name: Hyssop-leaved Bugseed.
Habitat: Sandy soil, including sandy beaches along Lake Michigan; disturbed soil.
Range: Native of Europe and Asia; probably adventive in the United States.
Illinois Distribution: Known from Carroll, Cook, Lake, Macon, Mason, Menard, St. Clair, and Whiteside counties.

There is disagreement among botanists as to whether this species is native or not in the United States.

Corispermum hyssopifolium differs from *C. nitidum* in its larger fruits and its lowest bracts that are as broad as or broader than the fruit. It differs from the strikingly similar *C. orientale* var. *emarginatum* by the presence of a wing on the seed.

This species flowers from July to September.

56. **Corispermum hyssopifolium**
a. Habit
b. Fruiting branch
c. Seed, ventral view
d. Seed, dorsal view
e. Flower

2. Corispermum orientale Lam. var. **emarginatum** (Rydb.) J. F. Macbr. Contr. Gray Herb. 53:13. 1918. Fig. 57.
Corispermum emarginatum Rydb. Bull. Torrey Club 31:404. 1904.

Annual herb from a slender taproot; stems slender, spreading, to 50 cm tall, much branched, glabrous or stellate-pubescent; leaves linear to linear-lanceolate, acute at the apex, cuneate at the sessile base, up 6 cm long, 1–3 mm broad, entire, glabrous or stellate-pubescent; spikes densely flowered, up to 7 cm long, 4–7 mm thick, the bracts usually lanceolate, up to 8 mm long, concealing the fruit; fruits 3.5–4.5 mm long, unwinged but usually emarginate, ovoid to ellipsoid, the styles persistent; seeds vertical, about 1 mm long, black.

Common Name: Emarginate Bugseed.
Habitat: Sandy soil.
Range: Native to Asia; sparingly adventive in the United States.
Illinois Distribution: St. Clair Co.: along the Mississippi River beneath the MacArthur Bridge, East St. Louis, August 18, 1949, J. O. Neill 2610.

This plant is very similar in appearance to *C. hyssopifolium*, differing by its unwinged seeds, although the seeds are usually emarginate.

This variety flowers from July to September.

3. Corispermum nitidum Kit. ex Schult. Oestr. Fl. ed. 2, 1:7. 1794. Fig. 58.
Corispermum hyssopifolium L. var. *microcarpum* S. Wats. Proc. Am. Acad. 8:123. 1874.

Annual herb from a slender taproot; stems slender, spreading, to 50 cm tall, much branched, glabrous or villous, usually reddish tinged; leaves narrowly linear, acute and cuspidate at the apex, cuneate or somewhat rounded at the sessile base, up to 6 cm long, 1–2 mm broad, entire, glabrous or villous; spikes open and loosely flowered, slender, 3–4 mm thick, the bracts linear to linear-lanceolate, up to 20 mm long, narrower and not concealing the fruits; fruit 2.0–3.5 mm long, narrowly winged, ovoid to ellipsoid, the styles persistent; seeds vertical, 0.8–1.0 mm long, black, shiny.

Common Name: Bugseed.
Habitat: Disturbed sandy soil.
Range: Native to Europe; adventive in several parts of North America, including the arctic.
Illinois Distribution: Known from Whiteside County.

Corispermum nitidum may be distinguished from *C. hyssopifolium* and *C. orientale* var. *emarginatum* by its smaller fruits and its narrower bracts that do not cover the fruits.

This species flowers from July to September.

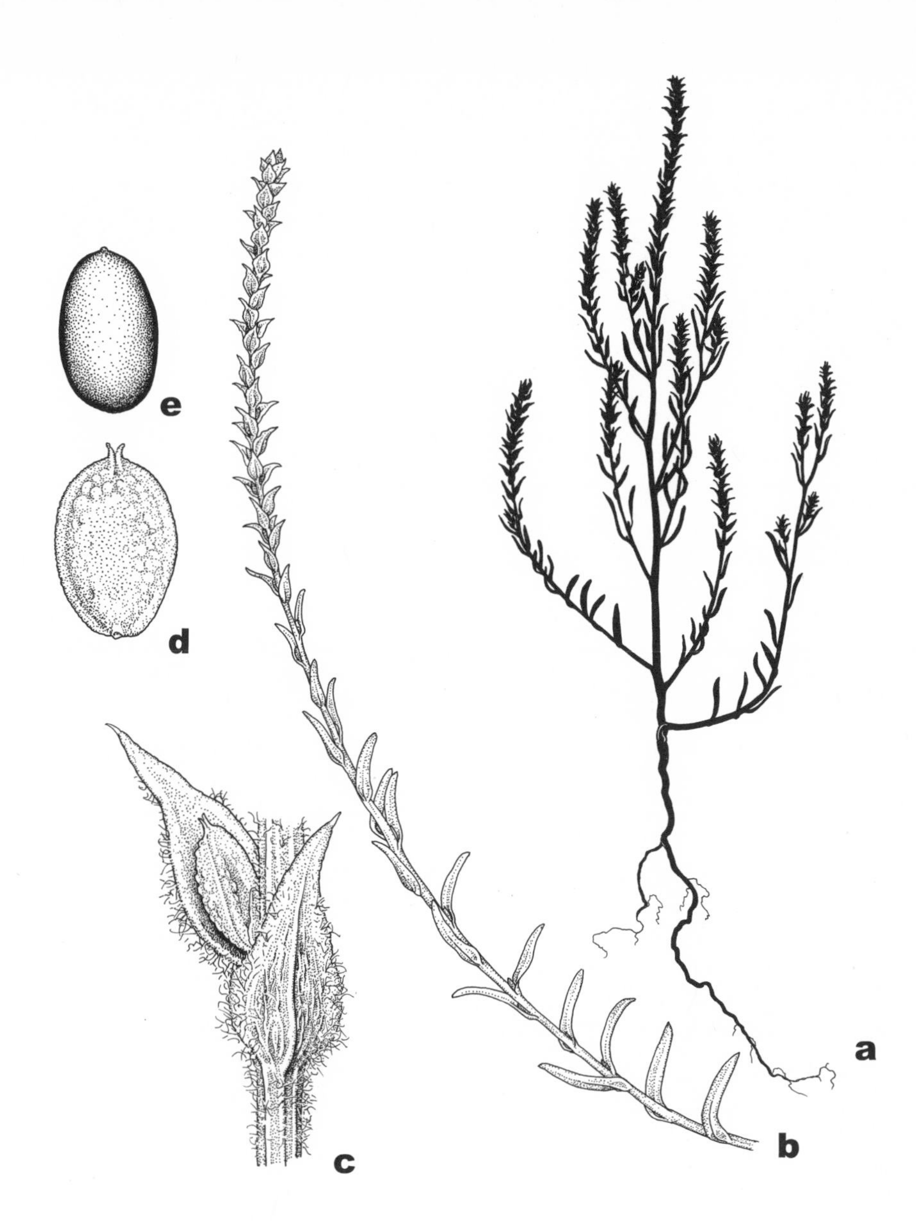

57. **Corispermum orientale** var. **emarginatum**
a. Habit
b. Habit
c. Fruiting node
d. Fruit
e. Seed

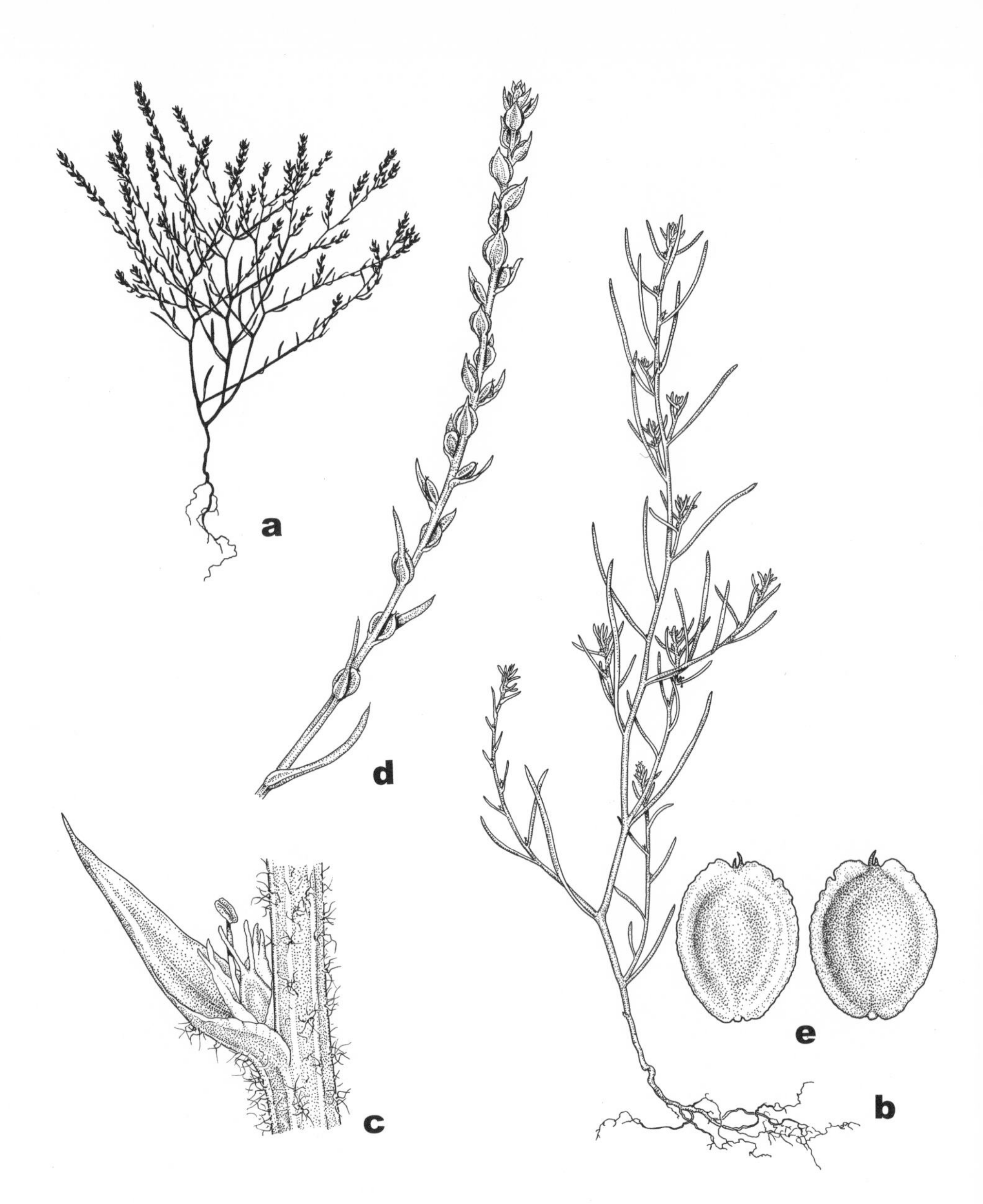

58. **Corispermum niditum**
a. Habit
b. Habit
c. Flowering node
d. Fruiting branch
e. Fruits

Amaranthaceae—Pigweed Family

Annual or perennial herbs; leaves alternate or opposite, simple, entire; flowers perfect or, if unisexual, the plants monoecious or dioecious, arranged in spikes or glomerules, the spikes often aggregated into panicles; calyx (3-) 5-parted, free or united below, sometimes united into a cup or tube, or calyx absent in some pistillate flowers; petals absent; stamens (1–) 5, the filaments sometimes united; styles 1–3; ovary superior, 1-locular; fruit dehiscent or indehiscent; seed 1.

There are about sixty-five genera and nine hundred species in this family, primarily in the tropics and subtropics.

Several species are grown as ornamentals or for their grains.

Key to the Genera of **Amaranthaceae** in Illinois

1. Leaves alternate.
   2. Flowers unisexual; utricles 1-seeded .... 1. *Amaranthus*
   2. Flowers perfect; utricles with 3 or more seeds .... 2. *Celosia*
1. All or most of the leaves opposite.
   2. Flowers borne in inconspicuous axillary clusters .... 3. *Tidestromia*
   2. Flowers in terminal and axillary spikes, the spikes sometimes arranged in panicles.
      3. Spikes silvery-white, not lanate or sericeous.
         4. Flowers unisexual, the plants dioecious; inflorescence a broad panicle of spikes .... 4. *Iresine*
         4. Flowers perfect; inflorescence of capitate spikes .... 5. *Alternanthera*
      3. Spikes whitish, lanate or sericeous .... 6. *Froelichia*

1. **Amaranthus** L.—Amaranth; Pigweed

Annual (in Illinois) monoecious or dioecious herbs from taproots; stems erect, ascending, or prostrate, branched or unbranched; leaves alternate, simple, entire, petiolate; flowers unisexual, less commonly perfect, borne in spikes or glomerules, the spikes often arranged in panicles, terminal and axillary; bracts subtending at least some of the flowers, sometimes rigid-tipped, longer or shorter than the sepals and fruits; staminate flowers with (3–) 5 sepals; pistillate flowers with (3–) 5 sepals, or sometimes reduced to 1 sepal and 1 rudimentary sepal; petals absent; stamens (3–) 5; styles (2–) 3; ovary superior, 1-locular; fruit a 1-seeded circumscissile utricle, in one species bursting irregularly at maturity, in one species indehiscent; seed 1, lenticular to orbicular, black or dark red-brown.

Amaranthus is a genus of about sixty species found worldwide. In tropical regions, particularly, some species are grown for their seeds, which have value as food. Other species may be used as potherbs. Some species, particularly those with red or purple flowers, are grown as ornamentals. *Celosia,* the cock's-comb, is an important ornamental.

Some botanists segregate the genus *Acnida* from *Amaranthus* based on the reduced sepals in the pistillate flowers. I am somewhat reluctantly following Sauer (1955) in the treatment of most of the dioecious amaranths.

Key to the Species of **Amaranthus** in Illinois

1. Stems with pairs of short spines at some or all the nodes .... 1. *A. spinosus*
1. Stems without spines.

2. Plants prostrate.
   3. Plants monoecious; stamens 3; sepals 4–5 in the pistillate flowers . . 3. *A. blitoides*
   3. Plants dioecious; stamens 5; sepals 1–2, at least one of them rudimentary in the pistillate flowers . . . . 13. *A. tuberculatus*
2. Plants ascending to erect.
   4. Plants monoecious.
      5. Plants ascending, diffusely branched; flowers in small, axillary clusters; stamens 2–3 . . . . 2. *A. albus*
      5. Plants erect, branched or unbranched but rarely diffusely branched; flowers in spikes or paniculate spikes; stamens 5 (3 in *A. powellii* and *A. cruentus*).
         6. Margin of sepals of the pistillate flowers overlapping; flowers red, less commonly purple; terminal spike pendulous . . . . 4. *A. hypochondriacus*
         6. Margin of sepals of the pistillate flowers not overlapping; flowers usually purple or green; terminal spike rarely pendulous.
            7. Sepals of the pistillate flowers 1.5–2.0 mm long; bracts up to 4 mm long, as long as to about twice as long as the sepals.
               8. Bracts 1.5–2.0 mm long; stems, leaves, and flowers purplish or purple-tinged; fruit conspicuously longer than the sepals; stamens 3; panicle branches not stiff . . . . 5. *A. cruentus*
               8. Bracts 2–4 mm long; stems, leaves, and flowers rarely purplish-tinged; fruit about as long as the sepals; stamens 5; panicle branches stiff . . . . 6. *A. hybridus*
            7. Sepals of pistillate flowers usually 4 mm long or longer; bracts usually 2–3 times longer than the sepals.
               9. Terminal spicate panicle usually lobulate; some part of the stem usually lanate; sepals of the pistillate flowers obtuse; stamens 5–7 . . . . 7. *A. retroflexus*
               9. Terminal spicate panicle usually not lobulate; stems not lanate; sepals of pistillate flowers acute; stamens 3 . . . . 8. *A. powellii*
   4. Plants dioecious.
      10. Pistillate flowers with 5 well-developed sepals.
         11. Bracts 3–6 mm long, almost always longer than the sepals, usually awned . . . . 9. *A. palmeri*
         11. Bracts 1.5–2.5 mm long, about half as long to sometimes almost as long as the sepals, rarely awned.
            12. Fruits indehiscent; pistillate flowers borne in short, thick spikes . . . . 10. *A. ambigens*
            12. Fruits dehiscent; pistillate flowers borne in long, slender spikes . . . . 11. *A. arenicola*
      10. Pistillate flowers with 1–2 sepals, if 2, then 1 of them very rudimentary.
         13. Fruits regularly dehiscent, circumscissile; sepals of staminate flowers usually awn-tipped . . . . 12. *A. rudis*
         13. Fruits bursting irregularly upon dehiscence, not circumscissile; sepals of staminate flowers acute, rarely awn-tipped . . . . 13. *A. tuberculatus*

1. **Amaranthus spinosus** L. Sp. Pl. 991. 1753. Fig. 59.

Annual monoecious herb from a deep taproot; stems stout, erect, to 1.2 m tall, branched, glabrous to sparsely pubescent, bearing a pair of spines at most nodes, each spine up to 1 cm long; leaves broadly lanceolate to ovate, obtuse but mucronate at the apex, cuneate at the base, to 8 cm long, to 6 cm broad, entire, glabrous or sparsely pubescent, the petioles sometimes as long as the blade; flowers unisexual,

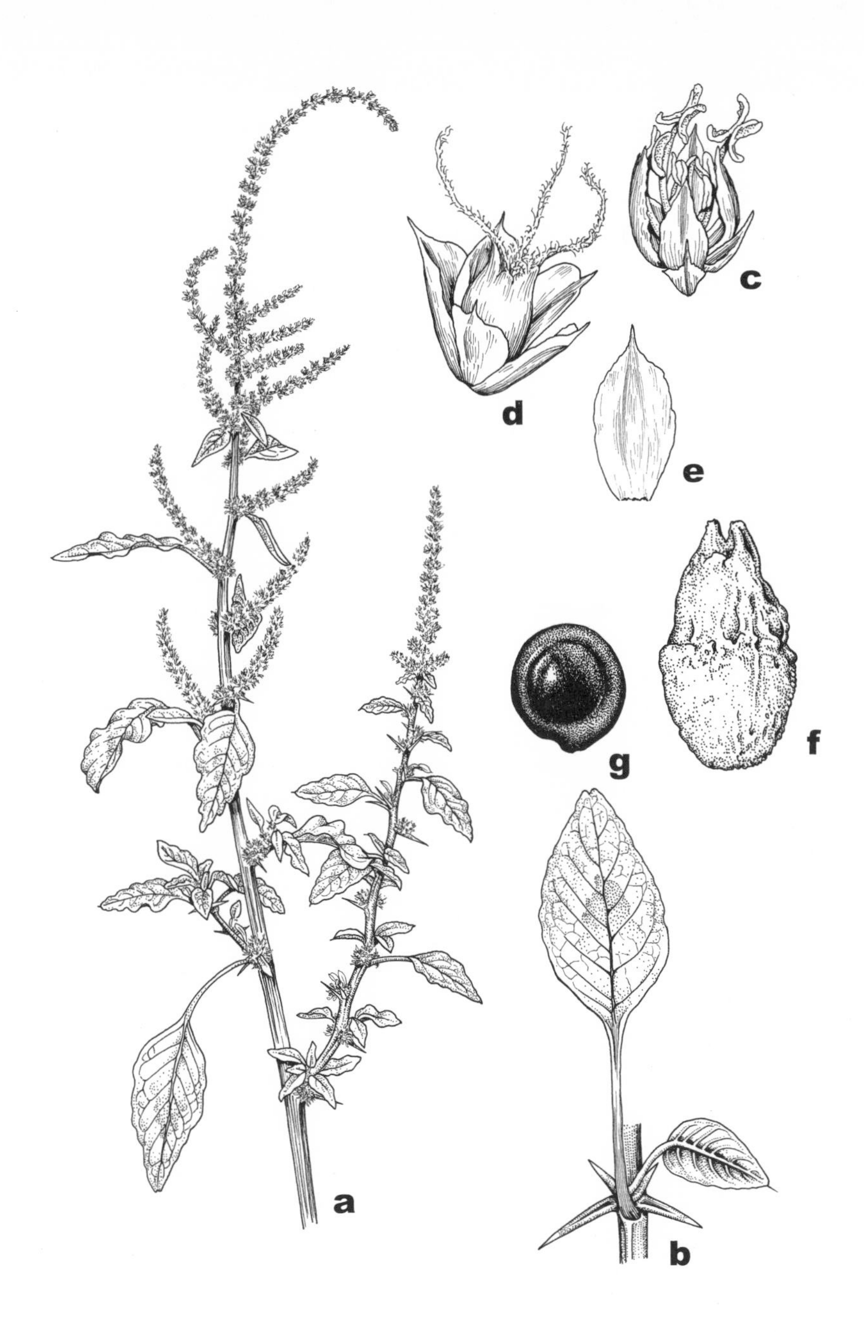

59. **Amaranthus spinosus**
a. Flowering branch
b. Node with leaves and spines
c. Staminate flower
d. Pistillate flower
e. Sepal
f. Fruit
g. Seed

less commonly perfect, the staminate in terminal spikes up to 15 cm long, up to 1 cm thick, the pistillate in shorter, capitate, axillary clusters; bracts up to 1.5 mm long, narrowly lanceolate to subulate; staminate flowers with 5 sepals, the sepals narrowly oblong; pistillate flowers with 5 sepals, the sepals oblong, 1.0–1.5 mm long; stamens 5; styles 3; fruit a 1-seeded utricle, oblongoid, rugose, 1.5–2.0 mm long; seeds orbicular, 0.6–1.0 mm in diameter, shiny, black.

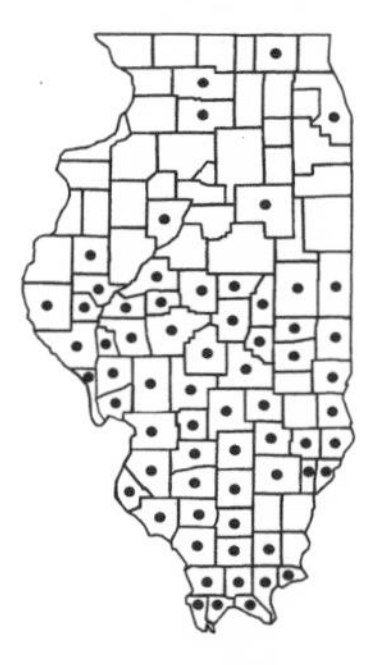

Common Name: Spiny Pigweed.
Habitat: Disturbed soil.
Range: Native to tropical America; adventive throughout much of the United States.
Illinois Distribution: Common in the southern two-thirds of Illinois, rare elsewhere.

The pair of sharp spines at most of the nodes readily distinguishes this species from all other species of *Amaranthus* in Illinois.

Amaranthus spinosus flowers from July to October.

2. **Amaranthus albus** L. Syst. ed. 10, 2:1268. 1759. Fig. 60.
Amaranthus graecizans L. Sp. Pl. 990. 1753, *nomen ambiguum.*

Annual monoecious herb from a strong taproot; stems stout, erect, to 1.2 m tall, much branched, glabrous or sparsely pubescent, whitish to pale green; leaves elliptic to obovate, obtuse but usually mucronate at the apex, cuneate at the base, to 7 cm long, to 2 cm broad, glabrous or sparsely pubescent, entire, white-veiny, with slender petioles up to half as long as the blade; flowers unisexual, less commonly perfect, in small axillary clusters shorter than the leaves; bracts to 4 mm long, oblong to lanceolate, stiff and sharp-pointed; staminate flowers with 3 sepals, the sepals oblong, scarious; pistillate flowers with 3 sepals, the sepals linear to oblong, green along the midvein, sometimes reddish; stamens 3; styles 3; fruit a 1-seeded utricle, subglobose, rugose, longer than the sepals; seeds lenticular, 0.6–0.8 mm long, shiny, deep red-brown.

Common Name: Tumbleweed; White Amaranth.
Habitat: Fields; disturbed soil.
Range: Quebec to British Columbia, south to California, Texas, and Florida.
Illinois Distribution: Common throughout the state.

Although this species occurs in disturbed areas, it is native to the United States.

Several botanists in Illinois called this species *A. graecizans* between 1922 and 1952, but that binomial is considered a *nomen ambiguum* and is therefore not available for this species.

This species differs from the similar *A. blitoides* in its erect growth form, its sharp-tipped bracts, its rugose fruits, and its shiny, deep red-brown seeds.

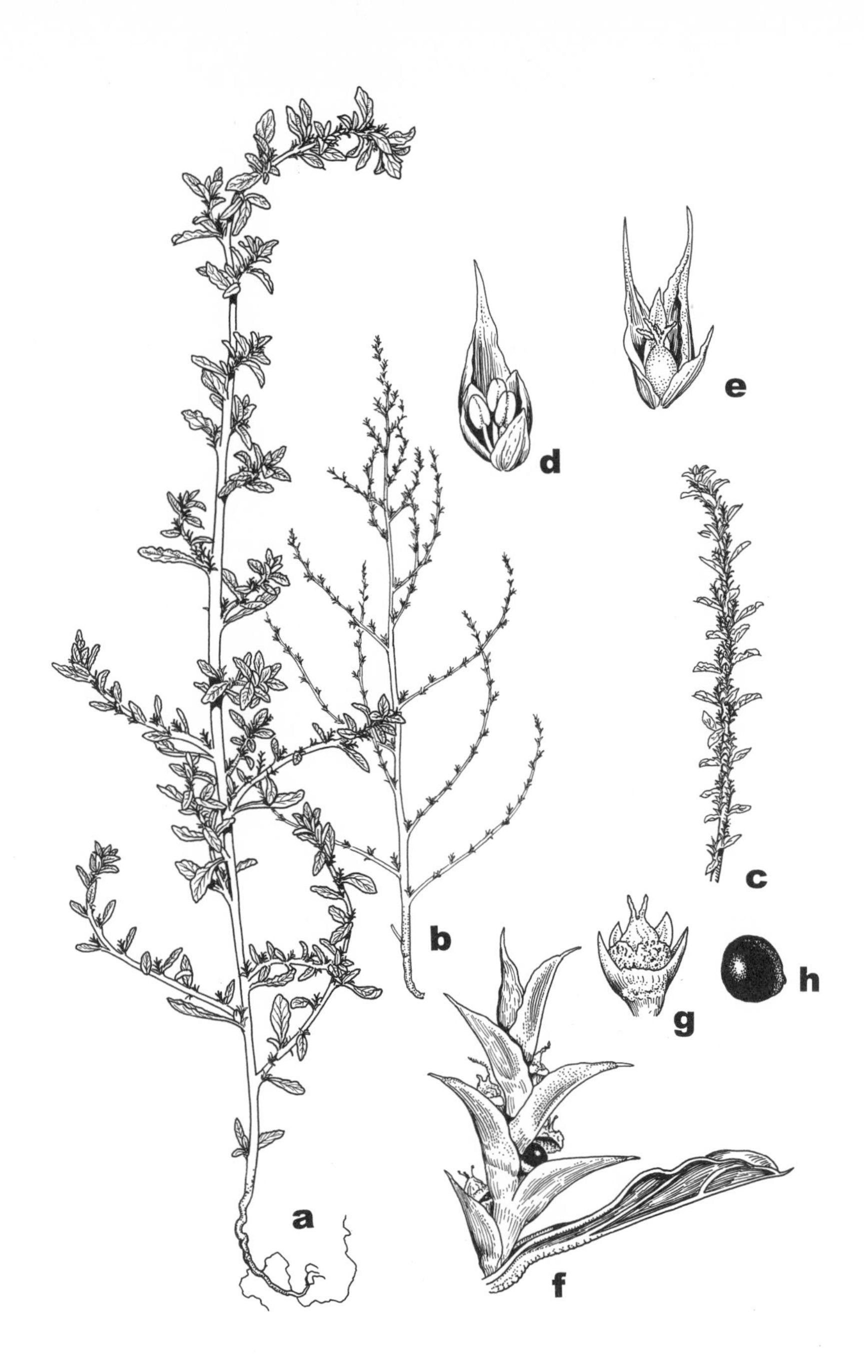

60. **Amaranthus albus**
a. Habit
b. Plant after it senesces
c. Stem variation
d. Staminate flower
e. Pistillate flower
f. Fruiting branch
g. Fruit
h. Seed

The entire plant is often uprooted and blown by the wind, tumbling across fields and often lodging in fence rows.

Amaranthus albus flowers from July to September.

3. **Amaranthus blitoides** S. Wats. Proc. Am. Acad. 12:273. 1877. Fig. 61.
Amaranthus graecizans L. Sp. Pl. 990. 1753, *nomen ambiguum.*

Annual monoecious herb from a strong taproot; stems stout, prostrate, to 60 cm long, much branched, glabrous or sparsely pubescent, whitish to pale green; leaves usually crowded, elliptic to obovate, obtuse to subacute at the apex, cuneate at the base, to 4 cm long, to 2 cm broad, entire, glabrous or sparsely pubescent, white-veiny, on stout petioles up to half as long as the blade; flowers unisexual, less commonly perfect, in small axillary clusters shorter than the leaves; bracts to 4 mm long, oblong to lanceolate, short-acuminate but not usually spine-tipped; staminate flowers usually with 4–5 sepals, the sepals oblong, scarious; pistillate flowers usually with 4–5 sepals, the sepals oblong, green along the midvein; stamens 3; styles 3; fruit a 1-seeded utricle, subglobose, smooth, as long as or slightly longer than the sepals; seeds orbicular, 1.2–1.5 mm in diameter, dull, black.

Common Name: Prostrate Pigweed.
Habitat: Disturbed soil.
Range: Native to the western United States; adventive over much of the eastern United States.
Illinois Distribution: Common throughout the state.

The most obvious difference between this species and *A. albus* is its prostrate growth form. *Amaranthus blitoides* also lacks spine-tipped bracts and has larger, dull seeds and smooth fruits.

This species has been erroneously called *A. graecizans* during the last half of the twentieth century. It should be called *A. blitoides,* a binomial used by Illinois botanists from 1882 to 1952 for this species and then abandoned.

Amaranthus blitoides flowers from July to September.

4. **Amaranthus hypochondriacus** L. Sp. Pl. 991. 1753. Fig. 62.
Amaranthus caudatus L. Sp. Pl. 990. 1753, misapplied.

Annual monoecious herb from a strong taproot; stems stout, erect, to 1.8 m tall, branched, glabrous; leaves elliptic to ovate, acute at the apex, tapering and decurrent at the base, to 20 cm long, to 15 cm broad, glabrous, with petioles about half as long as the blade; flowers unisexual, less commonly perfect, red, in large terminal, pendulous spikes and axillary spikes, the terminal up to 30 cm long, up to 3 cm thick; bracts lanceolate; staminate flower with 5 sepals, the sepals lanceolate to lance-ovate, acute; pistillate flowers with 5 sepals, the sepals obovate, with overlapping margins, 2.3–2.5 mm long; stamens 5; styles 3; fruit a 1-seeded utricle, about as long as to longer than the sepals; seeds orbicular, 1.0–1.2 mm in diameter, white to brown, shiny.

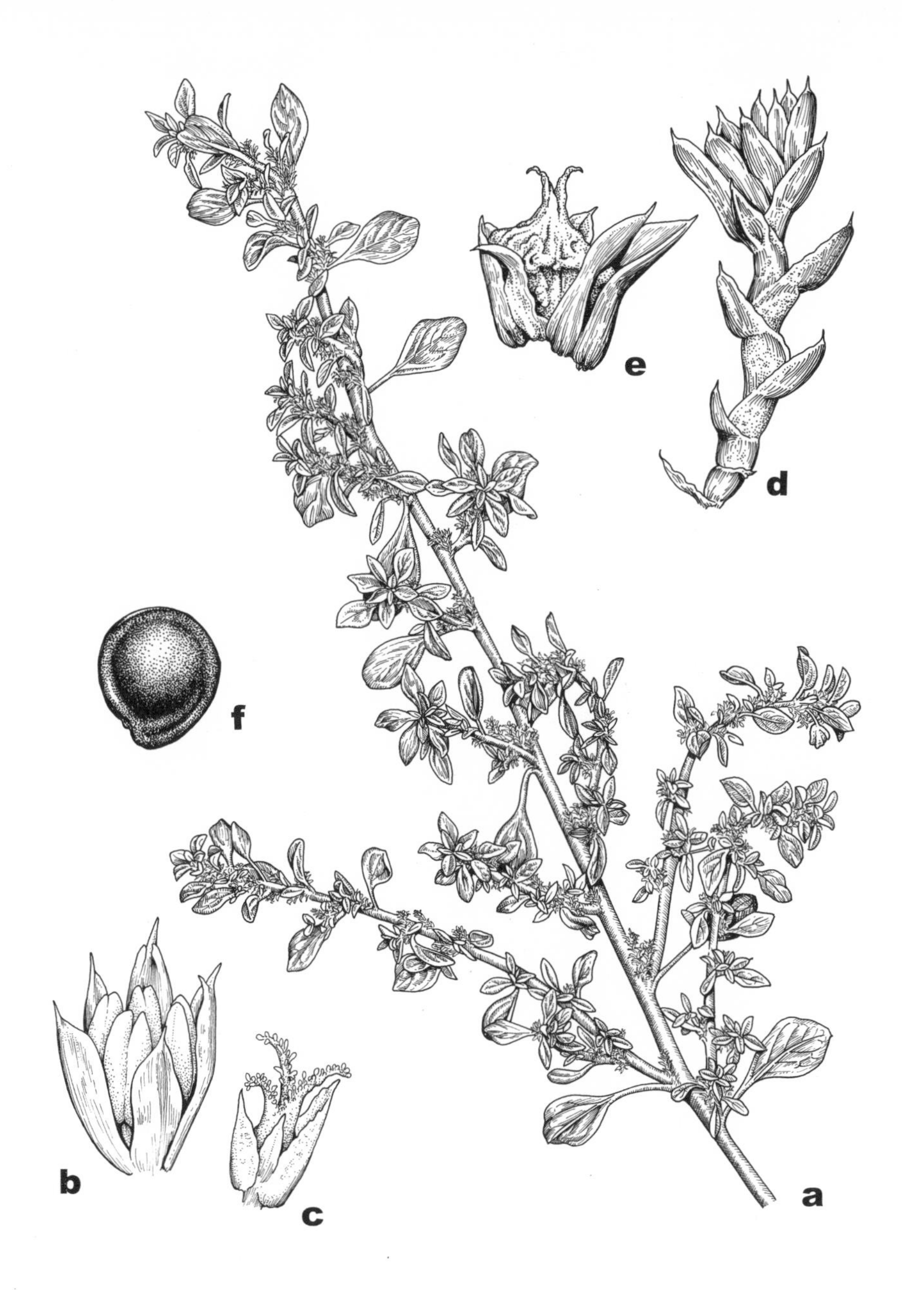

61. **Amaranthus blitoides**
a. Flowering branch
b. Staminate flower
c. Pistillate flower
d. Fruiting branch
e. Fruit
f. Seed

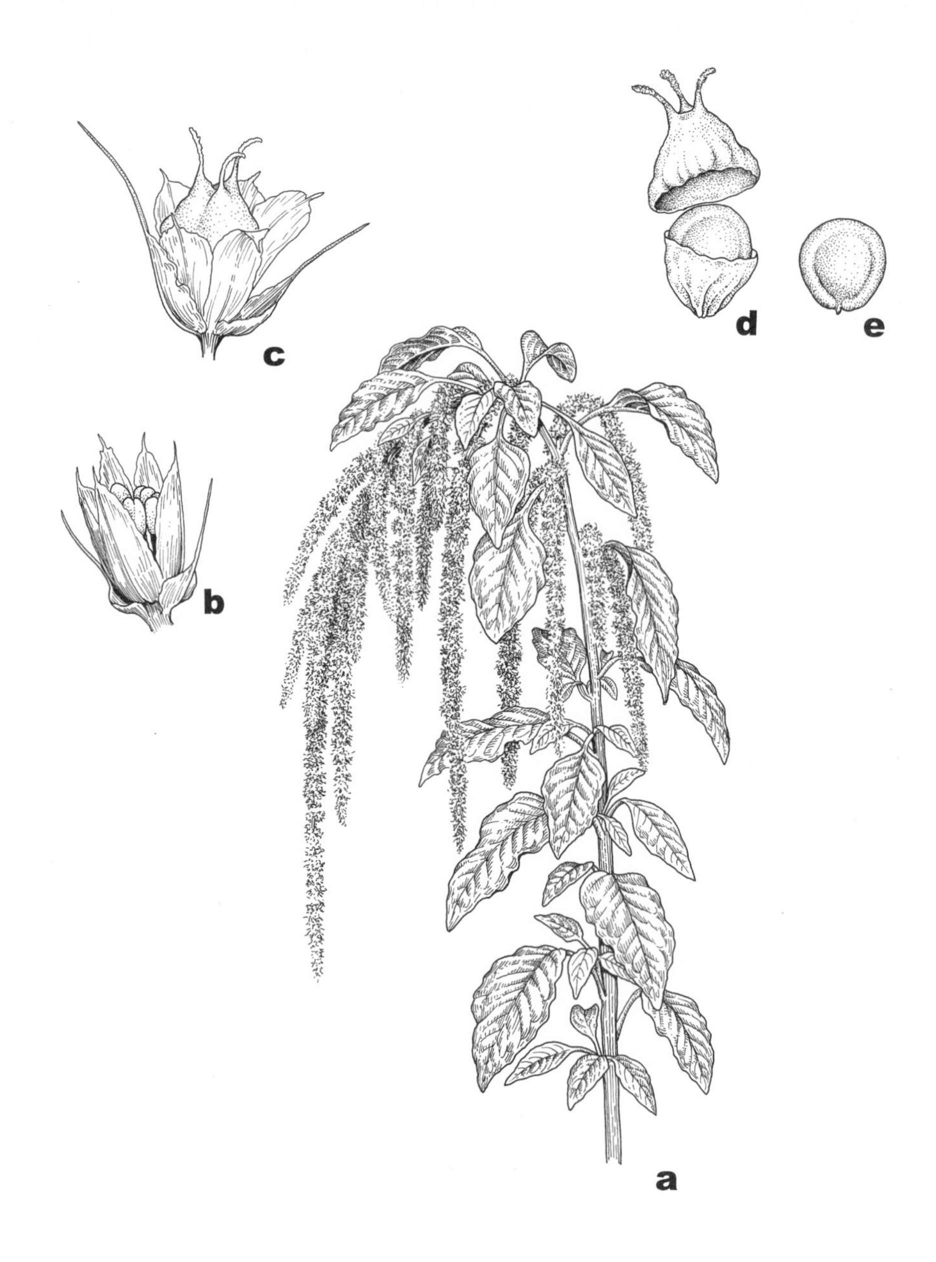

62. **Amaranthus hypochondriacus**
a. Habit, upper part of plant
b. Staminate flower
c. Pistillate flower
d. Fruit
e. Seed

Common Name: Prince's Feather.
Habitat: Disturbed soil.
Range: Native to tropical America; rarely escaped from cultivation in the United States.
Illinois Distribution: Known from Cook, Fayette, Jackson, Lawrence, and Wabash counties.

The thick, pendulous, red spikes are very attractive and are the reason for its being grown as a garden ornamental.

In addition to the large, pendulous, red spikes, this species differs in its overlapping sepals in the pistillate flowers.

Some American botanists have called this plant *A. caudatus*, but that binomial applies to a different species.

Amaranthus hypochondriacus flowers from July to October.

5. Amaranthus cruentus L. Syst. Veg. ed. 10, 1269. 1759. Fig. 63.

Annual monoecious herb from a strong taproot; stems rather stout, erect, to 2 m tall, branched or unbranched, glabrous in the lower half, usually pubescent in the upper half; leaves elliptic to ovate, obtuse to acute at the apex, cuneate or rounded at the base, to 20 cm long, to 15 cm broad, glabrous or sparsely pubescent, sometimes reddish or purplish, with petioles at least half as long as the blade; flowers unisexual, red or purple, crowded into spikes that are arranged in panicles, to 10 cm long, to 1.2 cm thick; bracts 1.5–2.0 mm long, lanceolate, spine-tipped, red or purple; staminate flowers with 5 sepals, the sepals oblong to ovate, acute, excurrent; pistillate flowers with 5 sepals, the sepals 1.5–2.0 mm long, oblong, obtuse to subacute, green or sometimes purplish, about 1.5 mm long; stamens 3; styles 3; fruit a 1-seeded utricle, globose, much longer than the sepals; seeds orbicular, about 1 mm in diameter, shiny, black or brown.

Common Name: Purple Amaranth.
Habitat: Disturbed soil.
Range: Native to Asia; occasionally adventive in the United States.
Illinois Distribution: Known sparingly from the northern half of the state; also St. Clair and Wabash counties.

This is a garden ornamental by virtue of its red or usually purple inflorescence. It rarely escapes from cultivation. Most Illinois collections were made during the nineteenth century when this plant was more popular as an ornamental.

Amaranthus cruentus differs from *A. hypochondriacus* in its erect rather than pendulous terminal spike. It differs from *A. hybridus* and *A. powellii* in its obtuse to subacute sepals of the pistillate flowers and its bracts that are not more than 1.5 times longer than the fruit.

This species flowers from July to October.

63. **Amaranthus cruentus**
a. Habit
b. Fruit
c. Seed

6. Amaranthus hybridus L. Sp. Pl. 990. 1753. Fig. 64.
Amaranthus chlorostachys Willd. Amaranth. 34, pl. 10, t. 19. 1790.

Annual monoecious herb from a strong taproot; stems stout, erect, to 2.5 m tall, branched, glabrous to rough-hairy in the lower half, more or less villous in the upper half; leaves lanceolate to ovate, obtuse to acute at the apex, cuneate or rounded at the base, to 15 cm long, to 10 cm broad, glabrous or variously pubescent, entire, usually somewhat paler on the lower surface; flowers unisexual, less commonly perfect, sometimes reddish, in slender terminal and axillary spikes arranged in panicles, the individual spikes up to 8 cm long, up to 1.5 cm thick; bracts up to 4 mm long, narrowly lanceolate, awned; staminate flowers with 5 sepals, the sepals oblong to ovate, excurrent; pistillate flowers with 5 sepals, the sepals linear to oblong, excurrent, 1.5–2.0 mm long; stamens 5; styles 3; fruit a 1-seeded utricle, subglobose, longer than the sepals; seeds orbicular, 1.1–1.3 mm in diameter, shiny, black.

Common Name: Green Pigweed.
Habitat: Fields; disturbed soil.
Range: Native to tropical America; adventive in much of the United States.
Illinois Distribution: Common throughout the state.

This coarse weed is a common plant in disturbed soil. It differs from the similar *A. retroflexus* in its shorter sepals that are up to 2 mm long and its bracts that are not more than twice as long as the sepals.

The flowers of this species are sometimes reddish.

Some Illinois botanists in the nineteenth century called this plant *A. hypochondriacus,* but that binomial belongs to a different species. Willdenow's *A. chlorostachys,* used by a few early Illinois botanists, is the same species as *A. hybridus.*

Amaranthus hybridus flowers from August to October.

7. Amaranthus retroflexus L. Sp. Pl. 991. 1753. Fig. 65.

Annual monoecious herb from a strong taproot; stems stout, erect, to 40 cm tall, branched or unbranched, rough-hairy, at least in the lower part of the stem, usually lanate in the upper part, usually white-striate; leaves lanceolate to ovate, acute to acuminate at the apex, cuneate at the base, to 15 cm long, to 10 cm broad, pubescent throughout or only on the veins, entire, with slender petioles sometimes as long as the blade; flowers unisexual, less commonly perfect, crowded into erect, terminal and often lobed as well as axillary spikes up to 20 cm long, up to 2 cm thick; bracts to 6 mm long, ovate, subulate-tipped; staminate flowers with 5 sepals, the sepals lanceolate to ovate, excurrent, scarious; pistillate flowers with 5 sepals, the sepals linear to oblong, obtuse, scarious but with a white midvein; stamens 5; styles 3; fruit a 1-seeded utricle, subglobose, more or less rugulose, shorter than the sepals; seeds orbicular, 0.9–1.1 mm in diameter, shiny, deep red-brown.

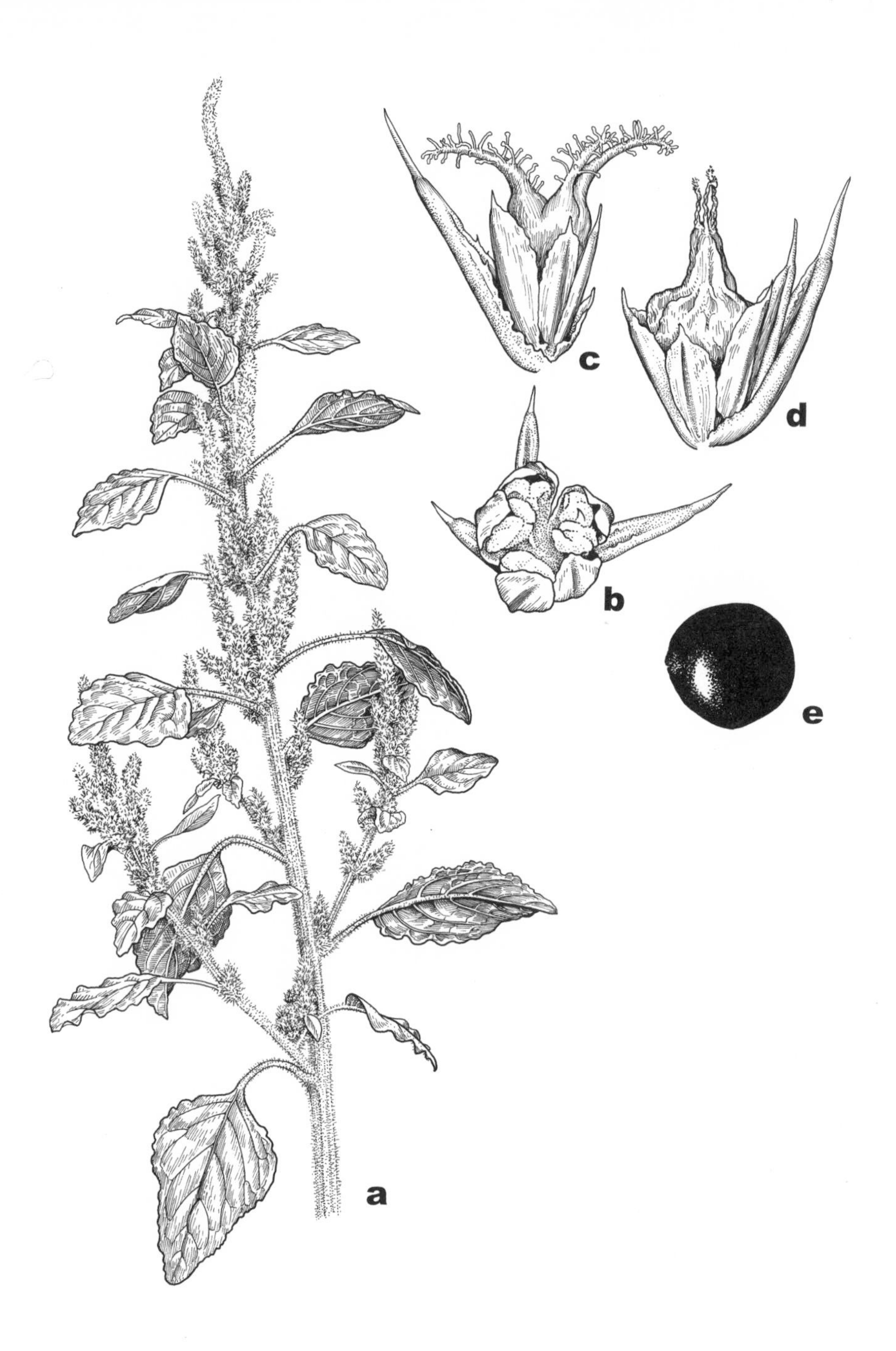

64. **Amaranthus hybridus**
a. Flowering branch
b. Staminate flower
c. Pistillate flower
d. Fruit
e. Seed

65. **Amaranthus retroflexus**
a. Flowering branch
b. Leaf
c. Staminate flower
d. Pistillate flower
e. Seed

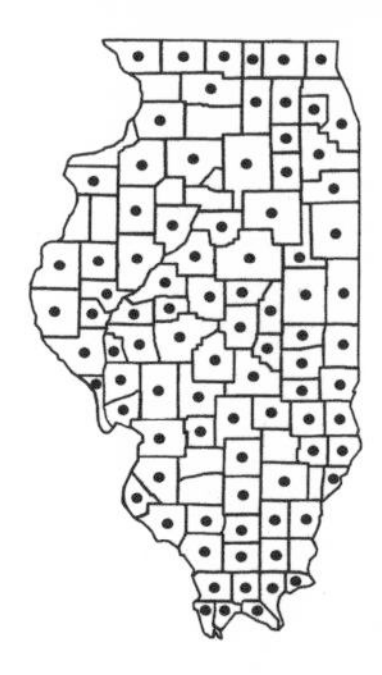

Common Name: Rough Pigweed.
Habitat: Fields; disturbed soil.
Range: Native to tropical America; adventive throughout much of North America.
Illinois Distribution: Common throughout the state.

This common, robust species of weedy habitats is distinguished by its fruits that are shorter than the sepals, the obtuse sepals of the pistillate flowers, and the often lobed terminal spike. It differs further from *A. powellii* in having five rather than three stamens.

Amaranthus retroflexus flowers from August to October.

8. Amaranthus powellii S. Wats. Proc. Am. Acad. 10:347. 1875. Fig. 66.

Annual monoecious herb from a strong taproot; stems stout, erect, to 3 m tall, branched or unbranched, pubescent in the upper half, sometimes red-tinged; leaves lanceolate to ovate, obtuse to acute at the apex, cuneate at the base, to 7.5 cm long, to 4 cm broad, glabrous or sparsely pubescent, entire, with slender petioles up to half as long as the blade; flowers unisexual, less commonly perfect, crowded in dense terminal and axillary spikes, the spikes up to 35 cm long, up to 2.5 cm thick; bracts 5–9 mm long, lanceolate to ovate, stiff and short-pointed; staminate flowers with 5 sepals, the sepals lanceolate to oblong, scarious; pistillate flowers with 5 sepals, the sepals lanceolate to oblong, acute; stamens 3; styles 3; fruit a 1-seeded utricle, subglobose, more or less rugose, usually slightly shorter than the sepals; seeds oval, 1.0–1.2 mm long, shiny, black.

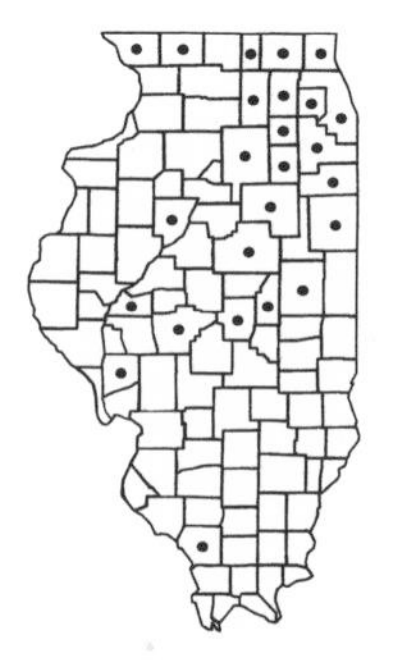

Common Name: Tall Amaranth.
Habitat: Cultivated fields; disturbed soil.
Range: Native to the southwestern United States and Mexico; adventive in much of the eastern United States.
Illinois Distribution: Occasional throughout the state except in the south.

This is a robust species that occurs commonly in corn and soybean fields. It has spread very rapidly in Illinois during the last three decades of the twentieth century.

Amaranthus powellii may be distinguished from other species of *Amaranthus* in its large, thick spikes; its sepals that are slightly longer than the utricle; and its bracts that are at least 5 mm long.

This species flowers from July to October.

9. Amaranthus palmeri S. Wats. Proc. Am. Acad. 12:274. 1876. Fig. 67.

Annual dioecious herb from a long taproot; stems rather slender, erect, to 1 m tall, branched, glabrous or pubescent; leaves broadly lanceolate to oblong to ovate, acute to obtuse at the apex, cuneate or rounded at the base, to 10 cm long, to 7 cm broad, entire, glabrous or pubescent, with petioles half as long as to slightly longer than the blade; flowers unisexual, in slender or somewhat thickened, erect or pendulous

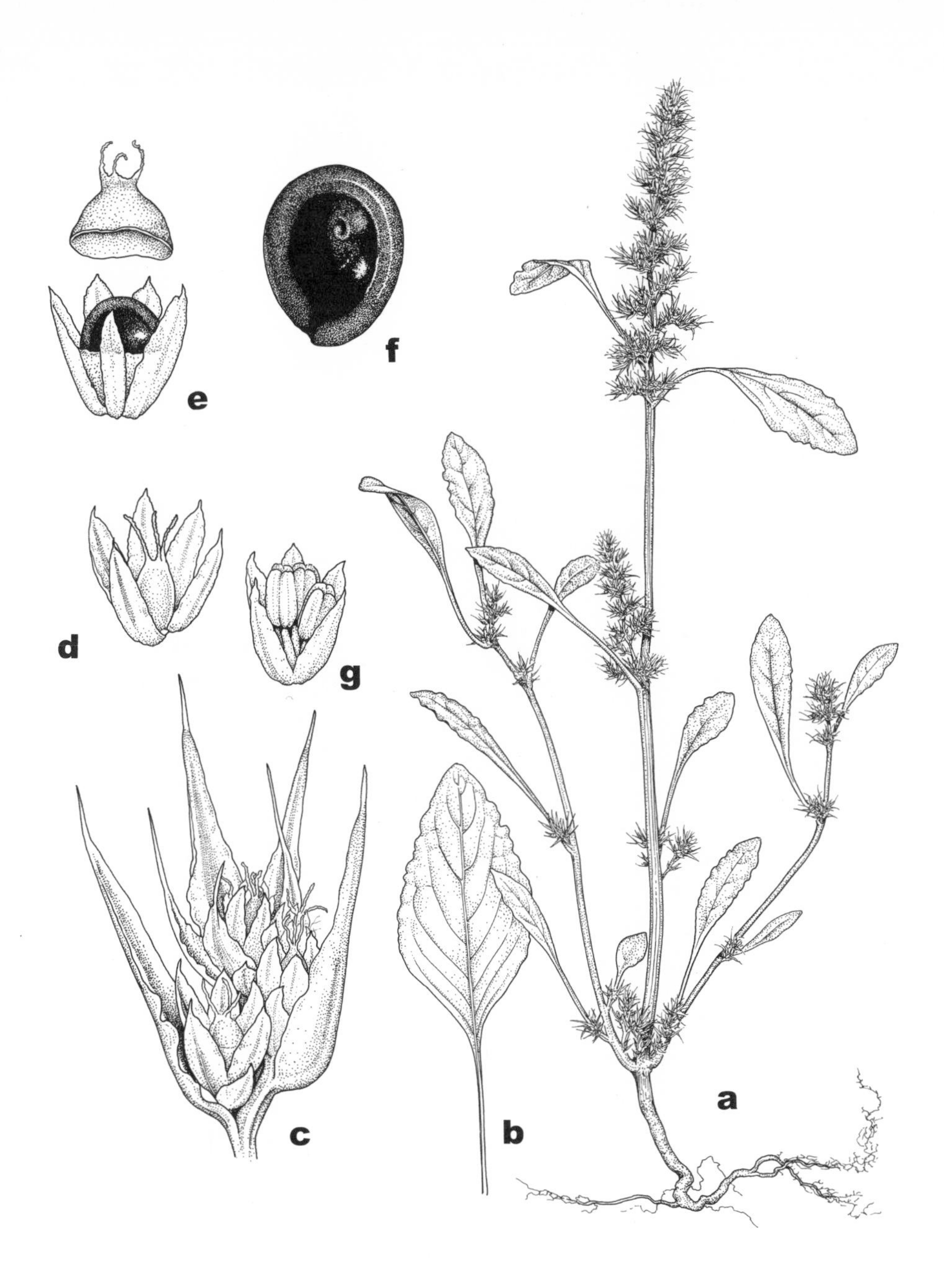

66. **Amaranthus powellii**
a. Habit, pistillate plant
b. Leaf
c. Pistillate flowering cluster
d. Pistillate flower
e. Fruit
f. Seed
g. Staminate flower

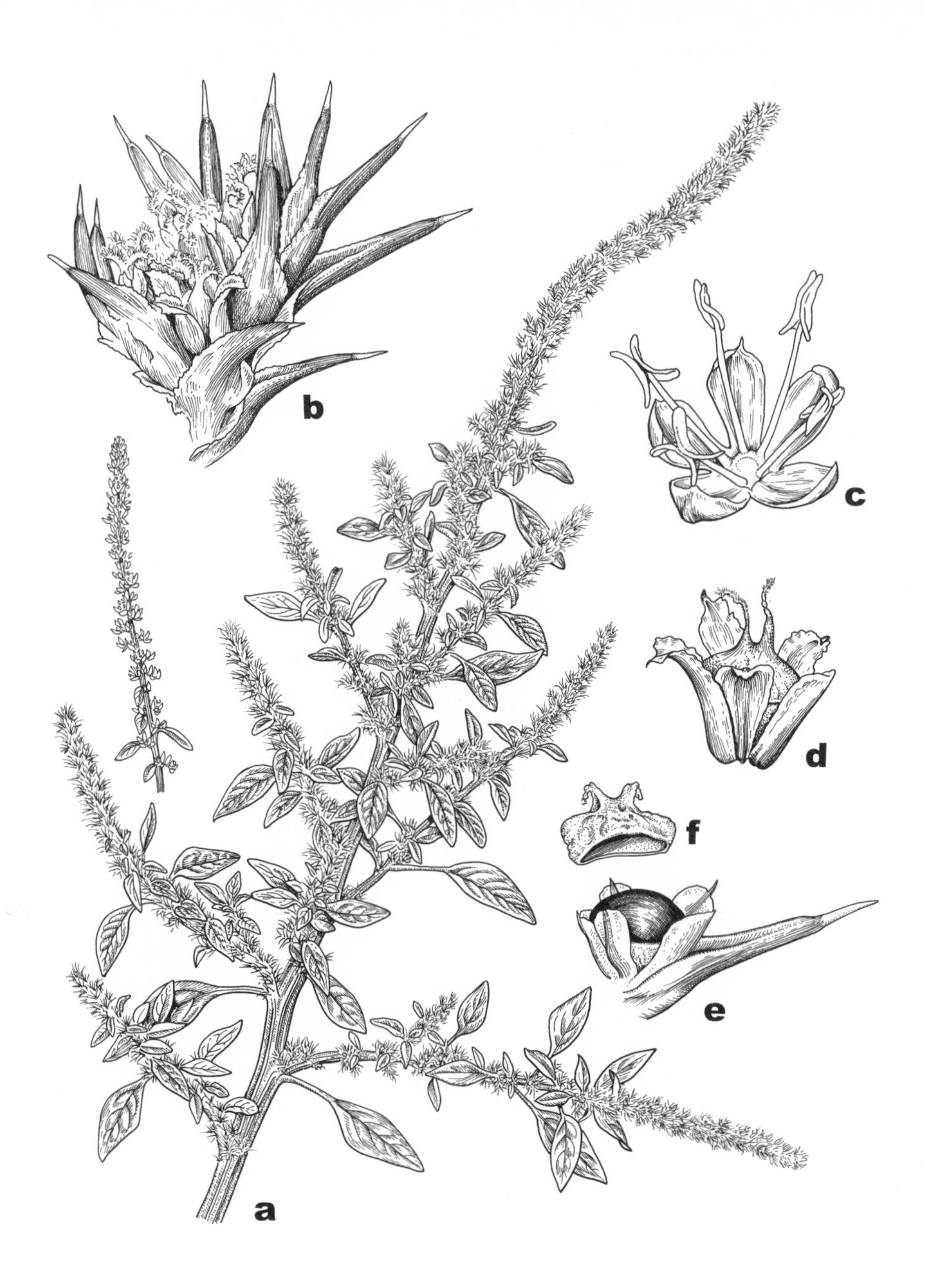

67. **Amaranthus palmeri**
a. Flowering branch
b. Cluster of flowers
c. Staminate flower
d. Pistillate flower
e. Fruit
f. Top of fruit

spikes; bracts 3–6 mm long, those of the staminate flowers thin, deltate, awned, scarious along the margins only along the lower half, those of the pistillate flowers subulate, awned, keeled, scarious along only the lower half of the margins, longer than the flowers; staminate flowers with 5 sepals, the sepals lanceolate to narrowly ovate, acuminate, 2–3 mm long; pistillate flowers with 5 sepals, the sepals obovate, acute and cuspidate, 2–3 mm long; stamens 5; styles 2 (–3); fruit a 1-seeded utricle, subglobose, more or less rugose, shorter than the sepals; seeds lenticular, 1.0–1.3 mm long, shiny, dark red-brown.

Common Name: Palmer's Amaranth.
Habitat: Disturbed soil.
Range: Native to the southwestern United States; sparingly adventive in the eastern United States.
Illinois Distribution: Scattered throughout the state.

This is one of three species of *Amaranthus* in Illinois that is dioecious and with five sepals in the pistillate flowers. It differs from *A. ambigens,* which has shorter, thicker spikes, and from *A. arenicola,* which has bracts only about as long as the flowers and bracts scarious along the margins all the way to the tip.

Amaranthus palmeri flowers from August to October.

10. **Amaranthus ambigens** Standl. N. Am. Fl. 21:106. 1917. Fig. 68.

Annual dioecious herb from an elongated taproot; stems rather stout, erect, to 1.5 m tall, mostly unbranched, glabrous; leaves oblong to oval, obtuse at the apex, cuneate at the base, to 7 cm long, to 2.5 cm broad, entire, glabrous, with slender petioles about half as long as the blade; flowers unisexual, the staminate in terminal paniculate spikes up to 5 cm long and in shorter, more densely crowded axillary spikes, the pistillate in simple or paniculate pendulous spikes up to 15 cm long; bracts lanceolate to deltate, 1.5–2.0 mm long, about half as long as the sepals; staminate flowers with 5 sepals, the sepals 2–3 mm long, broadly lanceolate, acute to acuminate; pistillate flower with 5 sepals, the sepals 2–3 mm long, broadly lanceolate, acute to acuminate; stamens 5; styles 3; fruit an indehiscent utricle, subglobose, 2–3 mm long; seeds orbicular, 1.0–1.1 mm in diameter.

Common Name: Water Hemp.
Habitat: Moist soil.
Range: Northwestern Illinois to southeastern Minnesota.
Illinois Distribution: Known only from Winnebago County.

This is presumably a very enigmatic plant, in part because there are so very few collections of it. Some botanists believe it is an aberrant form of *A. tuberculatus,* but I reject this view.

Amaranthus ambigens is also enigmatic because it superficially looks like some plants of *A. tuberculatus,* but it has a calyx in the pistillate flowers of five well developed sepals, whereas the sepals in the pistillate flowers of *A. tuberculatus* are reduced to one or two rudiments or even absent.

68. **Amaranthus ambigens**
a. Habit, staminate plant
b. Habit, pistillate plant
c. Staminate flower, unopened
d. Staminate flower
e. Pistillate flower
f. Ovary
g. Seed

The fact that *A. ambigens* differs in characters of the pistillate sepals, the characters that are deemed significant in the taxonomy of the genus, proves the case for recognizing *A. ambigens. Amaranthus arenicola* also is enigmatic in that it, too, appears more like the asepalous *A. tuberculatus* than it does other species of *Amaranthus* that have sepals.

The reason for accepting *A. arenicola* without question seems to be that it is a common plant, whereas *A. ambigens* is rare and, therefore, an anomaly.

Amaranthus ambigens differs from *A. tuberculatus* and *A. rudis* in having pistillate flowers with five sepals. It differs from *A. palmeri* and *A. arenicola,* the other two species of *Amaranthus* that are dioecious and that have five sepals in the pistillate flower, in that its fruits are indehiscent.

The type collection of this species was made by M. S. Bebb at Fountaindale, Winnebago County. The only other specimen I have seen of this species was collected in southeastern Minnesota.

Amaranthus ambigens flowers from July to October.

11. **Amaranthus arenicola** I. M. Johnst. Journ. Arn. Arb. 29:193. 1948. Fig. 69.

Annual dioecious herb from a long taproot; stems erect, to nearly 2 m tall, branched or unbranched, usually glabrous, whitish; leaves narrowly oblong to ovate, obtuse to subacute and mucronate at the apex, rounded to cuneate at the base, to 8 cm long, to 5 cm broad, glabrous, usually yellow-green, conspicuously reticulate, at least on the lower surface, with petioles sometimes slightly longer than the blade; flowers unisexual, in slender continuous or interrupted, terminal and axillary spikes, the spikes to 40 cm long, to 1 cm thick; bracts 1.5–2.5 mm long, lanceolate, acuminate, shorter than to about as long as the sepals, with scarious margins extending from the base of the bract to the tip; staminate flowers with 5 sepals, the sepals 2–3 mm long, oblong, acuminate, cuspidate; pistillate flowers with 5 sepals, the sepals about 2 mm long, obovate, obtuse and often mucronate at the apex; stamens 5; styles 2; fruit a 1-seeded utricle, subglobose, about 1.5 mm long; seeds lenticular, 1.0–1.3 mm long, shiny, dark red-brown.

Common Name: Sandhills Amaranth.
Habitat: Disturbed, sandy soil.
Range: Native to the western United States; occasionally adventive in the eastern United States.
Illinois Distribution: Known from Cook, Crawford, DuPage, Fayette, Jackson, Mason, Morgan, Vermilion, and White counties.

This species of *Amaranthus* is common in sandhills and on river sandbars in the western United States. In Illinois, it has been found in disturbed, sandy soil.

Amaranthus arenicola differs from the other dioecious species of *Amaranthus* as follows: from *A. palmeri* by its more obtuse pistillate sepals and its bracts usually shorter than the flowers; from *A. ambigens* by its longer, more slender spikes

69. **Amaranthus arenicola**
a. Habit, pistillate plant
b. Pistillate flower
c. Pistillate bract
d. Fruit
e. Seed
f. Habit, staminate plant
g. Staminate bract
h. Staminate flower

and its dehiscent fruit; from both *A. tuberculatus* and *A. rudis* by possessing five sepals in each pistillate flower. It is often impossible to distinguish staminate plants of *A. arenicola* from *A. tuberculatus.*

Prior to 1948, and then sometimes since then, botanists have called this species *A. torreyi*, but that binomial belongs to a different species.

Amaranthus arenicola flowers from July to October.

12. Amaranthus rudis J. Sauer, Madrono 21:428. 1972. Fig. 70.
Amaranthus tamariscinus Nutt. Trans. Am. Phil. Soc. II 5:165. 1837, misapplied.
Acnida tamariscina (Nutt.) Wood, Bot. & Fl. 289. 1873.

Annual dioecious herb from an elongated taproot; stems slender or more commonly stout, erect, to 2 m tall, glabrous, branched or unbranched; leaves lanceolate to oblong to narrowly ovate, obtuse and sometimes emarginate at the apex, cuneate at the base, to 10 cm long, to 3.5 cm broad, glabrous, entire, with slender petioles up to half as long as the blade; flowers unisexual, the staminate crowded into slender terminal spikes up to 20 cm long and up to 1 cm thick, or crowded into a short panicle of spikes, the pistillate crowded into a solitary or paniculate group of very slender spikes; bracts 1.5–2.0 mm long, excurrent, shorter than the staminate sepals but longer than the fruits; staminate flowers with 5 sepals, the sepals 2–3 mm long, acuminate with a stiff subulate tip; pistillate flowers with 1–2 sepals, if 2, then 1 of them very rudimentary; stamens 5; styles 3; fruit a dehiscent, circumscissile utricle, ovoid, 1.3–1.5 mm long; seeds lenticular, about 1 mm long, dark red-brown.

Common Name: Water Hemp.
Habitat: Moist disturbed soil.
Range: Michigan to South Dakota, south to New Mexico, Texas, and Arkansas; adventive in the eastern United States.
Illinois Distribution: Scattered throughout the state.

Swink and Wilhelm (1994) believe this species has been introduced into the Chicago region. Some plants from the western counties of Illinois seem to be native as they grow in moist sand along rivers.

Amaranthus rudis and *A. tuberculatus* are the only species of *Amaranthus* in Illinois with reduced sepals in the pistillate flowers. As a result, these two species are sometimes segregated into the genus *Acnida.* Sauer (1955) has given reasons for combining *Acnida* with *Amaranthus.* Until 1961, most botanists called this plant either *Acnida tamariscina* or *Amaranthus tamariscinus.*

This species differs from *A. tuberculatus* in its regularly circumscissile utricle that is shorter than the subtending bract and in its stiff, subulate-tipped sepals in the staminate flowers.

Robertson (1999), however, believes that the differences separating *A. rudis* from *A. tuberculatus* overlap to such an extent that only one species should be recognized.

Amaranthus rudis flowers from July to October.

70. **Amaranthus rudis**
a. Habit
b. Staminate flowering branch
c. Pistillate flowering branch
d. Pistillate flower
e. Staminate flower
f. Bract
g. Fruit
h. Seed

13. Amaranthus tuberculatus (Moq.) J. Sauer, Madrono 13:18. 1955.
Acnida tuberculata Moq. in DC. Prodr. 13 (2):277. 1849.

Annual dioecious herb from a slender taproot; stems slender to stout, prostrate to ascending to erect, to 2.5 m tall, glabrous, branched or unbranched; leaves oblanceolate to lanceolate to obovate, obtuse at the apex, less commonly acute, cuneate at the base, to 15 cm long, to 8 cm broad, entire, glabrous, with slender petioles up to half as long as the blade; flowers unisexual, the staminate in continuous, slender, axillary spikes, or the spikes arranged in a terminal panicle, the pistillate in continuous or interrupted, slender or globose spikes, both axillary and terminal; bracts lanceolate, acute and sharp-tipped, 3.5–4.5 mm long, longer than the sepals and the fruits; staminate flowers with 5 sepals, the sepals 2–3 mm long, acute to acuminate, often cuspidate; pistillate flowers with 1–2 sepals, if 2, then one of them very rudimentary; stamens 5; styles 3; fruit a 1-seeded utricle that bursts irregularly, not circumscissile; seeds lenticular, 0.7–1.0 mm long, dark red-brown. This is a very characteristic species of moist, open soils.

Because of the reduced sepals in the pistillate flower, this species is sometimes segregated into the genus *Acnida.* Two other Illinois species are similar in appearance: *A. ambigens* has well developed sepals in the pistillate flowers; *A. rudis* has a fruit that dehisces circumscissily, and the fruits are shorter than the bracts.

Three varieties, differing in growth form and disposition of the pistillate flowers, can usually be differentiated in Illinois. They may be separated by the following key:

1. Plants erect; leaves up to 15 cm long; pistillate flowers crowded into slender spikes .... 13a. *A. tuberculatus* var. *tuberculatus*
1. Plants prostrate to ascending; leaves usually no more than 8 cm long; pistillate flowers arranged in dense, many-flowered glomerules or loose, few-flowered glomerules.
   2. Plants prostrate to ascending; pistillate flowers aggregated into dense, many-flowered, globular glomerules .... 13b. *A. tuberculatus* var. *subnudus*
   2. Plants entirely prostrate; pistillate flowers arranged in loose, few-flowered glomerules ... . 13c. *A. tuberculatus* var. *prostratus*

13a. Amaranthus tuberculatus (Moq.) J. Sauer var. **tuberculatus** Fig. 71.
Amaranthus altissimus Riddell, Syn. Fl. W. States 41. 1835.
Acnida altissima (Riddell) Riddell ex Moq. in DC. Prodr. 13 (2)278. 1849.

Plants erect; leaves up to 15 cm long; pistillate flowers crowded into slender spikes.

Common Name: Water Hemp.
Habitat: Moist soil, particularly stream banks, pond margins, and sandbars; also in moist, disturbed soil.
Range: Ontario to North Dakota, south to Colorado, Missouri, and Tennessee; adventive at locations east of this range.
Illinois Distribution: Scattered throughout the state.

This variety is apparently a little more common than the other two varieties in Illinois. Its erect growth form and its slender pistillate spikes are distinctive.

Amaranthus tuberculatus var. *tuberculatus* flowers from August to October.

71. **Amaranthus tuberculatus**
a. Habit, pistillate plant
b. Pistillate flower
c. Pistillate flower
d. Seed
e. Habit, staminate plant
f. Staminate flower
g. Bract of staminate flower

13b. Amaranthus tuberculatus (Moq.) J. Sauer var. **subnudus** (S. Wats.) Mohlenbr., comb. nov.
Acnida tuberculata var. *subnuda* S. Wats. in A. Gray, Man. Bot. ed. 6, 429.1889.
Acnida altissima (Riddell) Riddell var. *subnuda* (S. Wats.) Fern. Rhodora 43:288. 1941.

Plants prostrate to ascending; leaves usually no more than 8 cm long; pistillate flowers aggregated into dense, many-flowered, globular glomerules.

Common Name: Water Hemp.
Habitat: Moist sandy soil.
Range: Quebec to North Dakota, south to Kansas, Missouri, and Kentucky.
Illinois Distribution: Scattered in Illinois.

This variety flowers from August to October.

13c. Amaranthus tuberculatus (Moq.) J. Sauer var. **prostratus** (Uline & Bray) Mohlenbr., comb. nov.
Amaranthus tamariscinus Nutt. Trans. Am Phil. Soc. II 5:165. 1837.
Acnida tamariscina Riddell var. *prostrata* Uline & Bray, Bot. Gaz. 20:158. 1895.
Acnida tuberculata Moq. var. *prostrata* (Uline & Bray) B. L. Robins. Rhodora 43:288. 1941.

Common Name: Prostrate Water Hemp.
Habitat: Disturbed moist sandy soil.
Range: Ontario to South Dakota, south to Colorado, New Mexico, and Louisiana; adventive in the eastern United States, including Illinois.
Illinois Distribution: Not common in the state.

This entirely prostrate variety flowers from August to October.

2. **Celosia** L.—Cockscomb

Annual herb (in Illinois); stems erect; leaves alternate, petiolate; inflorescence spicate or paniculate, terminal and sometimes axillary, many-flowered; flowers perfect, subtended by a bract and 2 bracteoles; sepals 5, free, glabrous, scarious, concealing the fruit; stamens 5, the filaments connate below to form a cup; stigmas 2–3, the styles very short to elongate, persistent; ovules 1–many; utricle circumscissile, with 3–many seeds; seeds black.

There are about sixty species in the genus, many of them native to the tropics. Several species are common garden ornamentals.

Only the following species has been found in Illinois.

1. Celosia argentea L. Sp. Pl. 1:205. 1753. Fig. 72.

Annual herb; stems erect, glabrous, to 1 m tall; leaves ovate to lanceolate to narrowly linear, acute to acuminate at the apex, tapering to the base, to 15 cm long, to 7 cm broad, the petioles to 2 cm long; inflorescence a dense spike or abnormally

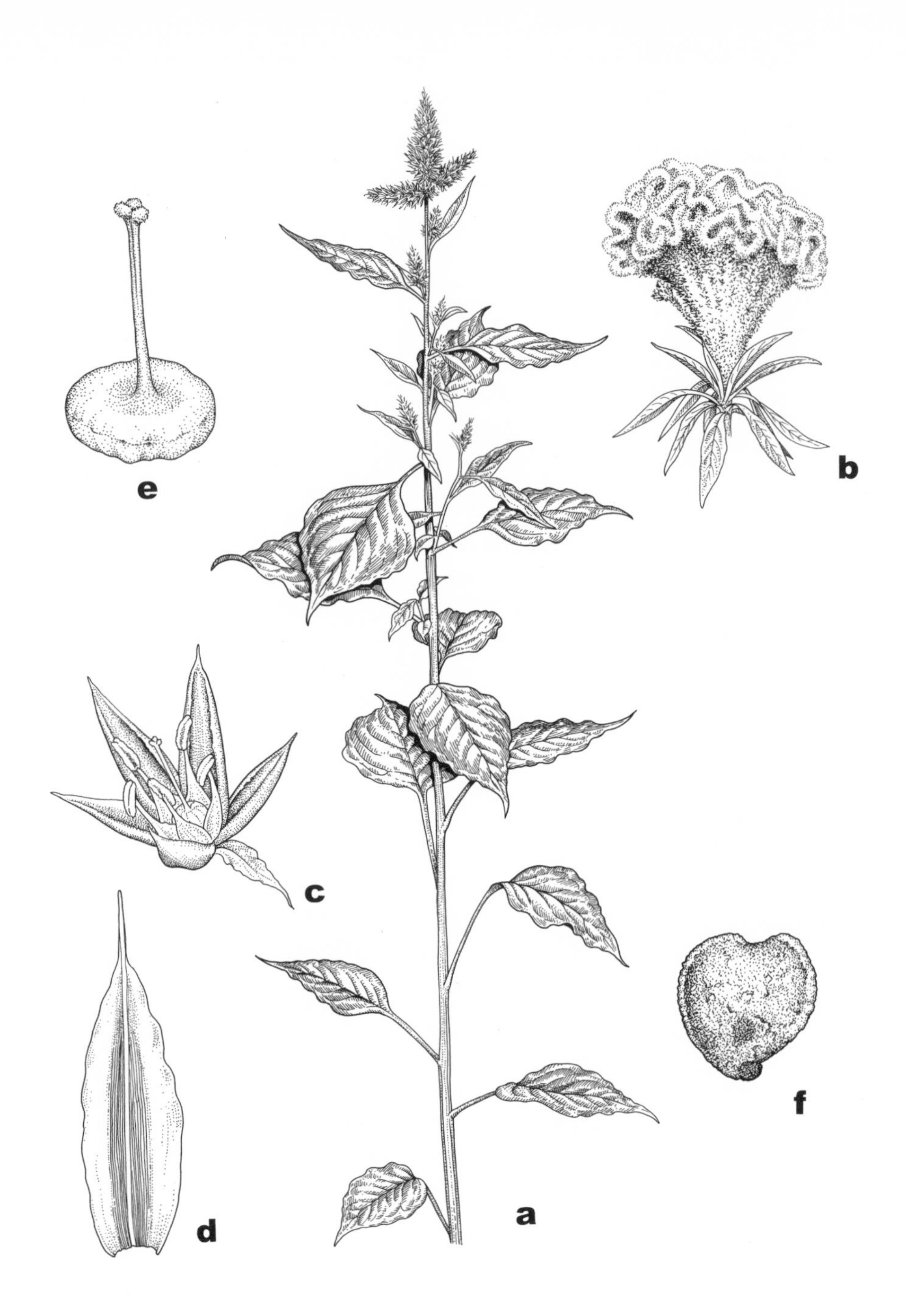

72. **Celosia argentea**
a. Habit
b. Crested flowering head
c. Flower
d. Sepal
e. Portion of pistillate flower
f. Seed

crested; sepals silvery-white to pinkish (red or yellow in cultivated varieties), scarious, 6–8 mm long, much longer than the bracts; stamens 5; styles 3–4 mm long, persistent; utricle up to 4 mm long, with 3–10 seeds; seeds smooth, shiny, black, about 1.5 mm in diameter.

Common Name: Celosia; Cockscomb.
Habitat: Disturbed soil.
Range: Native to the tropics; rarely adventive in the United States; occasionally escaped from cultivation.
Illinois Distribution: Known from Peoria and Vermilion counties.

The plant that was collected in Vermilion County is an escaped cockscomb and has a red crested inflorescence. It may be known as var. *cristata* (L.) O. Kuntze. It is sometimes considered a distinct species known as *C. cristata* L. The Peoria County specimen has a silvery, noncrested inflorescence.

This species flowers from July until October.

3. **Tidestromia** Standl.—Tidestromia

Annual herb (in Illinois); leaves alternate or opposite; flowers perfect, 1–5 in axillary clusters, enclosed by an involucre of hardened bractlike leaves; bracts present; sepals 5, more or less free; petals 0; stamens 5, the filaments united below to form a cup, with 5 staminodia; style 1; ovary superior, 1-locular; fruit a 1-seeded, indehiscent utricle.

The seven species that comprise this genus inhabit the deserts of western North America.

Only the following species occurs in Illinois.

1. **Tidestromia lanuginosa** (Nutt.) Standl. Journ. Wash. Acad. 6:70. 1916. Fig. 73.
Achyranthes lanuginosa Nutt. Trans. Am. Phil. Soc. II 5:166. 1820.

Annual herb from a slender taproot; stems prostrate to spreading to ascending, much branched, to only 15 cm tall but the plant often much broader, gray-tomentose, at least when young; leaves opposite, obovate to broadly ovate, obtuse at the apex, cuneate or rounded at the base, to 3 cm long, to 2 cm broad, entire, gray-tomentose with stellate hairs, at least when young, with petioles sometimes as long as the blade; flowers few, in axillary clusters, subtended by small leaves; bracts scarious; sepals 5, more or less free, 1.8–2.5 mm long, reflexed; petals absent; stamens 5, united below; fruit globose, 5–6 mm in diameter; seed 1, obovoid, 4.5–5.0 mm long, brown.

Common Name: Tidestromia.
Habitat: Disturbed soil.
Range: South Dakota to California, east to Texas; Mexico; rarely adventive east of the native range.
Illinois Distribution: Known from Cook and St. Clair counties.

73. **Tidestromia lanuginosa**
a. Habit
b. Habit
c. Floral node
d. Flower

The only Illinois collections of this species were made by W. F. Moffitt and H. Eggert during the nineteenth century.

This species, with inconspicuous flowers and fruits, is distinguished by its gray-tomentose young leaves and stems, its five reflexed sepals, its five stamens that are united below to form a cup, and its five staminodia.

Tidestromia lanuginosa flowers from July to October.

4. **Iresine** P. Br.—Bloodleaf

Annual or perennial (in Illinois), monoecious or dioecious herbs; leaves opposite, simple, petiolate; flowers perfect or unisexual, crowded into spikes arranged in panicles; bracts present; calyx deeply 5-parted; petals 0; stamens 5, the filaments connate at base; styles 2–3, very short; ovary superior, 1-locular, compressed; fruit an indehiscent, 1-seeded utricle; seed inverted.

There are about eighty species in this genus found primarily in North America, Africa, and tropical America.

Only the following species occurs in Illinois.

1. **Iresine rhizomatosa** Standl. Proc. Biol. Soc. Wash. 28:172. 1915. Fig. 74.

Perennial dioecious herb from slender rhizomes; stems erect, to 1.5 m tall, usually unbranched, glabrous or sometimes pilose at the nodes; leaves opposite, simple, thin, lance-ovate to ovate, acute to acuminate at the apex, cuneate at the base, to 15 cm long, to 7 cm broad, entire, glabrous or sparsely pubescent on the veins above and on the surface beneath, bright green, decurrent onto a short petiole; flowers unisexual, borne in spikes aggregated into panicles, the panicles terminal and in the uppermost leaf axils, to 30 cm long; bracts ovate, shorter than the calyx; calyx 5-parted, the lobes 1.2–1.5 mm long, silvery-white, ovate-lanceolate, subtended by long hairs in the pistillate flowers; petals 0; stamens 5; fruit a 1-seeded, indehiscent utricle, globose, 2.0–2.5 mm in diameter; seed 1, suborbicular, about 0.5 mm in diameter, shiny, dark red.

Common Name: Bloodleaf.
Habitat: Wet woods.
Range: Maryland to Kansas, south to Texas and Virginia.
Illinois Distribution: Known from Crawford, Massac, Pulaski, and Wabash counties.

This rare species of wet woods may be extirpated from Crawford, Pulaski, and Wabash counties, but it still occurs in Massac County.

Iresine rhizomatosa is distinguished by its opposite leaves, its silvery-white unisexual inflorescences, and the long white hairs that subtend the sepals in the pistillate flowers.

This species flowers from August to October.

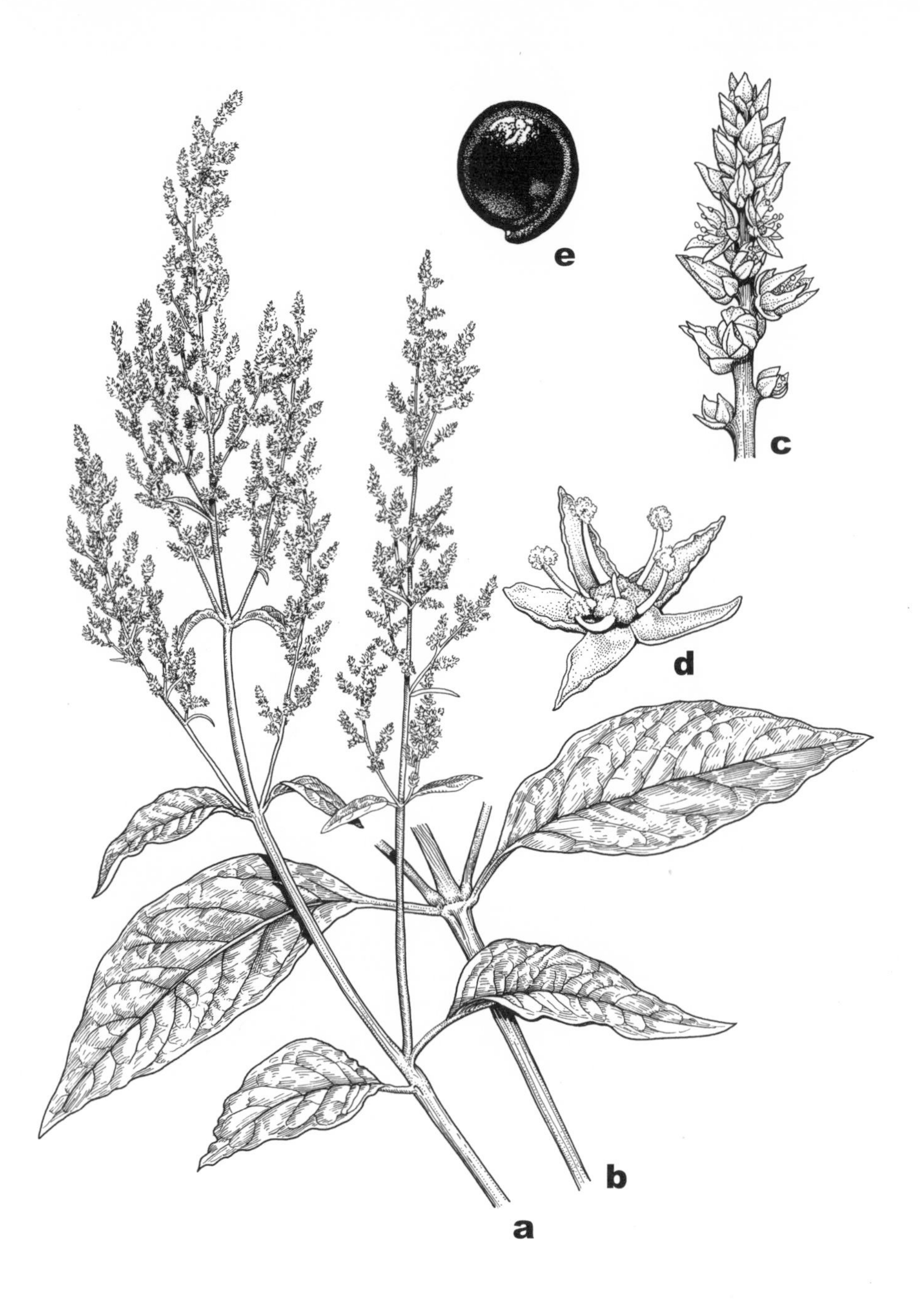

74. **Iresine rhizomatosa**
a. Flowering branch
b. Node with leaves
c. Staminate flowering branch
d. Staminate flower
e. Seed

5. **Alternanthera** Forsk.—Alligator-weed

Annual or perennial (in Illinois) herbs; leaves opposite, simple; flowers perfect, crowded into terminal and axillary, white or silvery, capitate spikes; sepals 5, free, white; petals 0; stamens usually 5, united near the base into a short tube, with 5 staminodia; style 1; ovary superior, 1-locular; fruit a 1-seeded, indehiscent utricle, compressed; seed 1, lenticular.

Approximately 170 species comprise this genus. Most of them occur naturally in tropical America.

Only the following species occurs in Illinois.

1. **Alternanthera philoxeroides** Griseb. Goett. Abh. 24:36. 1879. Fig. 75.

Perennial emergent aquatic or semiterrestrial herb; stems prostrate to decumbent, mat-forming, branched or unbranched, stoloniferous, to 1 m long, glabrous or with a pair of puberulent lines; leaves opposite, more or less fleshy, linear to narrowly obovate, acute and sometimes mucronate at the apex, cuneate at the base, to 10 cm long, to 2 cm broad, entire, glabrous; flowers perfect, borne in terminal and axillary capitate spikes; sepals 5, free, silvery-white, 5–6 mm long, lanceolate to narrowly ovate, glabrous; petals 0; stamens 5; style 1; fruit a 1-seeded, indehiscent, compressed utricle; seed 1.

Common Name: Alligator-weed.
Habitat: Disturbed soil along a river.
Range: North Carolina to Florida, west to Texas; adventive as far north as southern Illinois and as far west as California.
Illinois Distribution: Known only from Alexander County.

This is a very aggressive and noxious aquatic that tends to clog up waterways in the southeastern United States. It has been collected during the last decade of the twentieth century along the Mississippi River in Alexander County by John Schwegman.

Alternanthera philoxeroides is readily recognized by its opposite leaves and its silvery-white, capitate inflorescences.

This species flowers from June to August.

6. **Froelichia** Moench—Cottonweed; Snake-cotton

Annual, biennial, or perennial herbs from taproots; stems woolly, silky, or variously pubescent; leaves opposite, simple, entire; flowers perfect, borne in spikes, with scarious bracts; calyx tubular, lanate, 5-lobed, becoming tuberculate or crested in fruit; petals 0; stamens 5, the filaments united into a tube; style 1; ovary superior, 1-locular; fruit an indehiscent, 1-seeded utricle included within the filament tube; seed 1, inverted.

This genus is unique in its lanate calyx tube and in the fruits that are included within the filament tube.

75. **Alternanthera philoxeroides**

a. Habit
b. Inflorescence
c. Cluster of flowers
d. Fruit

Froelichia is a genus of about twenty species, all found in North America. Two taxa occur in Illinois.

Key to the Taxa of **Froelichia** in Illinois

1. Stems stout, usually more than 0.5 m tall; leaves 8–20 mm broad; calyx in fruit merely denticulate or entire . . . . 1. *F. floridana* var. *campestris*
1. Stems slender, rarely more than 0.5 m tall; leaves up to 9 mm broad; calyx in fruit spinose . . . . 2. *F. gracilis*

1. Froelichia floridana (Nutt.) Moq. var. **campestris** (Small) Fern. Rhodora 43:336. 1941. Fig. 76.
Froelichia campestris Small, Fl. S.E. U.S. 397. 1903.

Annual herb from an elongated taproot; stems rather stout, erect, up to 1.5 m tall, sparsely branched, at least the branches sericeous-tomentose; leaves opposite, spatulate to oblanceolate, obtuse to acute at the apex, cuneate at the base, to 10 cm long, 8–20 mm broad, entire, usually sericeous on both surfaces with fulvous hairs intermingled on the lower surface; flowers perfect, crowded into dense spikes, the spikes to 10 cm long, to 1.3 cm thick, whitish, on lanate peduncles; bracts acuminate, occasionally cuspidate, scarious, much shorter than the calyx; calyx tubular, lanate, to 6 mm long, usually with 2 tuberculate ridges, the 5 lobes oblong, obtuse; stamens 5; style 1; fruit an indehiscent, 1-seeded utricle, with the persistent calyx denticulate to entire; seed 1, lenticular, 1.3–1.5 mm long, brown.

Common Name: Cottonweed.
Habitat: Disturbed soil in sandy fields.
Range: Wisconsin to Minnesota, south to Colorado, Oklahoma, and central Illinois.
Illinois Distribution: Occasional in the northern two-thirds of Illinois; also Jackson County.

This plant differs from *F. gracilis* in its taller, stouter stems; its broader leaves; and the persistent calyx around the fruit that is merely denticulate or entire.

Froelichia floridana var. *campestris* flowers from June to September.

2. Froelichia gracilis (Hook.) Moq. in DC. Prodr. 13 (2):420. 1849. Fig. 77.
Oplotheca gracilis Hook. Icones Pl. sub pl. 256. 1840.

Annual herb from an elongated taproot; stems slender, erect, up to 50 cm tall, branched, usually villous-tomentose; leaves opposite, linear-lanceolate to elliptic to lanceolate, acute to acuminate at the apex, cuneate at the base, to 10 cm long, to 9 mm broad, entire, sericeous; flowers perfect, crowded into dense spikes, the spikes to 3 cm long, to 8 mm thick, whitish, on lanate peduncles; bracts acute to acuminate, stramineous to nearly black, sericeous, shorter than the calyx; calyx tubular, lanate, to 6 mm long, with 2 rows of spines, the 5 sepals linear to narrowly oblong, acute; petals 0; stamens 5; style 1; fruit an indehiscent, 1-seeded utricle with the persistent calyx spinose; seed 1, lenticular, 1.3–1.5 mm long, yellow-brown.

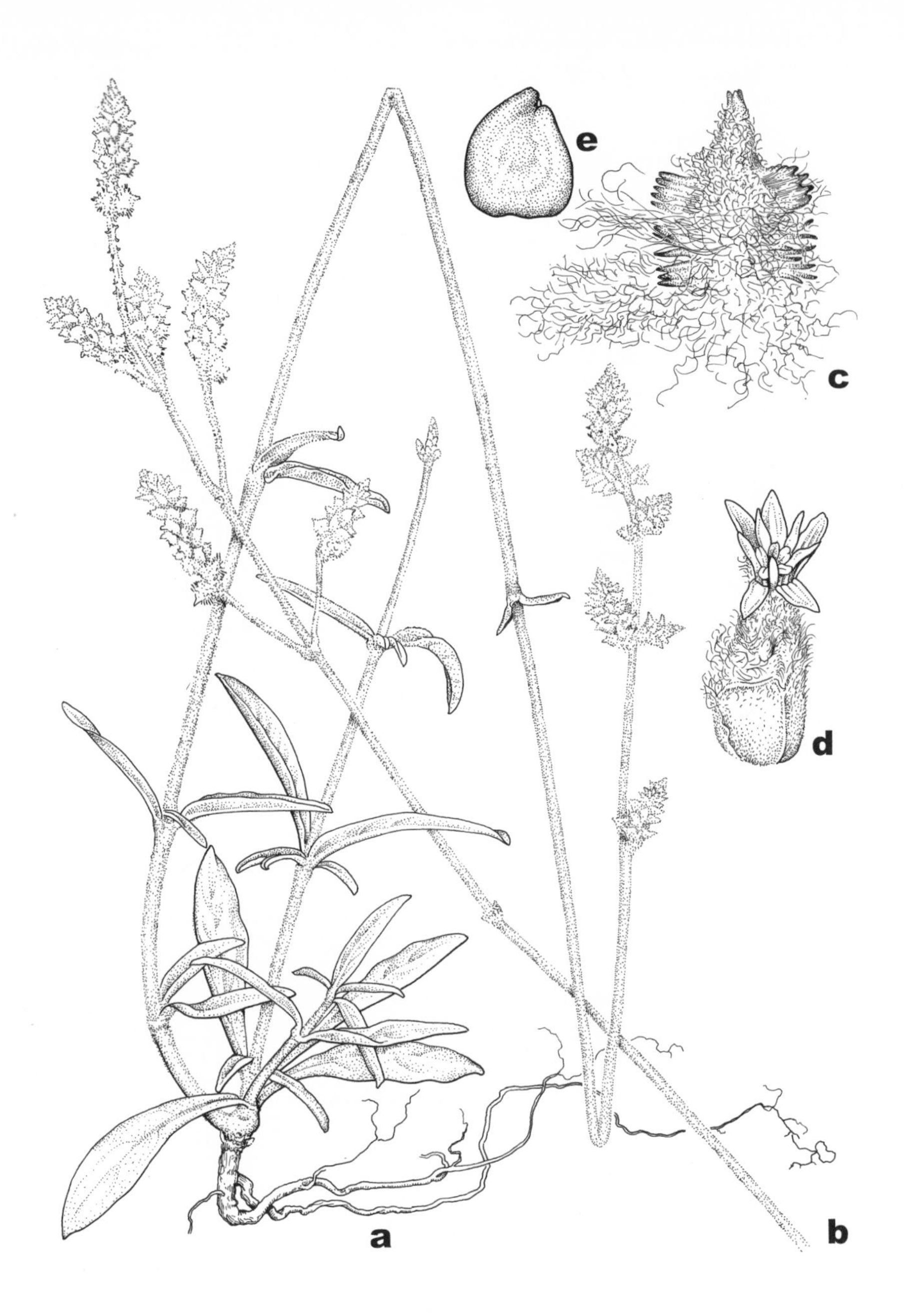

76. **Froelichia floridana** var. **campestris**

a. Habit with rare branching at base
b. Flowering branch
c. Fruit
d. Silky flower
e. Seed

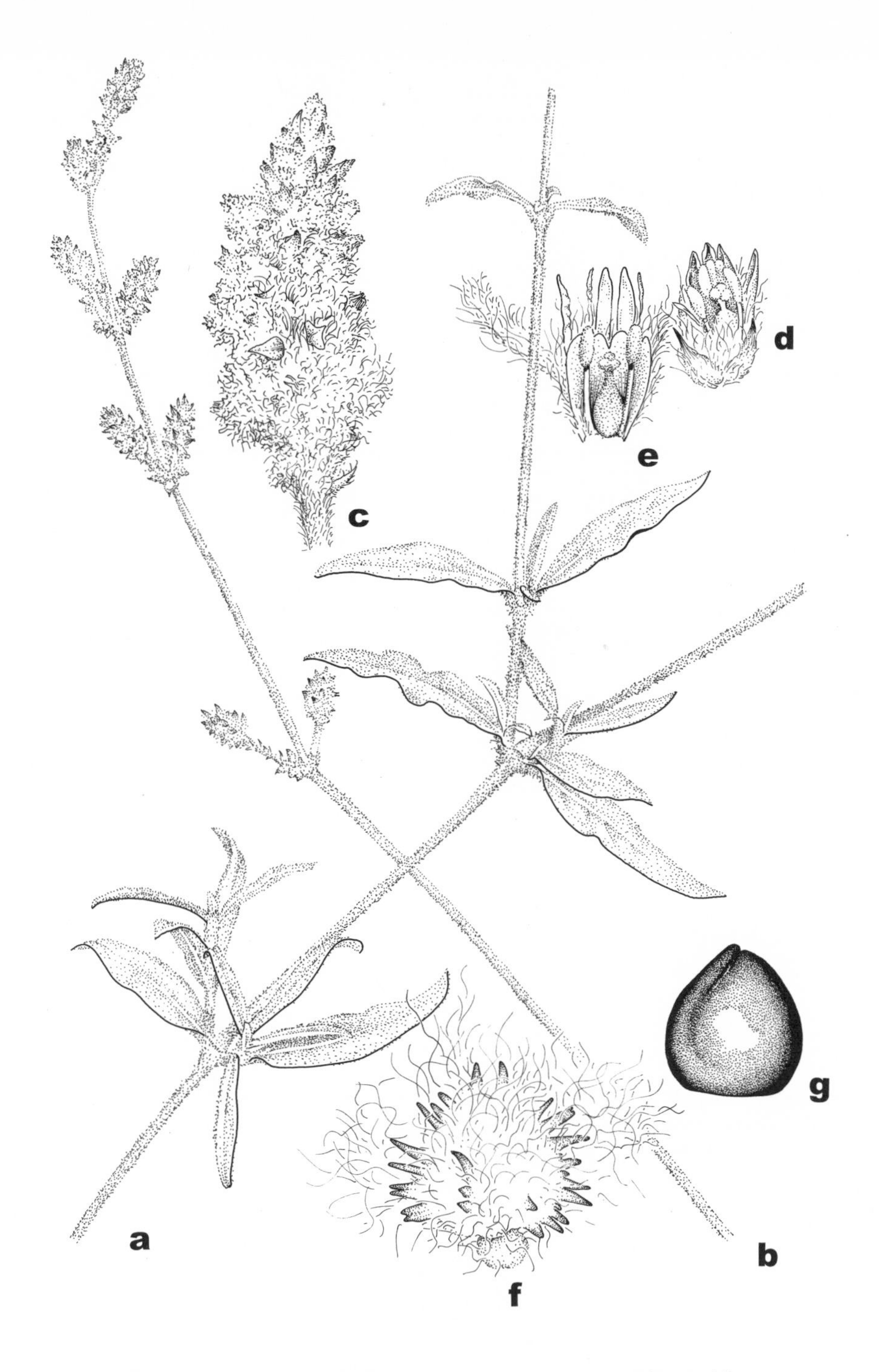

77. **Froelichia gracilis**
a. Leafy branch
b. Flowering branch
c. Spike
d. Open flower
e. Dissected flower
f. Fruit
g. Seed

Common Name: Cottonweed.
Habitat: Sandy or gravelly soil, particularly along railroads.
Range: Iowa to Colorado, south to Texas and Arkansas; Mexico; adventive in the eastern United States.
Illinois Distribution: Scattered throughout the state.

This species, which is probably not native in Illinois, is a frequent inhabitant of railroad rights-of-way.

Froelichia gracilis differs from *F. floridana* var. *campestris* in its shorter, more slender stature; its narrower leaves; and its spinose persistent calyx on the fruit.

This species flowers from May to September.

Caryophyllaceae—Pink Family

Annual, biennial, or perennial herbs; leaves simple, opposite or less commonly whorled, entire (rarely minutely serrulate), with or without stipules; inflorescence cymose or paniculate, terminal or axillary, or flower solitary in the leaf axils, with or without bracts; flowers actinomorphic, usually perfect; sepals (4–) 5, free or united into a tubular or urn-shaped calyx; petals (4–) 5, free, occasionally absent, sometimes with basal appendages; stamens 5 or 10, rarely 4 or 8; ovary superior, 1-locular; styles 2, 3 or 5, rarely 4; fruit a 1-seeded utricle or a capsule; seeds often tuberculate or roughened.

There are approximately eighty-five genera and twenty-four thousand species of Caryophyllaceae in the world, particularly in arctic and alpine areas of the northern hemisphere.

The members of this family are extremely uniform in characteristics. All are herbaceous. Almost all have opposite, entire, simple leaves and flowers that are 5-merous. Nonetheless, the family was split into three during the nineteenth century. The Alsinaceae included plants with the sepals more or less free from each other, and the Caryophyllaceae included plants with the sepals united to form a tubular or urn-shaped calyx. In addition, some botanists segregated the Illecebraceae for those plants with no petals and a 1-seeded utricle.

Key to the Genera of **Caryophyllaceae** in Illinois

1. Leaves with stipules, although sometimes the stipules reduced to threadlike structures less than 1 mm wide.
    2. Leaves whorled . . . . 4. *Spergula*
    2. Leaves opposite.
        3. Petals present; styles 3; fruit a capsule, with several seeds . . . . 3. *Spergularia*
        3. Petals absent; styles 2; fruit a 1-seeded utricle . . . . 1. *Paronychia*
1. Leaves without stipules.
    4. Fruit a 1-seeded utricle; petals absent . . . . 2. *Scleranthus*
    4. Fruit a capsule, with several seeds; petals present (absent in *Stellaria pallida*, *Sagina apetala*, and occasionally absent in *Sagina procumbens*, *Silene antirrhina*, and on lateral branches of *Cerastium glomeratum*).
        5. Sepals free from each other; petals not long-clawed.

6. Each petal 2-cleft, usually deeply so, or jagged toothed.
7. Styles 3.
8. Petals jagged toothed at the apex .... 6. *Holosteum*
8. Petals 2-cleft at the apex.
9. Capsule curved, at least in the upper half ...... *Cerastium dubium*
9. Capsule straight .... 7. *Stellaria*
7. Styles 5.
10. Capsule curved, at least in the upper half; styles opposite the sepals .... 9. *Cerastium*
10. Capsule straight; styles alternate with the sepals .... 8. *Myosoton*
6. Each petal entire, or petals absent.
11. Petals absent.
12. Styles 3; stamens 1–3 .... *Stellaria pallida*
12. Styles 4–5; stamens 4, 5, 8, or 10.
13. Styles 4; stamens 4 or 8.
14. Pedicels and sepals glandular-pubescent .... 5. *Sagina*
14. Pedicels and sepals glabrous .... 10. *Moenchia*
13. Styles 5; stamens 5 or 10.
15. Plants usually matted, not viscid-pubescent; sepals 1 mm long .... 5. *Sagiina*
15. Plants upright, viscid-pubescent; sepals 3.5–5.5 mm long .... *Cerastium glomeratum*
11. Petals present.
16. Styles 4; stamens 4 or 8; sepals 4; petals 4 .... 10. *Moenchia*
16. Styles 5; stamens 5 or 10; sepals 5; petals 5.
17. Leaves subulate, setaceous, filiform, or linear ...... 11. *Minuartia*
17. Leaves oval, oblong, or ovate.
18. Perennial with slender rhizomes; leaves obtuse to subacute; sepals obtuse, 2–3 mm long .... 12. *Moehringia*
18. Annual; leaves acuminate; sepals acuminate, 3–4 mm long .... 13. *Arenaria*
5. Sepals united, tubular or urn-shaped; petals long-clawed.
19. Styles 3–5; teeth of valves of fruit 3, 5, 6, or 10.
20. Bracts 2–6.
21. Calyx lobes longer than the tube, usually at least 15 mm long .... 18. *Agrostemma*
21. Calyx lobes shorter than the tube, up to 13 mm long.
22. Plants densely white-tomentose; calyx lobes twisted .... 16. *Lychnis*
22. Plants variously pubescent or glabrous, not densely white-tomentose; calyx lobes not twisted.
23. Styles 5; flowers bisexual; teeth of valves of fruit usually 5 .... 16. *Lychnis*
23. Styles 3 or, if 5, the plant dioecious; teeth of valves of fruit 3, 5, 6, or 10 .... 17. *Silene*
20. Bracts absent.
24. Calyx up to 5 mm long, conspicuously nerved .... 19. *Gypsophila*
24. Calyx at least 8 mm long, obscurely nerved.
25. Each petal with an awl-shaped scale at the base; flowers 1.8–2.5 cm broad, usually pink or white; calyx 20-nerved, tubular-cylindric .... 20. *Saponaria*
25. Each petal without a scale at the base; flowers less than 1.5 cm broad, red; calyx 5-nerved, ovoid .... 21. *Vaccaria*

19. Styles 2; teeth of valves of fruit usually 4.
  26. Veins or ribs of calyx 20–40 .... 14. *Dianthus*
  26. Veins or ribs of calyx usually 5 .... 15. *Petrorhagia*

1. **Paronychia** Mill.—Forked Chickweed

Annual or perennial herbs; leaves opposite, entire, subtended by scarious stipules; inflorescence usually cymose, with many small flowers; flowers perfect, actinomorphic; sepals 5, free or united at the base; petals 0; stamens 2, 3, or 5, attached to the sepals, sometimes with 5 staminodia; style 2-cleft; ovary superior, l-locular; fruit a 1-seeded utricle; seeds smooth, dark.

Of the fifty species of this genus known from the temperate regions of the world, two are found in Illinois.

Paronychia and *Scleranthus* are unique among the Caryophyllaceae in that all plants are always apetalous and the fruits are 1-seeded utricles.

Key to the Species of **Paronychia** in Illinois

1. Stems glabrous; sepals obtuse to subacute, 1.0–1.5 mm long .... 1. *P. canadensis*
1. Stems pubescent; sepals mucronulate, 2–3 mm long .... 2. *P. fastigiata*

1. Paronychia canadensis (L.) Wood, Classbook Bot. 262. 1861. Fig. 78.
Queria canadensis L. Sp. Pl. 90. 1753.
Anychia dichotoma Michx. Fl. Bor. Am. 1:113. 1803.
Queria capillacea Nutt. Gen. 1:159. 1818.
Anychia canadensis (L.) Ell. Bot. S.C. & Ga. 1:307. 1824.
Anychia capillacea (Nutt.) DC. Prodr. 3:369. 1828.

Annual with slender roots; stems slender, erect, glabrous, repeatedly forked, to 30 cm tall; leaves elliptic to oval, obtuse to acute at the apex, tapering to the base, entire, glabrous, 1-nerved, usually punctate, to 15 mm long, to 8 mm broad, the petioles up to 2 mm long, the stipules small, scarious; inflorescence much branched and spreading, with slender or nearly capillary branches; flowers minute, on pedicels up to 1 mm long, subtended by scarious bracts; sepals 5, oblong-lanceolate, green except for the scarious margins, obtuse to subacute, united below, 1-nerved, 1.0–1.5 mm long; petals 0; stamens 5; styles 2, free except at the base; fruit a 1-seeded utricle, obovoid to subglobose, 1.5–2.5 mm long, longer than the calyx; seed 1, minute, very dark.

Common Name: Slender Forked Chickweed.
Habitat: Dry woods, sandy soils, in shade of trees.
Range: New Hampshire to Minnesota, south to Kansas, Arkansas, and Georgia.
Illinois Distribution: Occasional and scattered throughout the state.

This species is a characteristic plant of dry, sandy woods, but it is often overlooked because of its inconspicuous nature. Swink and Wilhelm (1994) report this species to occur around the base of large trees where the soil is nearly free of litter and of other plants. They also cite a locality on sandy roadside cuts in the Kankakee River valley.

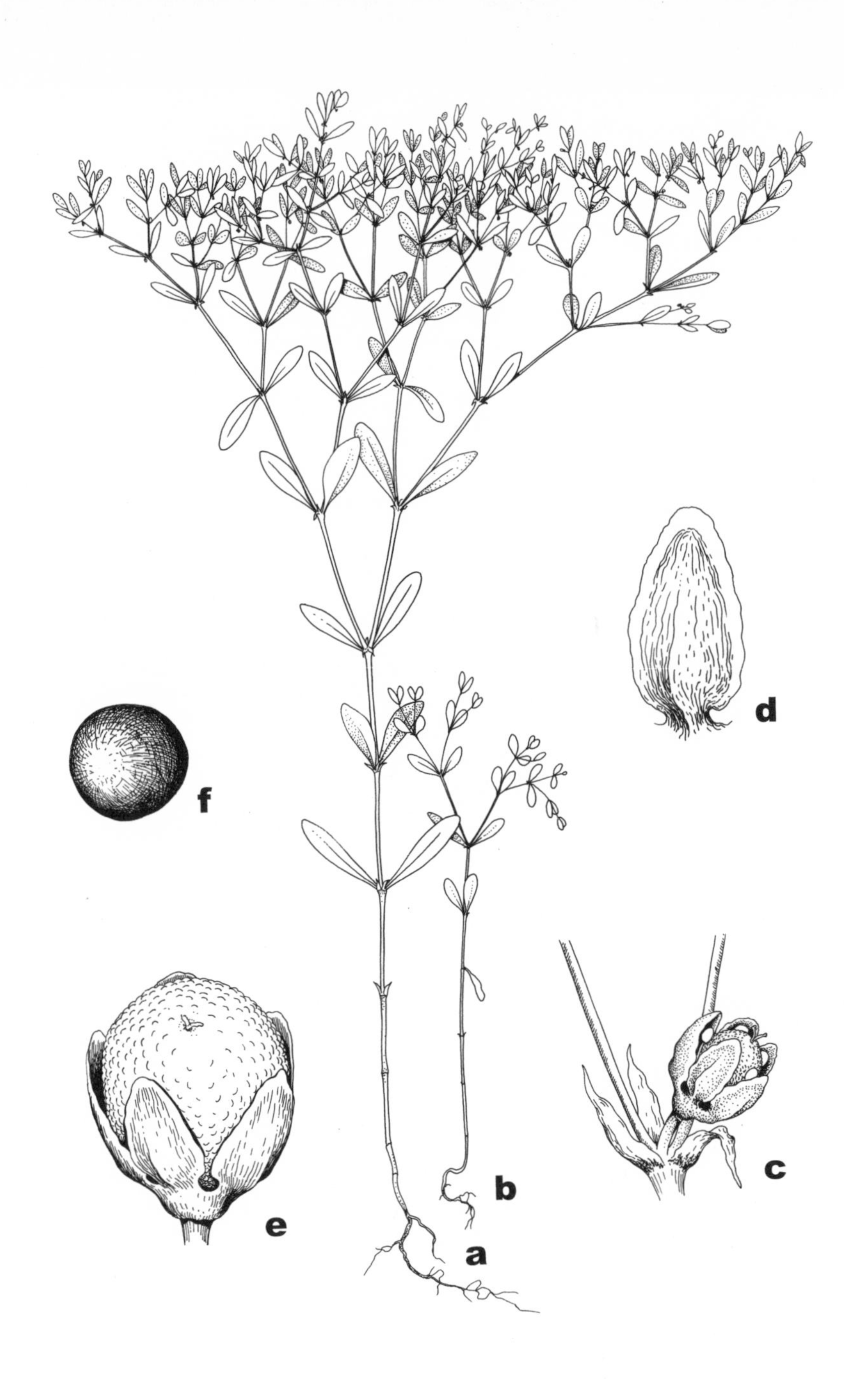

78. **Paronychia canadensis**
a. Habit
b. Habit
c. Flowering node
d. Sepal
e. Fruit
f. Seed

The slender forked chickweed shows a remarkable constancy of characters.

The earliest Illinois botanists referred to this species as *Anychia dichotoma.* Later workers, including Pepoon (1927), called it *Anychia canadensis.*

Paronychia canadensis flowers from June to October.

2. Paronychia fastigiata (Raf.) Fern. Rhodora 38:421. 1936. Fig. 79.
Anychia fastigiata Raf. Atl. Jour. 16. 1832.
Anychia polygonoides Raf. Atl. Jour. 16. 1832.
Paronychia fastigiata (Raf.) Fern. var. *paleacea* Fern. Rhodora 38:421. 1936.

Annual with slender roots; stems slender, wiry, spreading to erect, minutely pubescent, repeatedly forked, to 25 cm tall; leaves narrowly elliptic to oblanceolate, obtuse to acute at the apex, tapering to the base, entire, minutely pubescent, 1-nerved, usually punctate, to 12 mm long, to 2 mm broad, sessile or on petioles less than 1 mm long, the stipules small, scarious; inflorescence much branched and spreading, with slender, wiry branches; flowers minute, sessile or nearly so, subtended by scarious stipules; sepals 5, linear-lanceolate, green, apiculate to subulate and usually slightly cucullate at the apex, united below, strongly 1- to 3-nerved, 2–3 mm long; petals 0; stamens 2, 3, or 5; styles 2, free nearly to the base; fruit a utricle, obovoid, 2–3 mm long, about as long as the calyx; seed 1, less than 1 mm long, dark red.

Common Name: Forked Chickweed.
Habitat: Dry, often sandy or rocky woods in much of the state but in grassy areas in the Chicago region.
Range: Massachusetts to Minnesota, south to Florida and Texas.
Illinois Distribution: Known from the southern two-thirds of the state, plus Cook, Grundy, Kankakee, Putnam, and Will counties.

The forked chickweed is a regular component of dry, rocky woods in the southern part of Illinois.

Fernald (1950) and others recognize two varieties of this species for Illinois. The typical variety is said to have the bracts that subtend the flowers shorter than to equal in length to the calyx. Variety *paleacea* is supposed to have floral bracts equal to or usually longer than the calyx. Because of the presence of many intergrading specimens, I am not recognizing variety *paleacea* as distinct.

Paronychia fastigiata flowers from June to October.

2. **Scleranthus** L.—Knawel

Annuals; leaves opposite, connate at the base, entire, without stipules; inflorescence cymose or reduced; flowers perfect, actinomorphic; sepals 5, united below into a short, cuplike tube; petals 0; stamens usually 5 or 10, attached to the sepals; styles 2; ovary superior, 1-locular; fruit a 1-seeded utricle; seeds minute, beaked.

All ten species that comprise this genus are native to Europe and Asia.

Only the following adventive species occurs in Illinois.

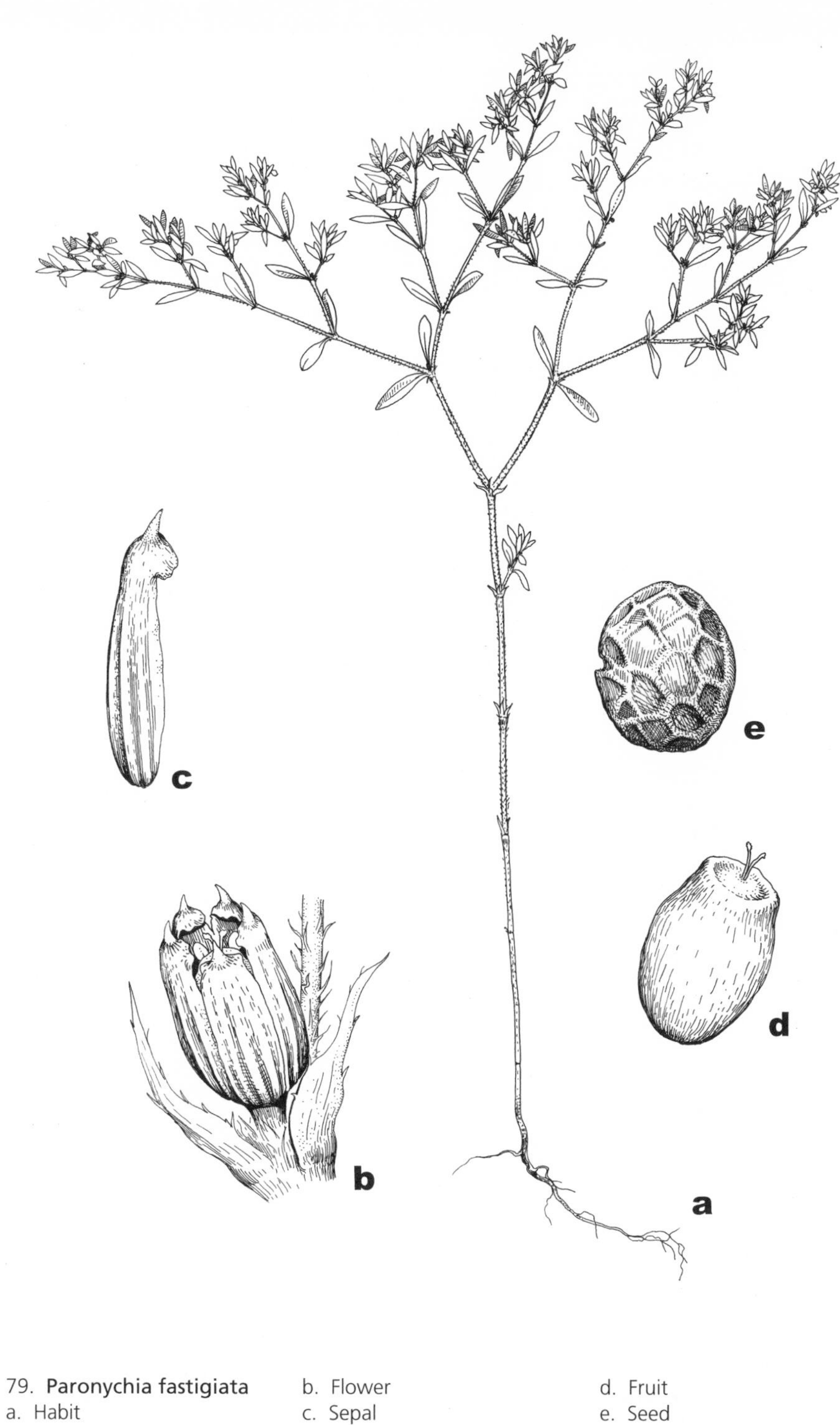

79. **Paronychia fastigiata**
a. Habit
b. Flower
c. Sepal
d. Fruit
e. Seed

1. **Scleranthus annuus** L. Sp. Pl. 406. 1753. Fig. 80.

Annual with long roots; stems slender, rather wiry, prostrate to spreading, glabrous or minutely hairy, repeatedly forked, to 12 cm long; leaves subulate, recurved, glabrous or minutely pubescent, to 15 mm long, without petioles and stipules; flowers minute, actinomorphic, not subtended by bracts; calyx 5-parted, the slenderly deltate lobes slightly shorter than the tube; petals 0; stamens 5 or 10; styles 2, free to the base; fruit a 1-seeded utricle, enclosed by the calyx; seed 1, about 1 mm long, stramineous.

Common Name: Knawel.
Habitat: Disturbed soil, particularly in sandy areas.
Range: Native to Europe and Asia; adventive in the eastern United States and in the Pacific States.
Illinois Distribution: Known from Clay, Cumberland, Kankakee, Lake, McHenry, and Winnebago counties.

This tiny, Eurasian species is a common weed in the eastern United States as far west as Indiana. West of Indiana, this species is very rare except in the Pacific States.

The first collection in Illinois was made in 1946 from Winnebago County by E. W. Fell.

Scleranthus annuus flowers during the summer.

3. **Spergularia** J. & C. Presl—Sand Spurrey

Annual or perennial herbs; leaves opposite, entire, sometimes fleshy, stipulate; inflorescence cymose, with several small flowers; flowers perfect, actinomorphic; sepals 5, united only at the base; petals 5, free; stamens 2–10; styles 3, free; ovary superior, 1-locular, with many ovules; fruit a 3-valved capsule; seeds globose or flattened, smooth or warty, with or without wings.

Spergularia is a genus of nearly fifty species found throughout the temperate world. It is particularly abundant in saline or alkaline areas.

Three species are known from Illinois, separated by the following key:

1. Seeds narrowly winged; sepals 4–6 mm long; petals white .... 1. *S. media*
1. Seeds wingless; sepals up to 4.3 mm long; petals usually pink.
   2. Stamens 6–10; leaves spinulose at the tip, scarcely fleshy .... 2. *S. rubra*
   2. Stamens 2–5; leaves obtuse to mucronate at the tip, more or less fleshy .... 3. *S. marina*

1. **Spergularia media** (L.) C. Presl ex Griseb. Spicil. Fl. Rumel. 1:213. Fig. 81.
Arenaria media L. Sp. Pl. ed. 2, 606. 1762.

Annual or perennial herb, somewhat fleshy; stems prostrate to erect, glabrous or minutely glandular-pubescent, branched, up to 30 cm tall; leaves rarely fascicled, subulate, obtuse to subacute at the apex, glabrous or minutely pubescent, to 4 (–5) cm long, 1.2–1.5 mm broad, sessile, the stipules deltate, to 6 mm long; flowers borne in reduced terminal cymes or solitary in the axils of the upper leaves, pedicellate, the pedicels glabrous or glandular-pubescent, becoming 1 cm or more long during

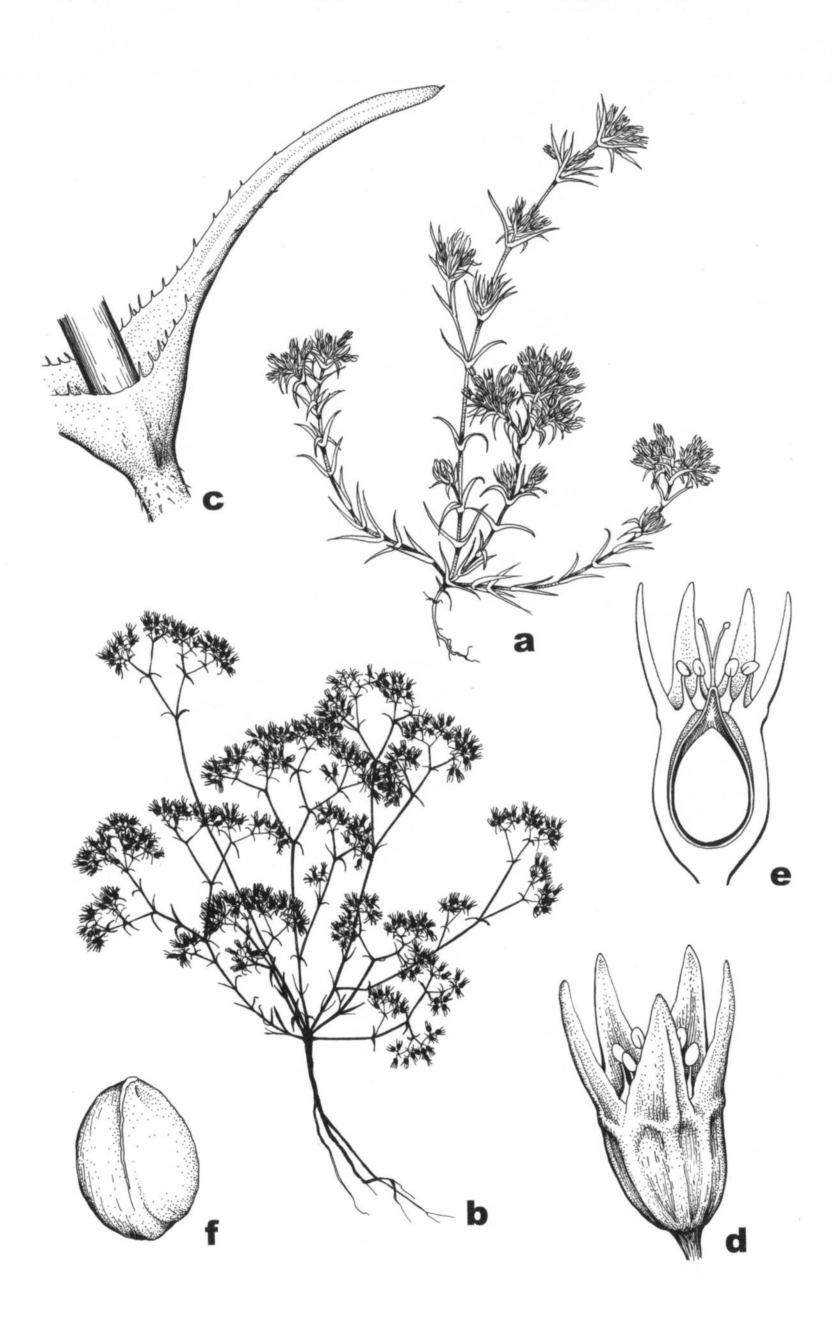

80. **Scleranthus annuus**
a. Habit in flower
b. Habit in fruit
c. Node
d. Flower
e. Cross-section showing fruit and seed enclosed by calyx tube
f. Seed

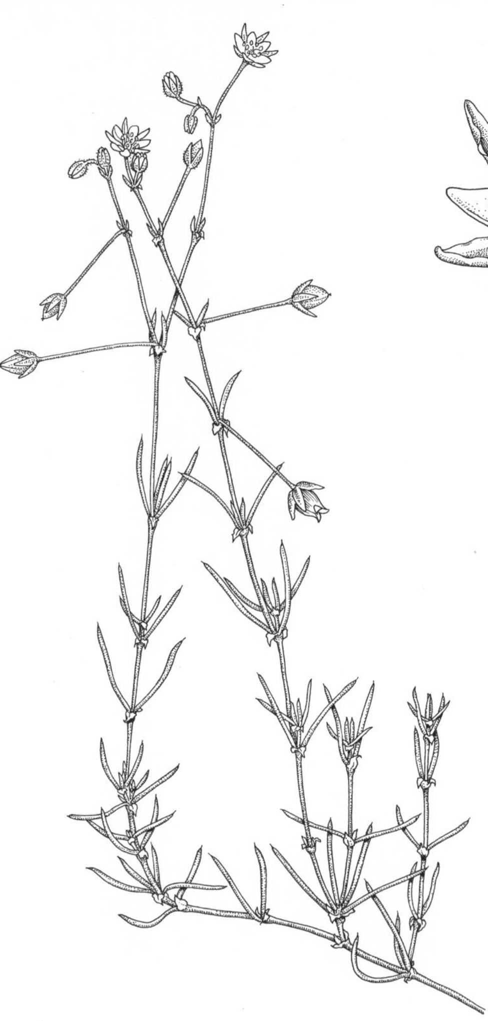

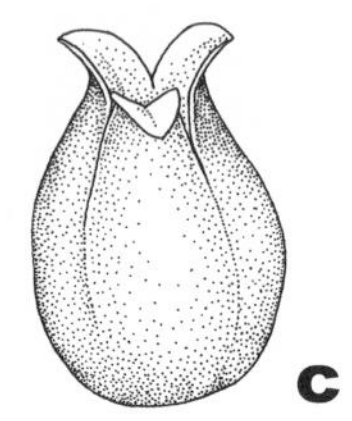

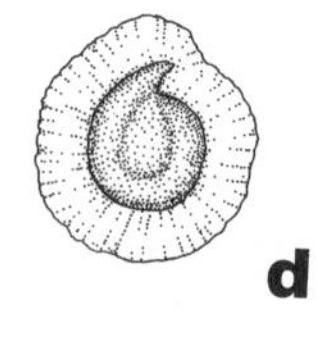

81. **Spergularia media**
a. Habit
b. Flower
c. Fruit
d. Seed

fruiting; sepals 5, barely united at the base, narrowly ovate, green, usually glabrous, 4–6 mm long, longer than the petals; petals 5, free, narrowly ovate, white; stamens 9–10; capsules 4.5–8.0 mm long, broadly ellipsoid; seeds 0.8–1.1 mm long, smooth, dark brown, narrowly winged.

Common Name: Sand Spurrey.
Habitat: Along roads, particularly where salt has been used during the winter for traffic control.
Range: Native to Europe; adventive in the United States, mostly along the coasts, much more rarely inland.
Illinois Distribution: Found thus far only in the extreme northeastern counties of the state; also Marion County.

This species, which grows in saline soils, has found a home along roads in northeastern Illinois where salt is applied during the winter months.

This sand spurrey differs from *S. rubra* and *S. marina* in having white flowers, winged seeds, and longer sepals.

Spergularia media flowers from July to September.

2. Spergularia rubra (L.) J. & C. Presl, Fl. Cech. 94. 1819. Fig. 82.
Arenaria rubra L. Sp. Pl. 423. 1753.

Annual or perennial herb, not fleshy; stems prostrate to ascending, glabrous or minutely glandular-pubescent, branched, up to 25 cm long; leaves fascicled, filiform, spinulose at the tip, glabrous or minutely pubescent, 2.0–2.5 cm long, 1.0–1.2 mm broad, sessile, the stipules narrowly triangular, white, to 5 mm long; flowers generally solitary from the upper axils, pedicellate, the glandular-pubescent pedicels becoming 1 cm or more long during fruiting; sepals 5, barely united at the base, lanceolate, green, usually pubescent, up to 4 mm long, longer than the petals; petals 5, free, lanceolate, pink; stamens 6–10; capsules 3.5–5.0 mm long, narrowly ellipsoid; seeds 0.4–0.6 mm long, warty, reddish to dark brown, wingless.

Common Name: Pink Sand Spurrey.
Habitat: Disturbed soil.
Range: Native to Europe; adventive throughout much of North America.
Illinois Distribution: Known from Kane and Kankakee counties.

This species has been collected from sandy nursery plots in Kankakee County. A Cook County collection by Moffatt, reported by various authors including Mohlenbrock (1986), is actually *S. marina*.

This pink sand spurrey blooms from July to September.

3. Spergularia marina (L.) Griseb. Spicil. Fl. Rumel. 1:213. 1843. Fig. 83.
Arenaria rubra L. var. *marina* L. Sp. Pl. 423. 1753.
Spergularia salina Presl, Fl. Cech. 95. 1819.
Tissa marina (L.) Britt. Bull. Torrey Club 16:126. 1889.

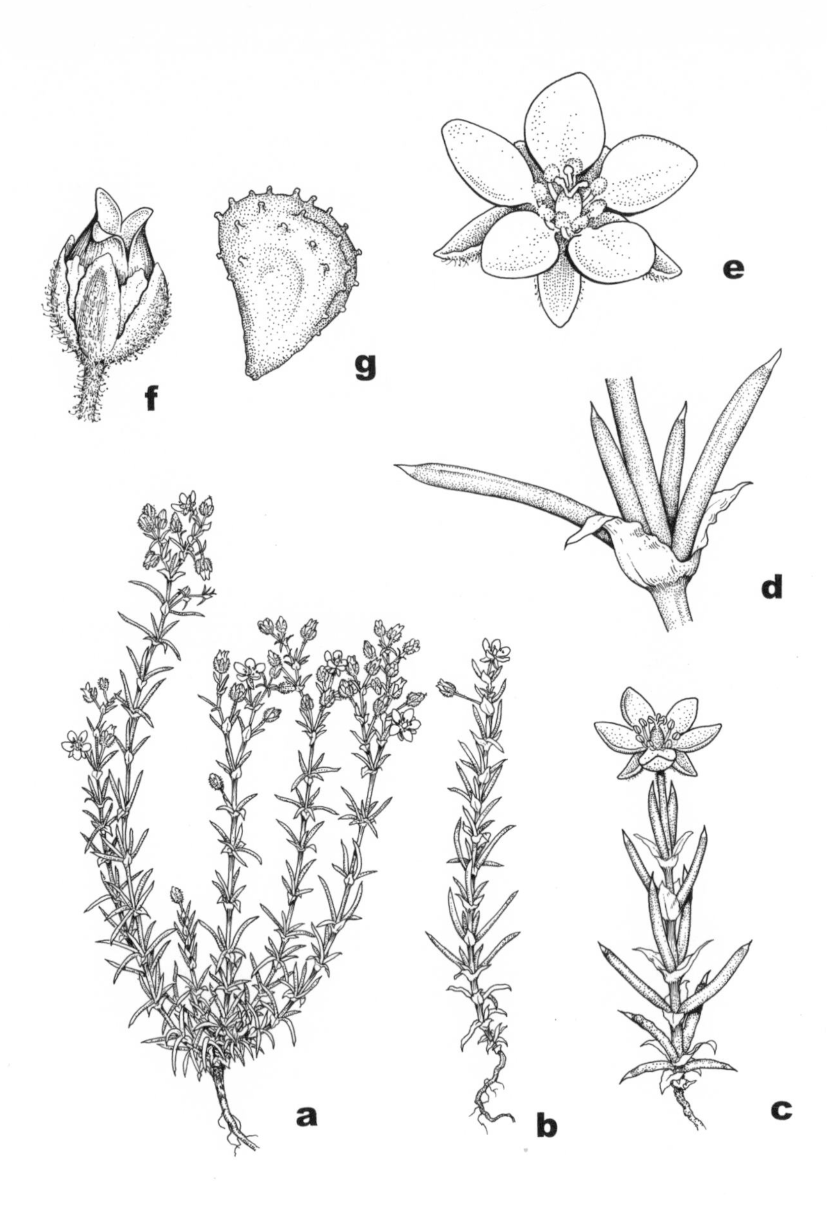

82. **Spergularia rubra**
a. Habit
b. Habit
c. Habit
d. Node
e. Flower
f. Fruit
g. Seed

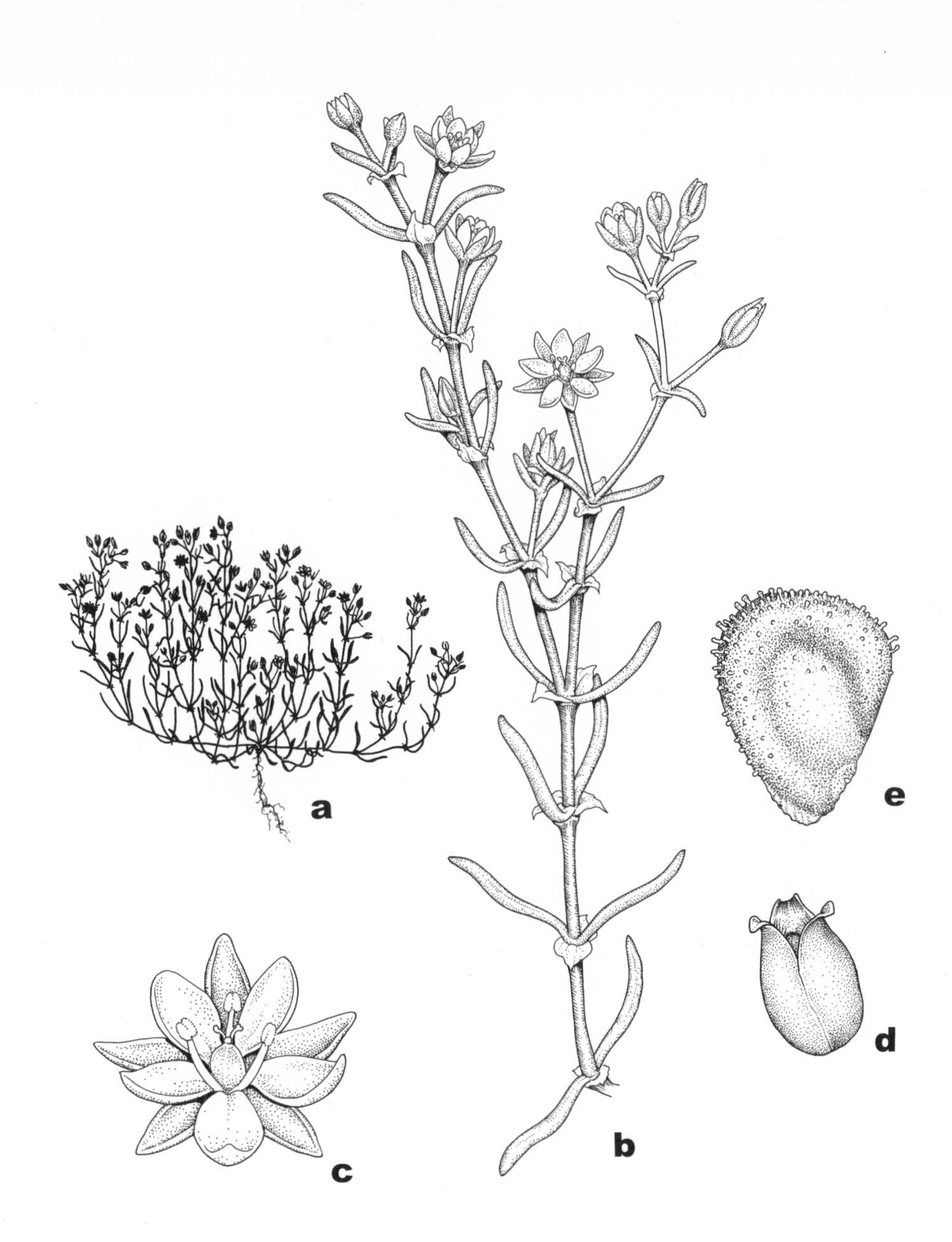

83. **Spergularia marina**
a. Habit
b. Habit
c. Flower
d. Fruit
e. Seed

Annual herb, fleshy, stems prostrate to ascending, glabrous or minutely glandular-pubescent, branched, up to 20 cm long; leaves not fascicled, filiform, obtuse to mucronate at the apex, glabrous or minutely pubescent, to 20 mm long, to 1 mm broad, sessile, the stipules broadly triangular, white, to 3 mm long; flower solitary in the axils of the leaves or few in reduced terminal cymes, pedicellate, the glandular-pubescent pedicels becoming 1 cm long during fruiting; sepals 5, barely united at the base, lanceolate, green, usually pubescent, up to 4.3 mm long, longer than the petals; petals 5, free, lanceolate, usually pink; stamens 2–5; capsules 3.0–6.5 mm long, narrowly ellipsoid; seeds 0.5–0.8 mm long, smooth or slightly warty, light brown to reddish brown, wingless.

Common Name: Coastal Sand Spurrey.
Habitat: Along highways that are heavily salted during the winter.
Range: Native of Europe; adventive in the United States, particularly along the coasts.
Illinois Distribution: Confined to the northeastern counties of the state.

Although Moffatt first collected this plant in Cook County in the last part of the nineteenth century, it has only been since 1974 that it has become an occasional plant of highway rights-of-way.

Spergularia marina differs from the very similar *S. rubra* in having fewer stamens and blunt-tipped, fleshy leaves. It differs from *S. media* in its usually pink flowers, its shorter sepals, and its wingless fruits.

This species flowers from late May to late September.

4. **Spergula** L.—Corn Spurrey

Annuals; leaves appearing whorled, entire, somewhat fleshy, stipulate; inflorescence cymose, with several small flowers; flowers perfect, actinomorphic; sepals 5, free; petals 5, free; stamens 5 or 10; styles 5, free; ovary superior, 1-locular, with many ovules; fruit a 5-valved capsule; seeds globose or flattened, more or less winged.

Five species native to Europe and Asia comprise this genus.

Only the following adventive species is known from Illinois.

1. **Spergula arvensis** L. Sp. Pl. 440. 1753. Fig. 84.

Annual from a taproot; stems ascending to erect, glabrous to minutely glandular-pubescent, branched, up to 45 cm tall; leaves appearing whorled, filiform to linear, often revolute, glabrous or minutely pubescent, to 4 (–5) cm long, with small, connate stipules, the stipules 1–2 mm long; flowers borne in terminal cymes, up to 6 mm across, pedicellate, the pedicels up to 2 cm long in fruit, with usually purplish bracts; sepals 5, free, ovate, obtuse to acute, green, 2–4 mm long, a little longer than the petals; petals 5, free, narrowly ovate to lanceolate, white, 1–3 mm long, persistent on the fruit; stamens 5 or 10; capsules ovoid, 4–6 mm long; seeds 1.0–1.5 mm long, black but with white papillae, very narrowly winged.

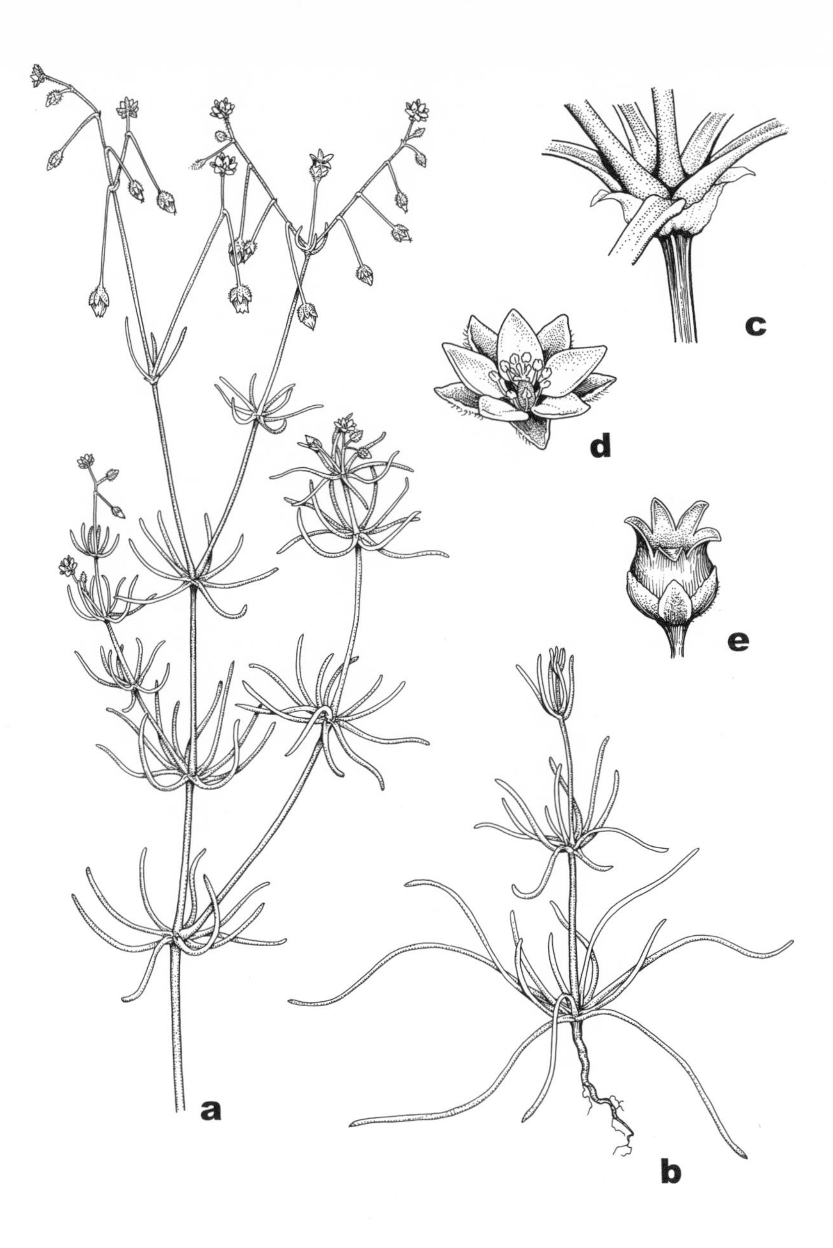

84. **Spergula arvensis**
a. Habit
b. Habit
c. Node
d. Flower
e. Fruit

Common Name: Corn Spurrey.
Habitat: Disturbed soil, sometimes in grain fields.
Range: Native to Europe; adventive throughout the United States but very uncommon in some areas.
Illinois Distribution: Known from Champaign, Cook, and Marion counties.

This weedy annual species is distinguished from all others in the family by its apparently whorled, filiform to linear leaves.

Spergula arvensis flowers from July to September.

5. **Sagina** L.—Pearlwort

Annual or perennial herbs; leaves opposite, entire, without stipules; inflorescence cymose, or reduced to a solitary flower in the upper leaf axils; flowers perfect, actinomorphic; sepals 4–5, free; petals 4 or 5, free, rarely absent; stamens 4, 5, or 10; styles 4–5, free; ovary superior, 1-locular, with many ovules; fruit a 4- to 5-valved capsule; seeds minute, usually dark, unwinged.

Sagina is a genus of about twenty-five low-growing, matted species native to the north temperate regions of the World. Crowe (1978) has revised the North American species.

Four species, distinguished by the following key, are known from Illinois.

1. Annuals without a persistent basal rosette; petals sometimes absent, or usually 5 and often poorly developed.
  2. Petals usually present; sepals usually 5.
    3. Pedicels without glandular-pubescence; seeds reddish brown, with slender ridges .... 1. *S. decumbens*
    3. At least the upper part of the pedicels with glandular-pubescence; seeds dark brown, tuberculate .... 2. *S. japonica*
  2. Petals absent; sepals 4 .... 3. *S. apetala*
1. Perennials with a persistent basal rosette; petals usually 4, well developed .... 4. *S. procumbens*

1. Sagina decumbens (Ell.) Torr. & Gray, Fl. N. Am. 1:177. 1838. Fig. 85.
Spergula decumbens Ell. Bot. S.C. & Ga. 1:523. 1817.
Sagina subulata Torr. & Gray var. *smithii* Gray, Man. Bot. ed. 5, 95. 1867.
Sagina decumbens (Ell.) Torr. & Gray var. *smithii* (Gray) S. Wats. Bibl. Index 1:105. 1878.

Dwarf annual from slender roots; stems usually matted, decumbent to ascending to erect, glabrous or minutely glandular-pubescent, branched or unbranched, up to 15 cm long; leaves linear to subulate, mucronate at the apex, connate at the base, glabrous or minutely glandular-pubescent, to 20 mm long, often with sterile shoots growing from the leaf axils; flowers up to 3 mm across, on ascending, filiform, eglandular pedicels up to 20 mm long, usually remaining straight after fruiting; sepals 5, elliptic to oblong, obtuse to subacute, green, 1.5–2.5 mm long; petals 5 (rarely 1–4 and poorly developed or absent), narrowly elliptic, white, more or less glandular-pubescent, 1–2 mm long; stamens 5, rarely 10; styles 5; capsules ovoid to oblongoid,

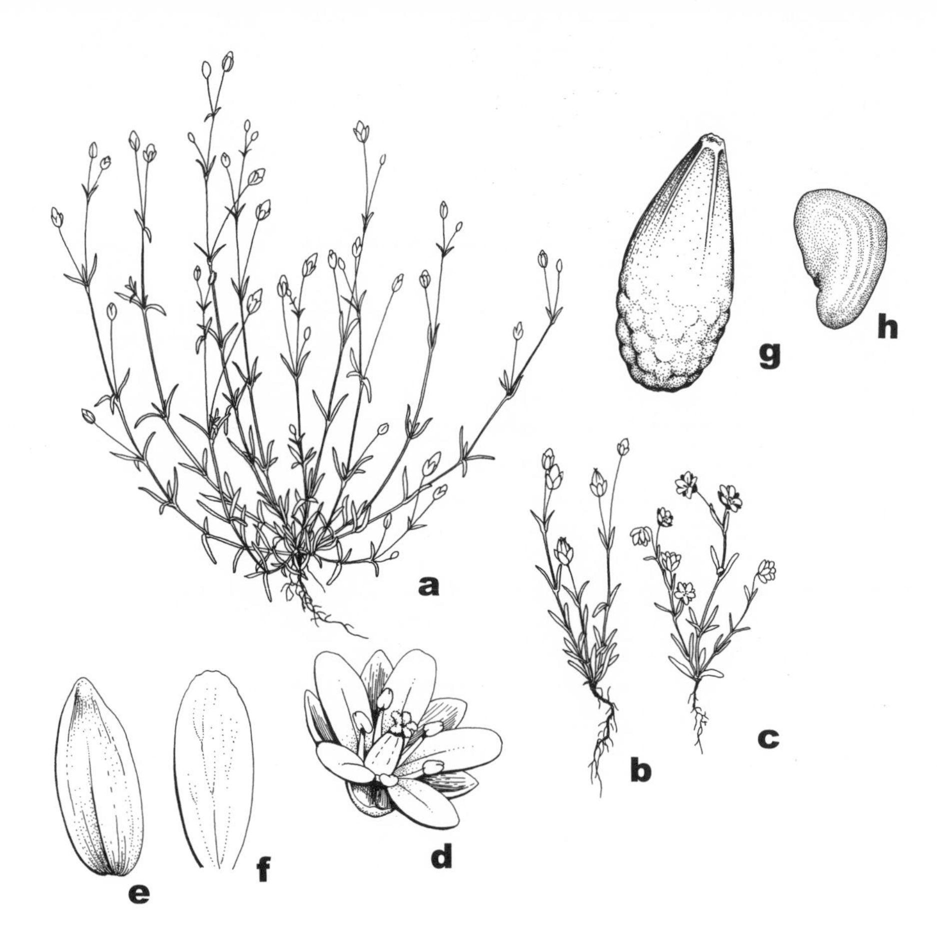

85. **Sagina decumbens**
a. Habit
b. Habit
c. Habit
d. Flower
e. Sepal
f. Petal
g. Fruit
h. Seed

2.0–3.5 mm long, 5-valved; seeds reddish brown, smooth but with slender ridges, 0.2–0.4 mm long.

Common Name: Annual Pearlwort.
Habitat: Moist or dry, often sandy soil; disturbed moist soil.
Range: Massachusetts across Illinois to Kansas, south to Texas and Florida.
Illinois Distribution: Occasional in the southern half of the state; also Champaign and Peoria counties.

This dwarf, matted pearlwort is so small and obscure that it is probably often overlooked.

Most Illinois specimens have 5 petals, although a few flowers with 1–4 poorly developed petals have been observed. Even some apetalous flowers have been seen in Illinois. These apetalous individuals may be called var. *smithii.* Early Illinois botanists called the apetalous plants *S. apetala,* but *S. apetala* has only four sepals per flower.

Although this species is native and occupies some obvious native habitats, it also occurs in weedy situations such as along paths and between the cracks in sidewalks.

This species differs from the very similar *S. procumbens* in its annual habit without a persistent basal rosette of leaves; its fruiting pedicels, which remain straight during fruiting; and its usually five petals. It differs from *S. japonica* by its eglandular pedicels and its ridged, reddish brown seeds.

Sagina decumbens flowers from April to June.

2. Sagina japonica (Sw.) Ohwi, Journ. Jap. Bot. 13:438. 1937. Fig. 86.
Spergula japonica Sw. Gesells. Naturforsch. Freulnde Berl. 3:164, t. 1, f. 2. 1801.

Dwarf annual from slender roots; stems decumbent to ascending to erect, filiform, glabrous or glandular-puberulent, branched or unbranched, up to 12 cm long; rosette leaves, if present, soon withering in early spring; leaves linear to subulate, mucronate at the apex, glabrous or glandular-puberulent, to 15 mm long, connate at base; flowers up to 3 mm across, on ascending, more or less erect, filiform pedicels, the pedicels glandular-puberulent at least in the upper half, up to 20 mm long, remaining straight after fruiting; sepals 5, elliptic to oblong, obtuse to subacute, green, glandular-pubescent, 2.0–2.5 mm long; petals usually 5, ovate to orbicular, white, 1–2 mm long; stamens usually 5; capsules ovoid to oblongoid to globose, 2.5–3.5 mm long, 5-valved; seeds dark brown or black, tuberculate, 0.4–0.5 mm long.

Common Name: Japanese Pearlwort.
Habitat: Disturbed soil.
Range: Asia; Hawaii; rarely adventive in the United States.
Illinois Distribution: The only Illinois collection is from Sangamon County.

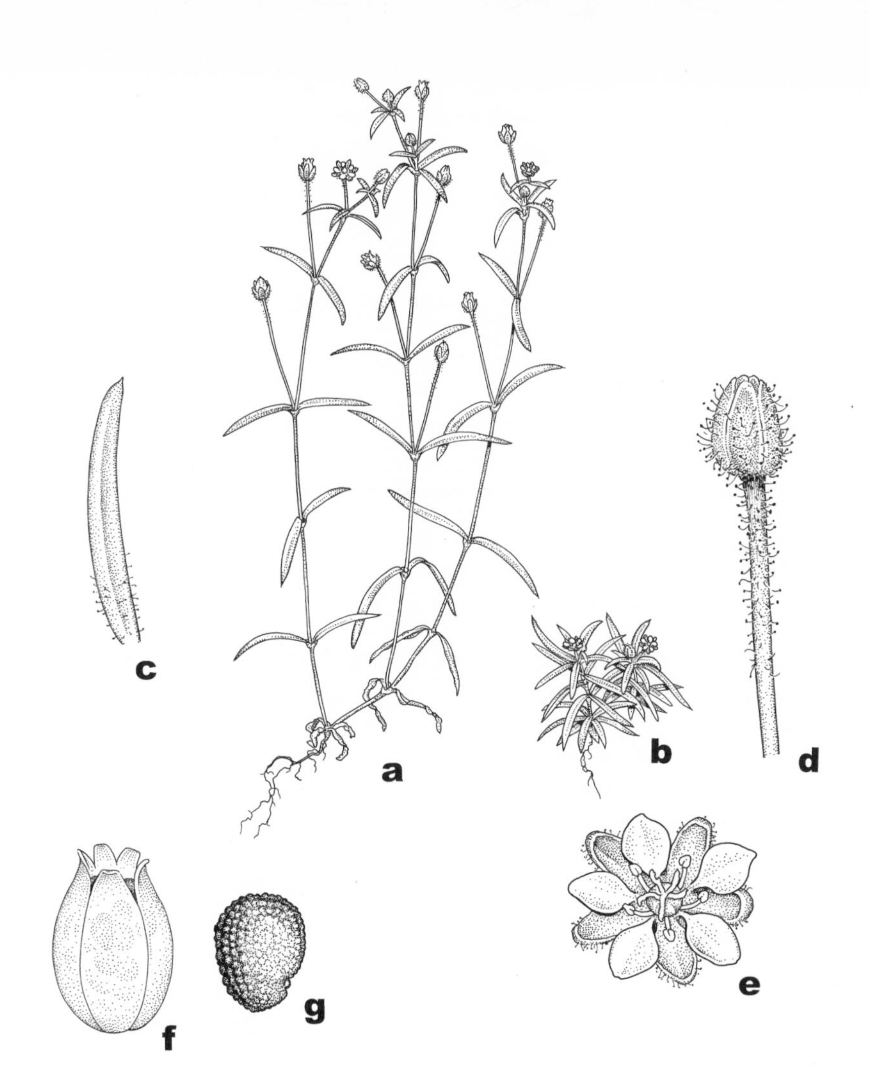

86. **Sagina japonica**
a. Habit
b. Habit
c. Leaf
d. Bud
e. Flower
f. Fruit
g. Seed

This dwarf species looks remarkably like *S. decumbens* but differs in having glandular-puberulent pedicels (at least in the upper half) and dark, tuberculate seeds. It differs from *S. apetala* by its five petals, five sepals, and five stamens.

Sagina japonica flowers during May and June.

3. Sagina apetala Ard. Animadv. Bot. Spec. alt. 2:22. 1764. Fig. 87.
Sagina apetala Ard. var. *barbata* Fenzl in Ledeb, Fl. Ross. 1:338. 1842.

Annual with slender roots; stems ascending or decumbent, branched, filiform, glabrous or glandular-pubescent (in the Illinois specimen); basal rosette, if present, withering early in the spring; leaves linear to subulate, up to 3 mm wide, usually with cilia at least near the base of each leaf; flowers borne on glandular pedicels up to 5 (–13) mm long; sepals 4, elliptic to ovate, 1.5–2.0 mm long, glandular-pubescent; petals absent; stamens 4; styles 4; capsules globose, 1.5–2.5 mm long, barely exceeding the sepals; seeds brown, smooth to papillose, 0.3–0.4 mm long.

Common Name: Apetalous Pearlwort.
Habitat: Apparently disturbed soil.
Range: Native to Europe; sparingly adventive in the Pacific States; rarely adventive in the eastern United States.
Illinois Distribution: Known from Union County.

This obscure little plant has been found a single time in Illinois in Union County by Stephen A. Forbes during the nineteenth century. The exact date of the collection is unknown, as is the precise locality and the habitat.

Sagina apetala is distinguished by its four sepals, no petals, four stamens, four styles, and annual habit. *Sagina procumbens* often has four sepals, but it possesses four petals and has a perennial habit.

This species flowers from April to June.

4. Sagina procumbens L. Sp. Pl. 128. 1753. Fig. 88.

Dwarf perennial, often with sterile basal rosettes present; stems decumbent to ascending, glabrous, branched or unbranched, up to 19 cm long; leaves linear, mucronate at the tip, connate at the base, glabrous, to 20 mm long; flowers up to 3 mm across, on ascending, filiform pedicels up to 25 mm long, usually becoming reflexed during fruiting; sepals 4, elliptic to oblong, obtuse to subacute, green, 1.5–2.0 mm

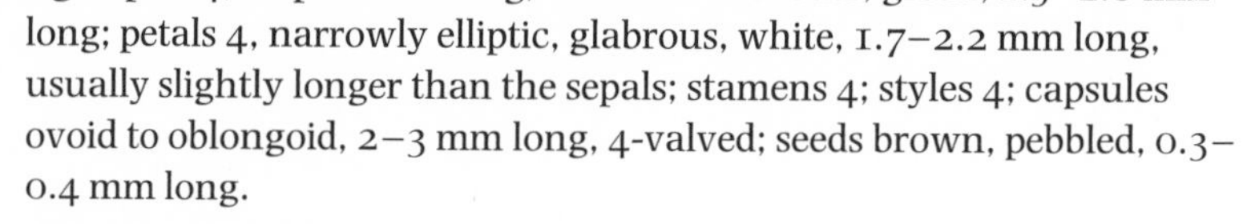

long; petals 4, narrowly elliptic, glabrous, white, 1.7–2.2 mm long, usually slightly longer than the sepals; stamens 4; styles 4; capsules ovoid to oblongoid, 2–3 mm long, 4-valved; seeds brown, pebbled, 0.3–0.4 mm long.

Common Name: Perennial Pearlwort.
Habitat: Moist lawns, paths, and between bricks and patio blocks.
Range: Newfoundland and Greenland to British Columbia, south to California, Missouri, northern Illinois, and Delaware.
Illinois Distribution: Apparently confined to the northern one-fourth of the state.

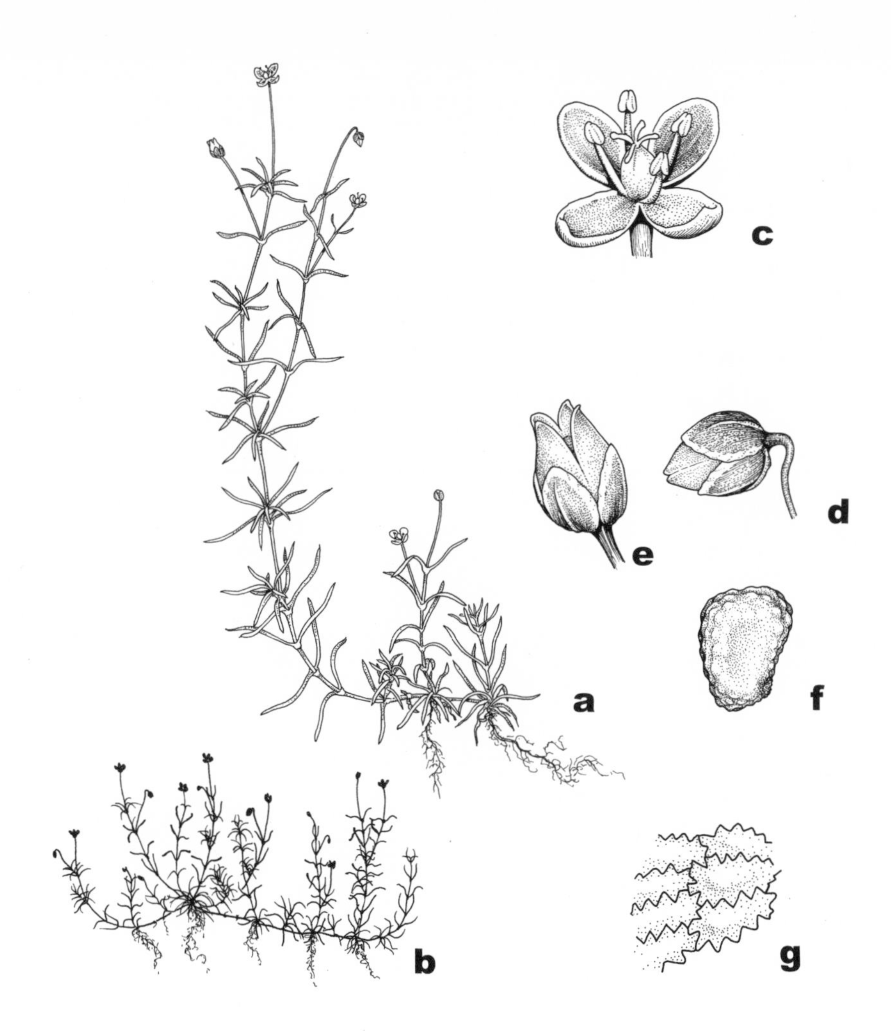

87. **Sagina apetala**
a. Habit
b. Habit
c. Flower
d. Fruit
e. Fruit
f. Seed
g. Surface of seed

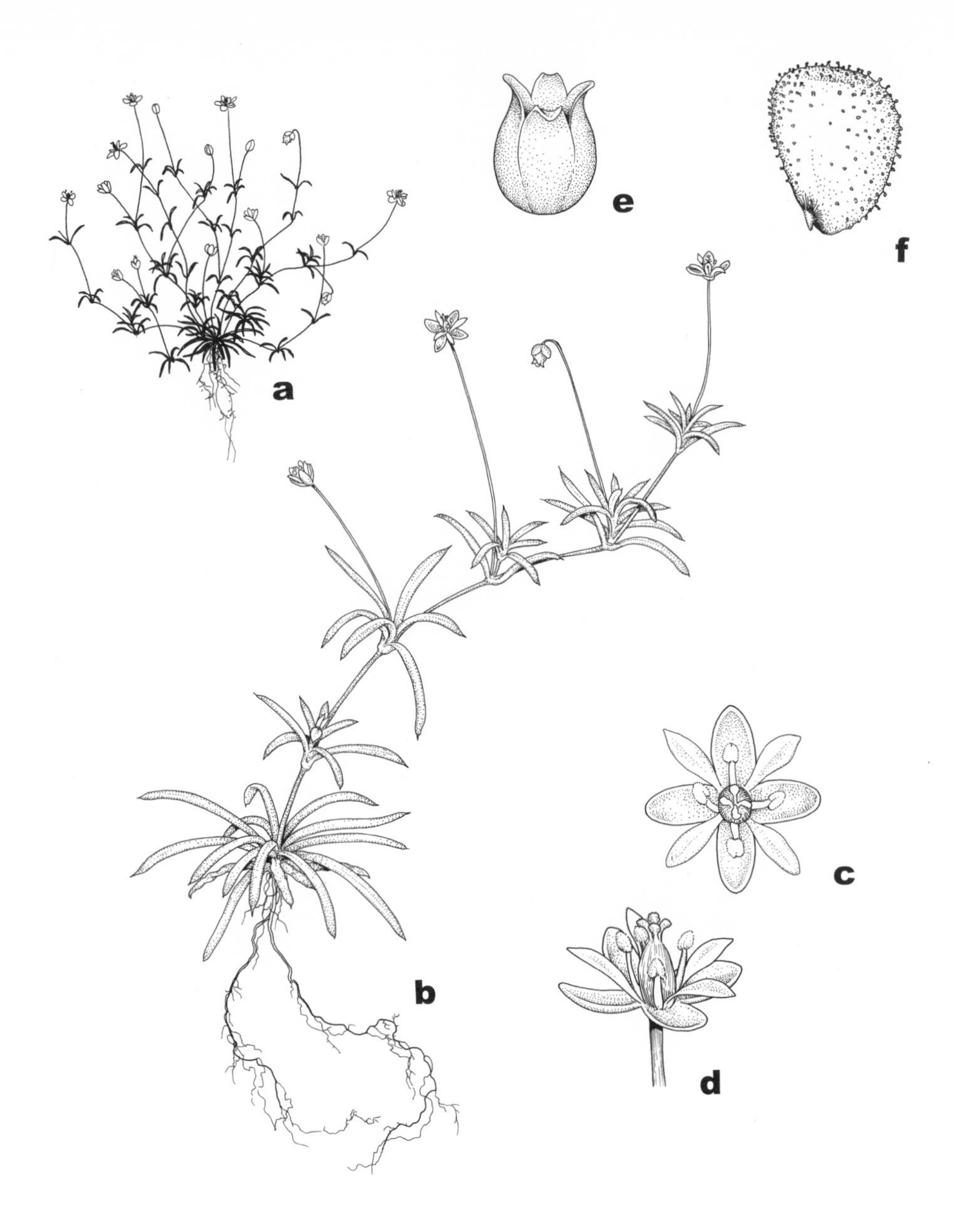

88. **Sagina procumbens**
a. Habit
b. Habit
c. Flower, face view
d. Flower, side view
e. Fruit
f. Seed

This species differs from the very similar *S. decumbens* in its perennial growth form with sterile basal rosettes of leaves usually present; its 4 sepals, 4 stamens, 4 styles; and its pebbled seeds. The fruiting stalks are usually reflexed.

Until 1994, all Illinois specimens of *Sagina* had been called *S. decumbens*. Examination of specimens collected in the northern one-fourth of Illinois, however, prove to be the more northern *S. procumbens*.

Sagina procumbens flowers from June to early August.

6. **Holosteum** L.—Jagged Chickweed

Annual herbs; leaves opposite, entire, without stipules; inflorescence an umbel-like cyme from a long peduncle; flowers perfect, actinomorphic; sepals 5, free or united at the base; petals 5, free; stamens 3–5 or 10; styles 3; ovary superior, 1-locular, with many ovules; fruit a 6-valved capsule; seeds compressed, minutely warty.

The genus *Holosteum* is comprised of three species native to Europe and Asia.

Only the following species has been found in Illinois.

1. Holosteum umbellatum L. Sp. Pl. 88. 1753. Fig. 89.

Annual from a slender taproot; stems tufted, erect, glandular-viscid, unbranched, up to 30 cm tall, quickly turning stramineous; basal leaves linear to oblong to oblanceolate, acute at the apex, tapering to a narrowed, petiolelike base, glandular-viscid, quickly turning stramineous; cauline leaves oblong to oblanceolate, acute to obtuse at the apex, tapering to a sessile base, glandular-viscid, to 2.0 (–2.5) cm long, quickly turning stramineous; inflorescence an umbel-like cyme, with 1–3 (–8) rays, each ray up to 3 cm long, 1-flowered; flowers up to 6 mm across, subtended by a minute bract at the base of the pedicel; sepals 5, lanceolate to ovate, acute to obtuse, green with scarious margins, 2.5–4.0 mm long; petals 5, free, narrowly elliptic to lanceolate, jagged at the apex, white, 3.0–4.5 mm long; stamens 3–5; styles 3; capsules ovoid, straight, 3–5 mm long, the pedicels reflexed during fruiting, with usually 6 recurved teeth; seeds compressed, 0.7–0.8 mm long, reddish brown, roughened.

Common Name: Jagged Chickweed.
Habitat: Disturbed soil, primarily along roads.
Range: Native to Europe, becoming widely distributed in the United States.
Illinois Distribution: Throughout the state except for the extreme northwestern counties.

The first collection of the jagged chickweed in Illinois was made in Johnson County on April 19, 1951. Since that time this species has spread very rapidly throughout most of the state. Shinners (1965) has discussed the rapid spread of this species in the United States.

Shortly after flowering, this plant turns stramineous to tawny and disappears. In the stramineous or tawny condition, it is readily observed along roadsides.

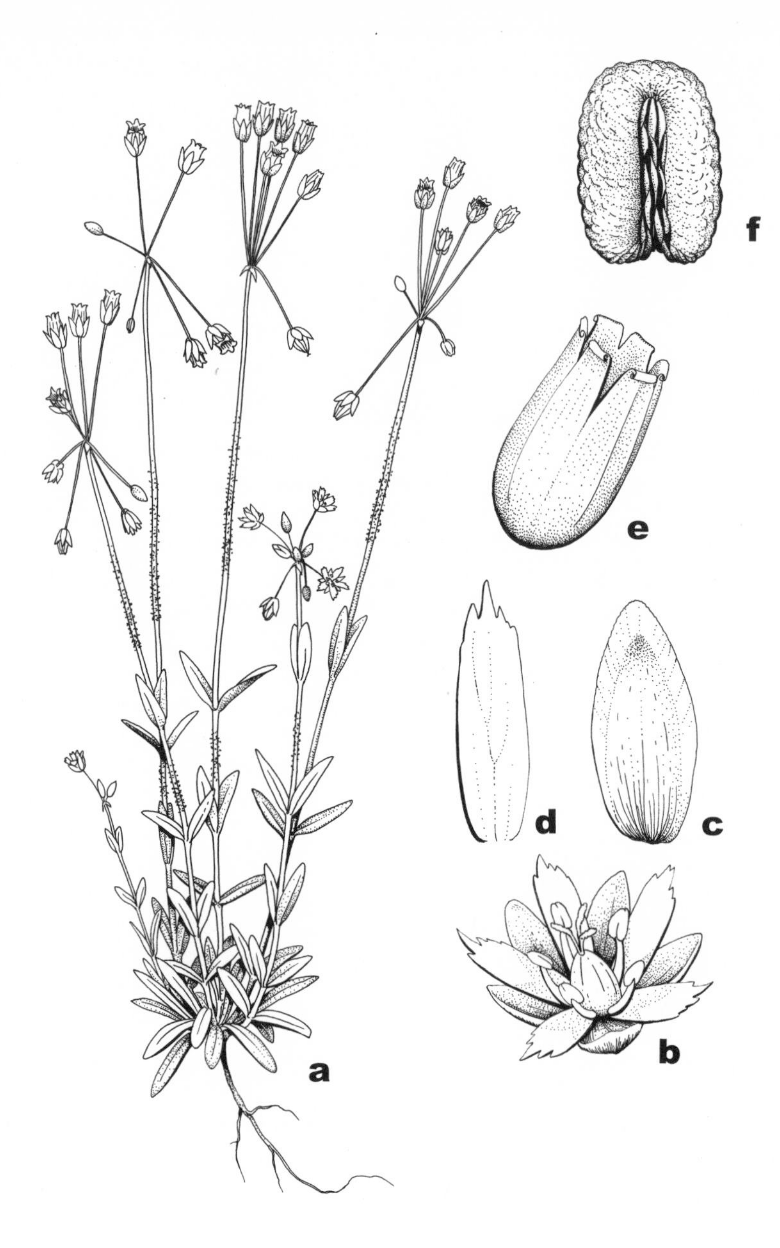

89. **Holosteum umbellatum**
a. Habit
b. Flower
c. Sepal
d. Petal
e. Fruit
f. Seed

The jagged apex of each white petal readily distinguishes this species from any other in the family.

Holosteum umbellatum flowers from mid-April through May.

7. **Stellaria** L.—Chickweed

Annual or perennial herbs; leaves opposite, entire, without stipules; inflorescence cymose or reduced to a single axillary flower; flowers perfect, actinomorphic; sepals (4–) 5, free; petals 5, free, rarely absent; stamens (2, 8) 10; styles 3 (–4); ovary superior, 1-locular, with many ovules; fruit a capsule, usually with 6 teeth; seeds compressed to globose, smooth or variously roughened.

About ninety species throughout most of the northern hemisphere comprise this genus.

Some species of *Stellaria* in Illinois are weedy and introduced from Europe, and some of the species are native. *Stellaria holostea*, the greater stitchwort, is sometimes grown as a garden ornamental.

Key to the Species of **Stellaria** in Illinois

1. Median leaves ovate, usually perfoliate; petals shorter than the sepals or absent; annuals from slender taproots.
  2. Petals 5; sepals 3–6 mm long; seeds 1.0–1.2 mm in diameter, usually dark reddish brown or purple-brown . . . . 1. *S. media*
  2. Petals absent; sepals 2–3 mm long; seeds 0.7–0.8 mm in diameter, light reddish brown . . . . 2. *S. pallida*
1. Median leaves linear or lanceolate or elliptic, sessile; petals longer than the sepals or at least as long as the sepals; perennials from slender rhizomes.
    3. Stems puberulent in lines, at least in the upper half of the stem; leaves elliptic to suborbicular . . . . 3. *S. pubera*
    3. Stems glabrous or essentially so; leaves lanceolate or linear.
        4. Leaves linear, up to 7 times longer than broad, completely glabrous; bracts without scarious margins, or sometimes with a very narrow scarious margin . . . . 4. *S. crassifolia*
        4. Leaves linear to lanceolate, at least some of them more than 8 times longer than broad, ciliate at base; bracts with obvious scarious margins.
            5. Bracts and sepals ciliate; seeds roughened; sepals strongly nerved; inflorescence terminal . . . . 5. *S. graminea*
            5. Bracts and sepals without cilia; seeds smooth; sepals nerveless or essentially so; inflorescence axillary . . . . 6. *S. longifolia*

1. **Stellaria media** (L.) Cyrillo, Char. Comm. 36. 1784. Fig. 90.
Alsine media L. Sp. Pl. 272. 1753.

Annual from a slender taproot; stems weak, decumbent to ascending, pubescent in lines, branched, up to 45 cm tall; leaves ovate to elliptic, acute at the apex, more or less rounded at the base, up to 3.0 (–4.5) cm long, glabrous, the upper leaves sessile, the median and lower leaves on ciliate petioles; flowers solitary or in few-flowered terminal cymes, up to 8 mm across, on often puberulent pedicels, leafy bracted; sepals 5, free, oblong to oblong-lanceolate, acute, more or less pubescent, at least at the base, green, 3–6 mm long; petals 5, free, 2-cleft at the apex, white,

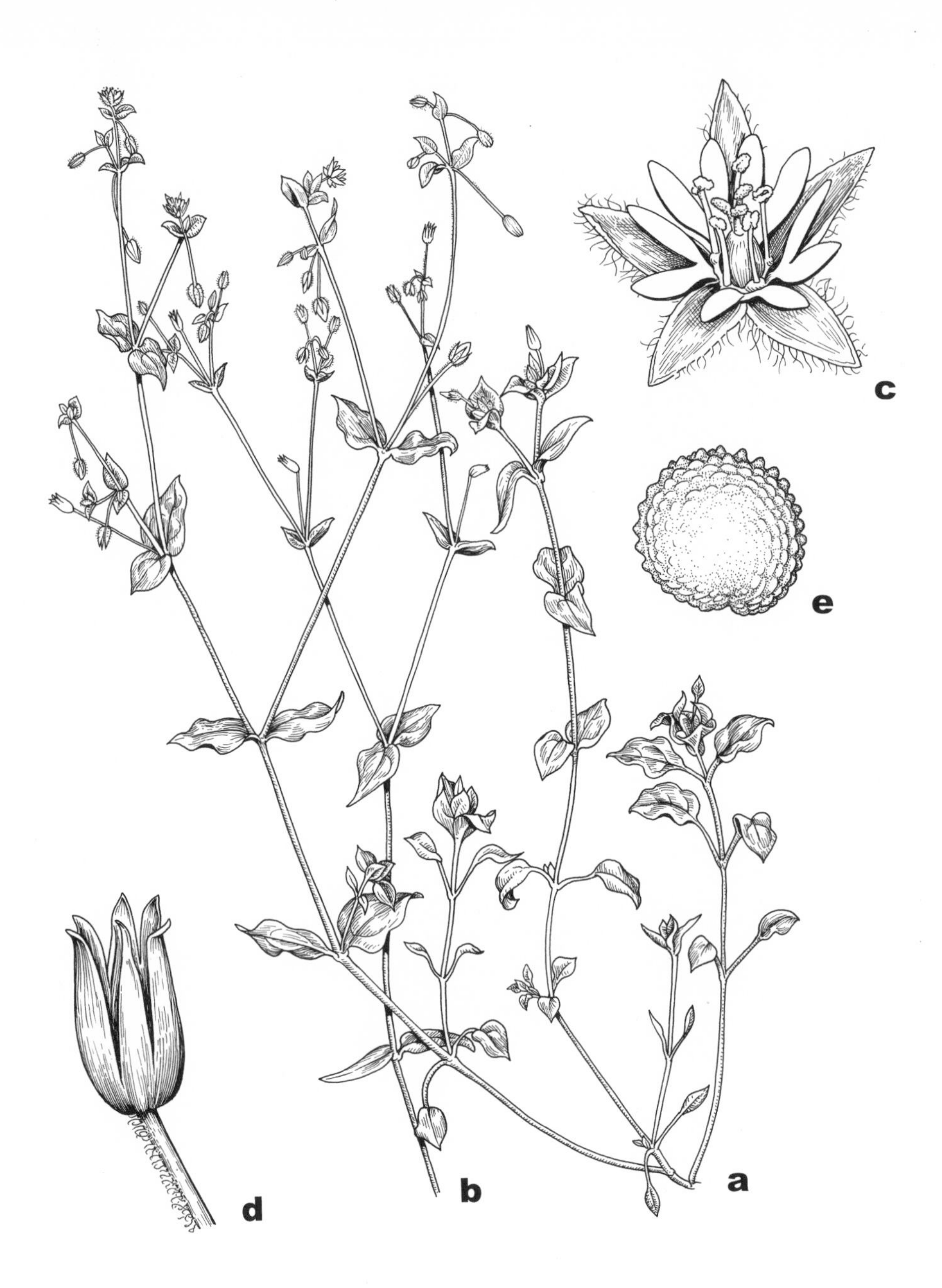

90. **Stellaria media**
a. Habit
b. Habit
c. Flower
d. Fruit
e. Seed

2.5–5.0 mm long, shorter than the sepals; stamens (2–) 10; capsules ovoid, 3–6 mm long, as long as or usually longer than the sepals, the pedicels curved to reflexed; seeds suborbicular, dark reddish brown to purple-brown, papillose, 1.0–1.2 mm in diameter.

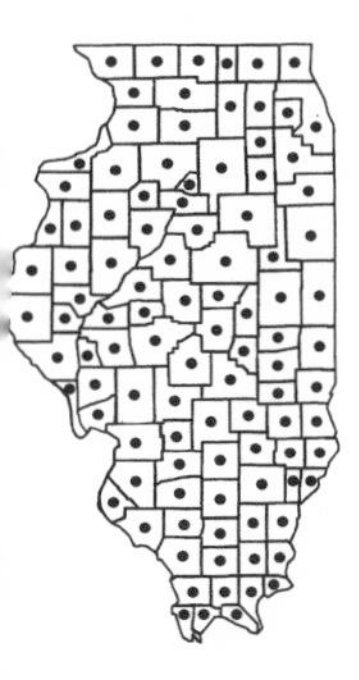

Common Name: Common Chickweed.
Habitat: Disturbed soils, particularly in lawns and gardens; infrequent in wooded floodplains where flooding has removed competition.
Range: Presumably native to Europe and Asia; adventive throughout all of the United States.
Illinois Distribution: Common throughout the state.

Common chickweed is a widespread weed throughout Illinois.

In addition to its prevalence in lawns, it may also be found in wooded floodplains that are periodically flooded.

Considerable variation is found within the species. Very small specimens occur, as do plants up to 45 cm tall. Stamen number varies from 2–10, but the prevailing number is 10. The only other annual species of *Stellaria* in Illinois is the apetalous *S. pallida.*

Stellaria media flowers from February to December.

2. Stellaria pallida (Dumort.) Pire, Bull. Soc. Bot. Belg. 2:43, 49. 1863. Fig. 91.
Alsine pallida Dumort. Fl. Belg. 109. 1827.

Annual from a slender taproot; stems weak, decumbent to erect, pubescent in lines, branched, to 50 cm tall; leaves ovate, acute at the apex, more or less rounded at the base, up to 4.5 cm long, glabrous, shiny, the upper leaves sessile, the median and lower leaves on ciliate petioles; flowers solitary or in few-flowered terminal and axillary cymes, up to 5 mm across, on often puberulent pedicels, leafy bracts; sepals 5, free, lanceolate to ovate, obtuse to acute, glabrous or glandular-pubescent, green, 2–3 mm long; petals 0; stamens 3–10; capsule ovoid, 2–4 mm long, as long as or usually longer than the sepals, the pedicels curved to reflexed; seeds 0.7–0.8 mm in diameter, light reddish brown.

Common Name: Apetalous Chickweed.
Habitat: Disturbed, often cultivated areas.
Range: Native to Europe; scattered as an adventive in the United States.
Illinois Distribution: Known from Cook, Grundy, Kane, and Will counties. It was first discovered in Illinois in 1985.

This annual chickweed is very similar to the widespread *S. media* but differs in its lack of petals, its shorter sepals, and its smaller seeds.

In the Chicago area, this species is usually found in sandy, shallow fields.

Stellaria pallida flowers from mid-April to mid-May.

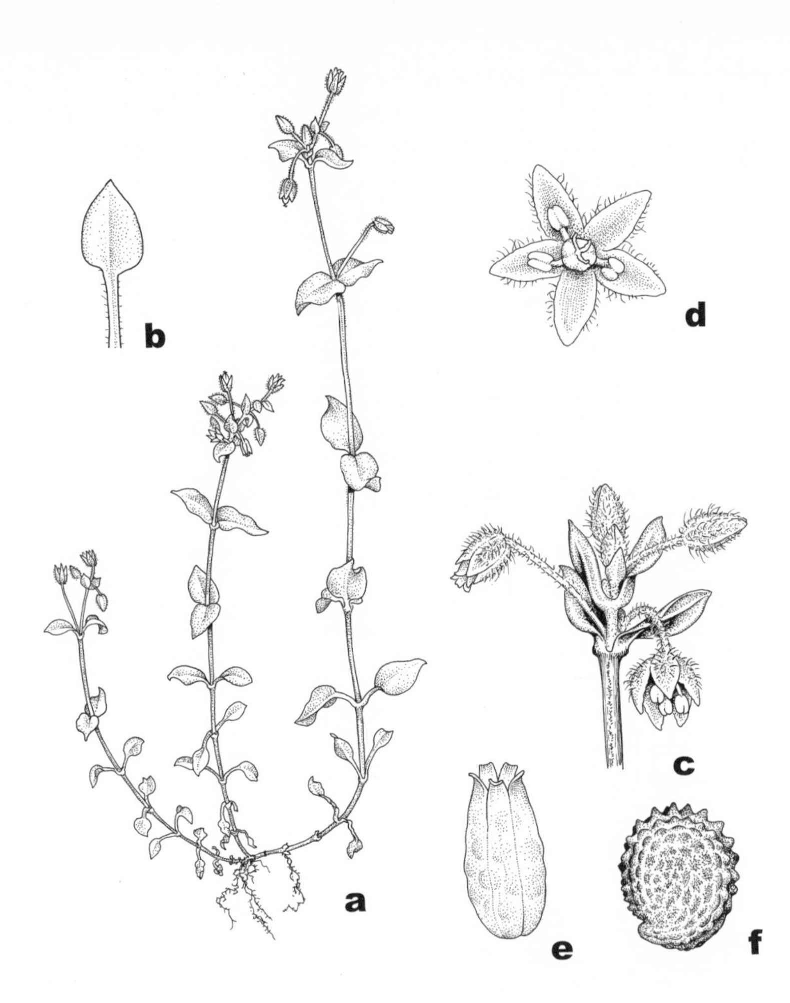

91. **Stellaria pallida**
a. Habit
b. Leaf
c. Buds and flower
d. Flower
e. Fruit
f. Seed

3. Stellaria pubera Michx. Fl. Bor. Am. 1:273. 1803. Fig. 92.
Alsine pubera (Michx.) Britt. Mem. Torrey Club 5:150. 1894.

Perennial herb from tough rhizomes; stems decumbent to erect, pubescent, 4-angled, sparingly branched, up to 30 cm tall; leaves elliptic to oblong to ovate-oblong, acute or obtuse at the apex, rounded or tapering at the base, up to 10 cm long, up to 4 cm broad, sparsely pubescent, ciliate, the upper leaves sessile or nearly so, the median and lower leaves petiolate; flowers up to 1.2 cm across, in terminal, leafy cymes, with pedicels up to 3 cm long; sepals 5, free, lanceolate to ovate, acute, more or less pubescent, green, 4–6 mm long; petals 5, free, 2-cleft at the apex, white, 3.5–5.5 mm long, a little shorter than the sepals; stamens usually 10; capsules subglobose to ovoid, 3–5 mm long, shorter than the sepals; seeds 1.0–1.2 mm long, reniform, minutely warty.

Common Name: Great Chickweed.
Habitat: Wooded cliffs.
Range: New Jersey to Illinois, south to Alabama and Florida.
Illinois Distribution: Known from Cook, DuPage, Pope, and possibly Will counties.

After flowering, this species usually sends up one or more vegetative stems that flower very sparingly, if at all.

The great chickweed is one of two species of *Stellaria* in Illinois that occupies a native habitat. At Burke Branch, in Pope County, this species occurs on rocky boulders on a wooded hillside. One location from a wooded bluff in Cook County may be an escape from a wildflower garden, according to Swink and Wilhelm (1994). Pech (1866) reported this species from Will County.

The flowers of this species are slightly larger than those of any other species of *Stellaria* in Illinois.

Stellaria pubera flowers from March to May.

4. Stellaria crassifolia Ehrh. Hannover. Mag. 8:116. 1784. Fig. 93.

Low, perennial herb with very slender rhizomes; stems matted to more or less erect, usually branched above, glabrous, to 20 cm long; leaves linear, up to 2.5 mm broad, not more than 7 times longer than broad, completely glabrous; flower solitary at the end of a branchlet, sometimes nodding, up to 6 mm across, the bracts not scarious, or with a very narrow scarious margin, not ciliate; sepals 5, free, green, narrowly lanceolate, acute at the apex, not scarious, not ciliate, 2.0–3.0 (–3.5) mm long; petals 5, free, white, 2-cleft at the apex, 2.5–4.0 mm long, usually longer than the sepals; stamens 10; capsules ovoid, 2.0–3.5 mm long, as long as or barely longer than the sepals, pale brown; seeds broadly oblong, 0.8–1.0 mm long, reddish brown, rugose.

Common Name: Matted Chickweed.
Habitat: Usually low, springy places. Habitat for the Illinois collection is unknown.

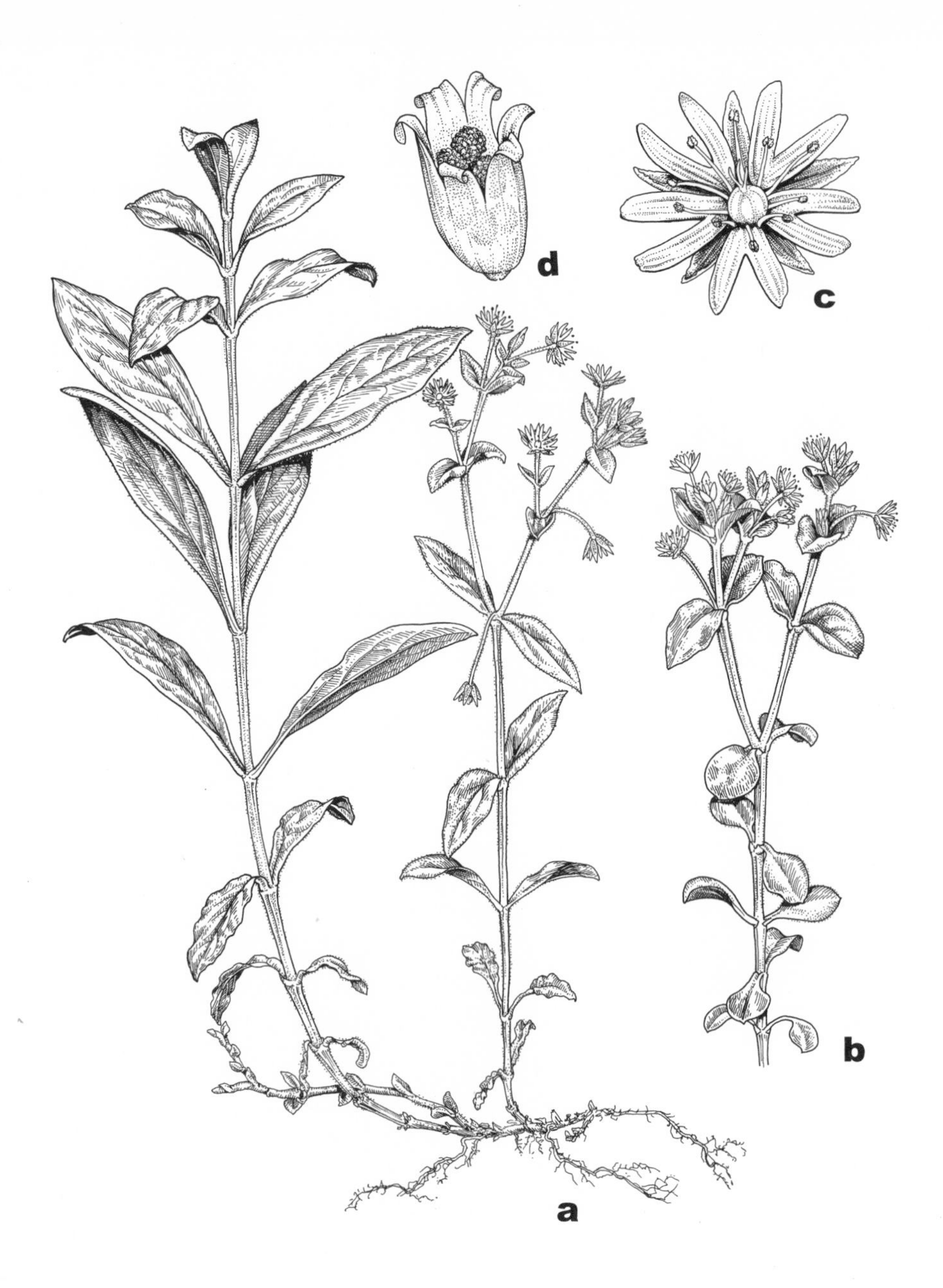

92. **Stellaria pubera**
a. Habit
b. Habit
c. Flower
d. Fruit

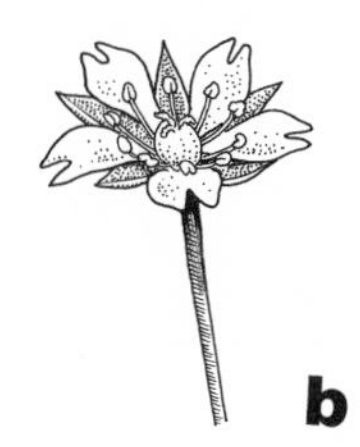

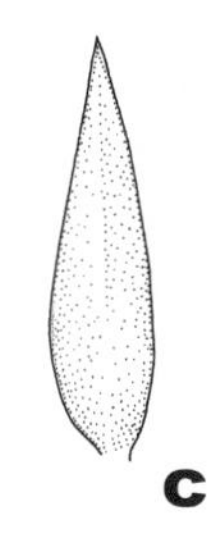

93. **Stellaria crassifolia**
a. Habit
b. Flower
c. Sepal

Range: Newfoundland to Alaska, south to Quebec, southern Michigan, northeastern Illinois, Colorado, and California.
Illinois Distribution: Known only from McHenry County.

This is another one of the enigmatic species collected by George Vasey during the nineteenth century. His label reads only Ringwood, a small village in McHenry County where Vasey resided for a while. There is speculation by some botanists that some of the Vasey collections labeled from Ringwood were actually made somewhere else. Since I have no proof of this, and since I have generally admitted to the Illinois flora any of the Vasey plants from Ringwood, I am concluding that *Stellaria crassifolia* was at one time collected in Illinois.

I have seen a few other Illinois collections labeled *Stellaria crassifolia,* but these are all misidentifications for *S. longifolia.* The Illinois specimen is very small but bears flowers and a few fruits. The leaves are completely glabrous and the seeds are rugose. The flower is borne solitary at the end of a branchlet. Some of the bracts of the Illinois specimen have a very narrow scarious margin.

There are no habitat data or time of collection of the Illinois specimen.

5. **Stellaria graminea** L. Sp. Pl. 422. 1753. Fig. 94.
Alsine graminea (L.) Britt. Mem. Torrey Club 5:150. 1894.

Perennial herb from slender, white rhizomes; stems decumbent to ascending, weak, glabrous, 4-angled, branched above, sometimes scabrous on the angles, up to 60 cm tall; leaves linear to lanceolate, acute at the apex, tapering to the sessile base, glabrous except for the ciliate margins near the base, shiny, up to 4 cm long, up to 6 mm broad; flowers to 1 cm across, numerous in much branched cymes, with scarious, ciliate, lanceolate bracts up to 6 mm long, with spreading to erect pedicels; sepals 5, free, lanceolate, acute, ciliate, green with scarious margins, 3-nerved, 4.0–5.5 mm long; petals 5, free, 2-cleft at the apex, white, 4–6 mm long, as long as or a little longer than the sepals; stamens usually 10; capsules oblongoid, pale brown to stramineous, 4–6 mm long; seeds globose to oblongoid, 0.8–1.2 mm long, reddish brown to dark brown, minutely pebbled.

Common Name: Common Stitchwort.
Habitat: Moist, disturbed areas.
Range: Native to Europe; adventive throughout North America.
Illinois Distribution: Scattered throughout the state.

Common stitchwort is an adventive species found frequently in moist, grassy areas.

This species differs from the similar narrow-leaved *S. longifolia* in its 3-nerved sepals and its distinctly pebbled seeds. *Stellaria crassifolia* is similar but smaller, has completely glabrous leaves, and has bracts that are not scarious or only with a very narrow scarious margin.

Stellaria graminea flowers from May to August.

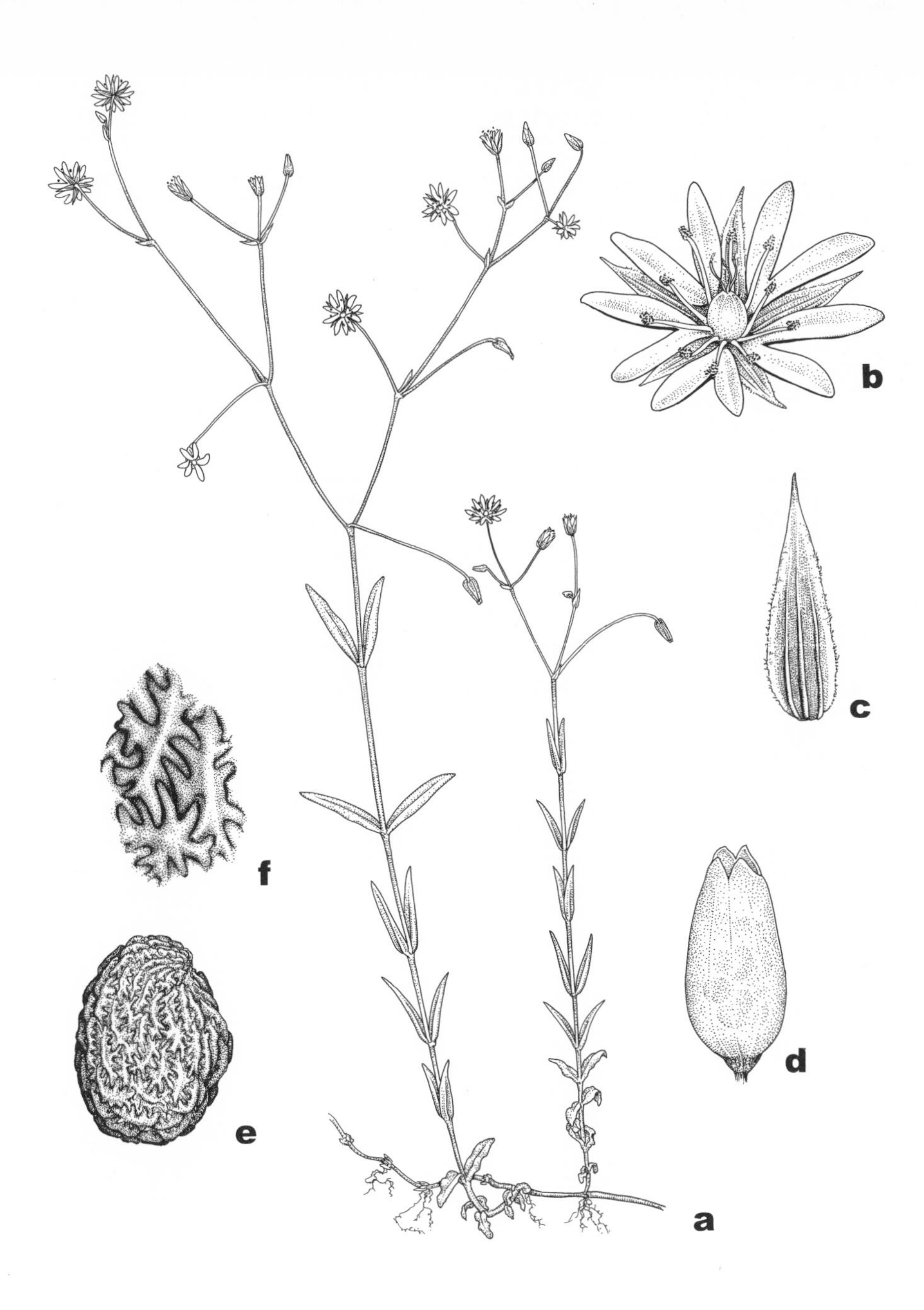

94. **Stellaria graminea**
a. Habit
b. Flower
c. Sepal
d. Fruit
e. Seed
f. Surface of seed

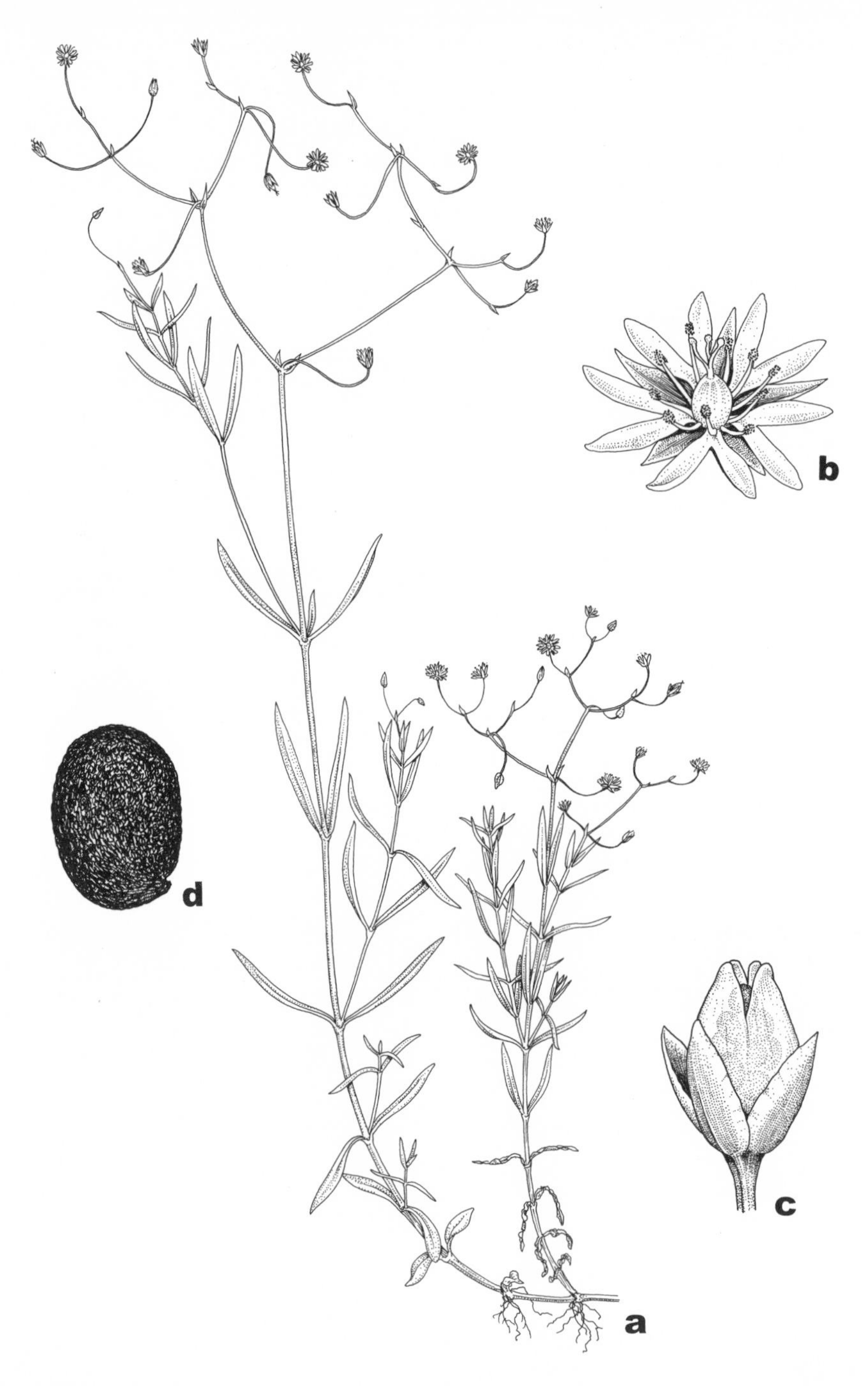

95. **Stellaria longifolia**
a. Habit
b. Flower
c. Fruit
d. Seed

6. Stellaria longifolia Muhl. ex Willd. Enum. Hort. Ber. 479. 1809. Fig. 95.
Alsine longifolia (Muhl.) Britt. Mem. Torrey Club 5:150. 1894.

Perennial herb from white rhizomes; stems decumbent to ascending, weak, glabrous, 4-angled, branched above, sometimes scabrous on the angles, up to 45 cm tall; leaves linear to linear-lanceolate, acute at the apex, tapering to the sessile base, glabrous except for ciliate margins near the base, up to 4 cm long, up to 5 mm broad; flowers to 8 mm across, numerous in much branched lateral cymes, with scarious, ciliate, lanceolate bracts up to 3 mm long, with spreading to ascending to erect pedicels; sepals 5, free, lanceolate to ovate-lanceolate, acute to obtuse, usually glabrous, green with somewhat scarious margins, without conspicuous nerves, 2.5–4.0 mm long; petals 5, free, 2-cleft at the apex, white, as long as or a little longer than the sepals; stamens usually 10; capsules ovoid to oblongoid, 5–8 mm long; seeds oblongoid to ovoid, 0.7–1.0 mm long, reddish brown, smooth or nearly so.

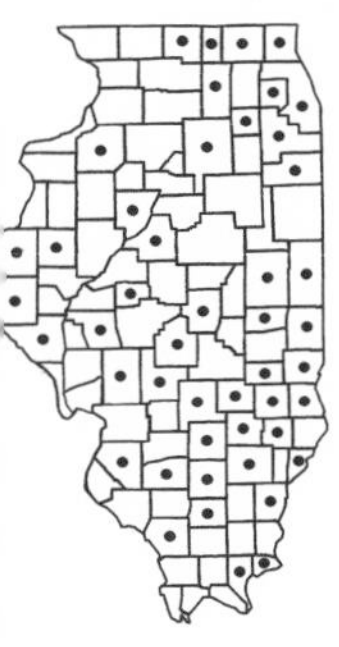

Common Name: Long-leaved Stitchwort.
Habitat: Moist ground, including bogs and floodplains.
Range: Newfoundland to Alaska, south to California, New Mexico, Nebraska, Kentucky, and Virginia; Europe; Asia.
Illinois Distribution: Scattered throughout the state but rare in the southern counties.

This species is similar to *S. graminea* from which it differs in its essentially nerveless sepals and its mostly smooth seeds. *Stellaria crassifolia* is also similar but has roughened seeds, completely glabrous leaves, and bracts that are not scarious or only very narrowly scarious along the margin.

Stellaria longifolia is a native species, occupying a variety of moist ground habitats. Swink and Wilhelm (1994) report it to be a species of shrub-carr communities.

This species flowers from May to July.

8. **Myosoton** Moench—Giant Chickweed

Perennial herb; leaves opposite, simple, entire, without stipules; inflorescence cymose or reduced to a single axillary flower; flowers perfect, actinomorphic; sepals 5, united below; petals 5, free; stamens 10; styles 5; ovary superior, 1-locular, with many ovules; fruit a capsule, 5-toothed; seeds not flattened, roughened.

The only species that comprises the genus *Myosoton* is often merged into the genus *Stellaria*. Reasons for segregating *Myosoton* from *Stellaria* are its five styles that are opposite the petals rather than three styles that are opposite the sepals in *Stellaria*.

Only the following adventive species occurs in Illinois:

1. Myosoton aquaticum (L.) Moench, Meth. 225. 1794. Fig. 96.
Cerastium aquaticum L. Sp. Pl. 439. 1753.
Stellaria aquatica (L.) Scop. Fl. Carn. ed. 2, 1:319. 1772.
Alsine aquatica (L.) Britt. Bull. Torrey Club 5:356. 1894.

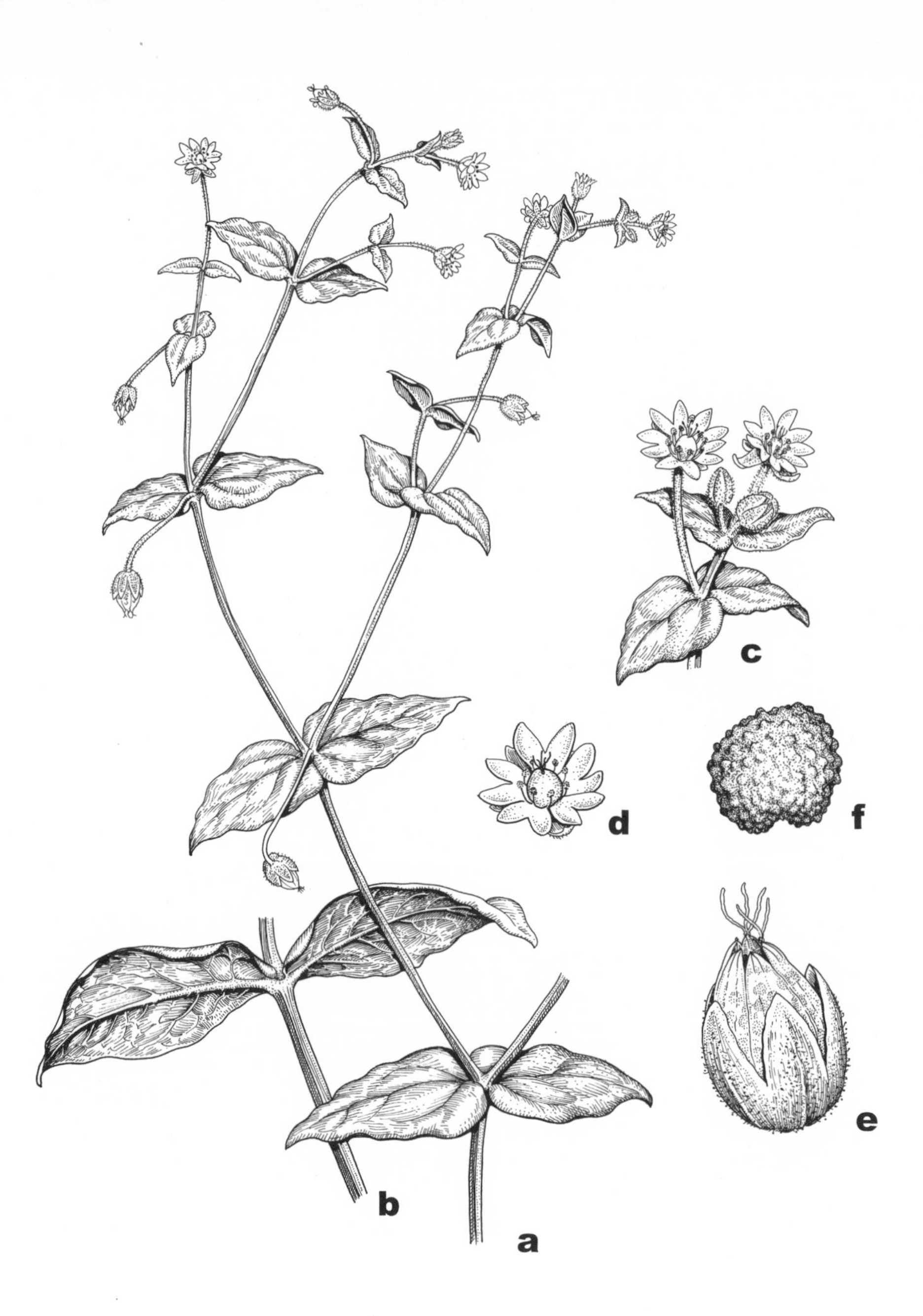

96. **Myosoton aquaticum**
a. Habit
b. Node
c. Flowering branch
d. Flower
e. Fruit
f. Seed

Perennial herb from extensive rhizomes; stems decumbent to ascending, glandular-puberulent, 4-angled, branched, up to 75 cm tall; leaves ovate-lanceolate to ovate, acute at the apex, rounded to subcordate at the base, usually puberulent, up to 5.5 (–7.0) cm long, the median and upper leaves sessile or barely clasping, the lowermost leaves usually short-petiolate; flowers to 1.0 (–1.2) cm across, in open cymes from the upper axils, with ovate-lanceolate, green bracts, with glandular-viscid pedicels up to 2.5 cm long, becoming deflexed in fruit; sepals 5, united below, ovate, acute to obtuse, glandular-puberulent, green, up to 9 mm long in fruit; petals 5, free, 2-cleft at the apex, longer than the sepals in flower; stamens 10; styles 5, free, opposite the petals; capsules ovoid to oblongoid, with five 2-cleft valves; seeds more or less orbicular, 0.7–0.8 mm in diameter, dark brown, roughened.

Common Name: Giant Chickweed.
Habitat: Moist soil along streams, sometimes in pastures.
Range: Native to Europe; adventive in the eastern United States.
Illinois Distribution: Occasional in the northern one-sixth of the state; also Clark, Cumberland, and Sangamon counties.

This coarse chickweed, although a native of Europe, often occupies moist soil along streams where it grows with native wetland species. It very much resembles the genus *Stellaria* except for the presence of five styles that are opposite the petals.

Myosoton aquaticum flowers from May to October.

9. **Cerastium** L.—Mouse-eared Chickweed

Annual or perennial herbs; leaves opposite, simple, entire, without stipules; inflorescence cymose or reduced to a single axillary flower; flowers perfect, actinomorphic; sepals (4–) 5, free; petals 5, free, deeply notched, rarely absent; stamens 4, 5, or 10; styles 5, rarely 3, free; ovary superior, 1-locular, with many ovules; fruit a capsule, usually curved, 10-toothed at the apex (6-toothed in *C. dubium*); seeds usually flattened, roughened.

Cerastium is composed of about one hundred species in temperate regions of the world.

This genus is distinguished from *Stellaria* by its five styles (except *C. dubium*), by its usually more elongated pubescence, and by its usually curved capsules that are 10-toothed (6-toothed in *C. dubium*) at the apex.

Key to the Species of **Cerastium** in Illinois

1. Styles 3; capsule 6-toothed at the apex .... 1. *C. dubium*
1. Styles 5; capsule 10-toothed at the apex.
    2. Some or all of the bracts with scarious margins or at least the upper part of the bracts with scarious margins.
        3. Petals usually 10 mm long or longer, at least twice as long as the sepals; stems not viscid .... 2. *C. arvense*
        3. Petals less than 10 mm long, about as long as the sepals; stems more or less viscid.
            4. Stamens 10; plants perennial, with basal offshoots .... 3. *C. fontanum*

4. Stamens 5; plants annual, without basal offshoots.
    5. Uppermost bracts with narrow scarious margins, the lower bracts green throughout; petals deeply 2-cleft, with branched veins; capsules on erect pedicels .... 4. *C. pumilum*
    5. Uppermost bracts as much as half composed of scarious margins; petals shallowly 2-cleft, with unbranched veins; capsules on deflexed pedicels .... 5. *C. semidecandrum*

2. All bracts entirely green, not scarious.
  6. Pedicels at least twice as long as the capsules.
    7. Stamens 4 or 5 .... 6. *C. diffusum*
    7. Stamens 10.
      8. Filaments glabrous; stems viscid throughout; sepals pubescent but not distinctly bearded at the apex .... 7. *C. nutans*
      8. Filaments ciliate; stems viscid only near the apex; sepals long-bearded at the apex .... 8. *C. brachypetalum*
  6. Pedicels up to twice as long as to shorter than the capsules.
    9. Sepals acute; capsules nearly twice as long as the sepals .... 9. *C. glomeratum*
    9. Sepals obtuse to subacute; capsules about three-fourths as long as the sepals .... 10. *C. brachypodum*

1. Cerastium dubium (Bast.) O. Schwarz, Mitt. Thuring Bot. Ges. 1:98. 1807. Fig. 97.
Stellaria dubia Bast. Fl. Maine-et-Loire, Suppl. 24. 1809.

Annual without basal offshoots; stems ascending to erect, glabrous or somewhat pubescent, occasionally branched, up to 20 cm tall; leaves elliptic to ovate, acute to obtuse at the apex, rounded or tapering to the nearly sessile base, usually pubescent, up to 2 cm long, less than 1 cm across; flowers up to 6 mm across, in cymes, with bracts with narrow scarious margins, the pedicels up to 1 cm long; sepals 5, free, lanceolate, acute, 2.5–4.0 mm long; petals 5, free, 2-cleft at the apex, white, 3.5–6.0 mm long, about 1.5 times longer than the sepals; stamens usually 10; styles 3; capsules cylindrical, 4–7 mm long, 6-toothed at the apex; seeds obovoid, 0.5–0.6 mm long, brown, minutely warty.

Common Name: Mouse-eared Chickweed.
Habitat: Disturbed soil.
Range: Native to Europe; rarely adventive in North America.
Illinois Distribution: Known from Effingham, Fayette, and Shelby counties.

This rare plant of weedy habitats is anomalous in the genus *Cerastium* in having only three instead of five styles and in possessing capsules that are six-toothed at the apex rather than ten-toothed.

Cerastium dubium flowers during April and May.

2. Cerastium arvense L. var. **villosum** Holl. & Britt. Bull. Torrey Club 14:49. 1887. Fig. 98.
Cerastium velutinum Raf. Med. Rep. II, 5:359. 1808.
Cerastium oblongifolium Torr. Fl. U. S. 460. 1824.

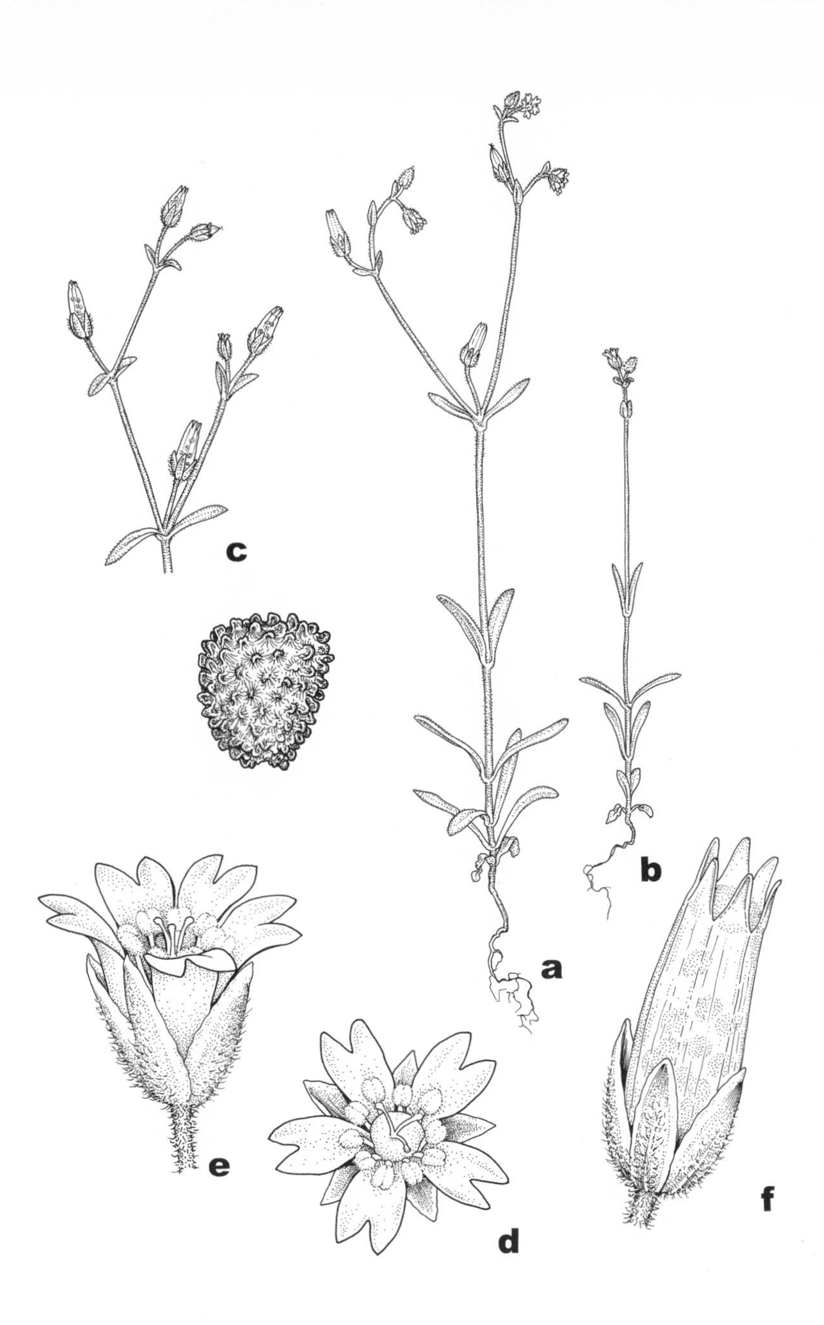

97. **Cerastium dubium**
a. Habit
b. Habit
c. Fruiting branch
d. Flower
e. Flower
f. Fruit

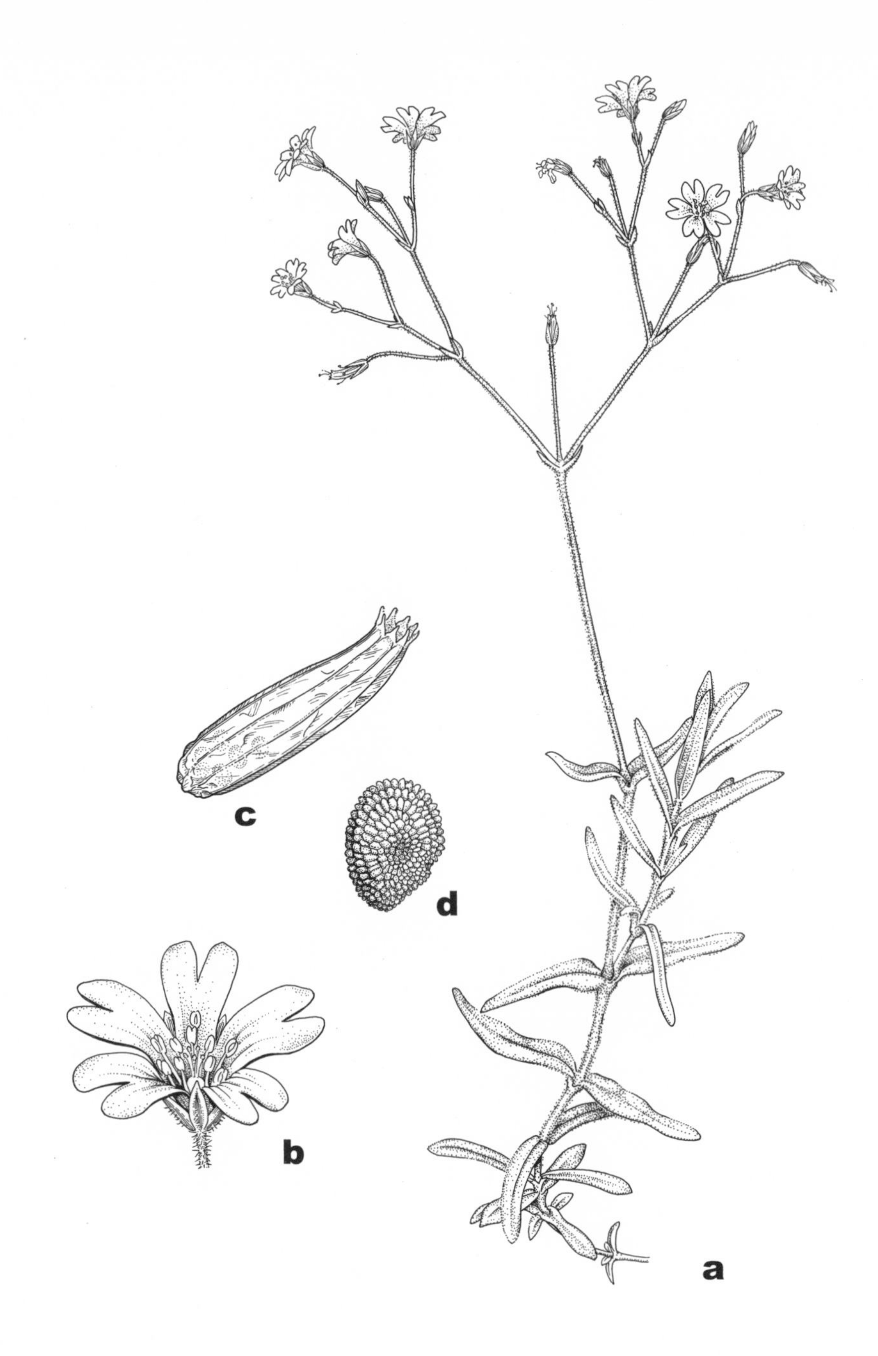

98. **Cerastium arvense** var. **villosum**

a. Habit
b. Flower
c. Fruit
d. Seed

Cerastium arvense L. var. *oblongifolium* (Torr.) Holl. & Britt. Bull. Torrey Club 14:49. 1887.
Cerastium arvense L. var. *maximum* Holl. & Britt. Bull. Torrey Club 14:47. 1887.
Cerastium arvense L. f. *oblongifolium* (Torr.) Pennell, Bartonia 12:10. 1931.

Perennial herb with leafy basal offshoots; stems tufted, spreading to erect, pilose, with or without glandular hairs, usually not viscid, branched from the base, up to 45 cm tall; leaves lanceolate to ovate-lanceolate, acute or obtuse at the apex, tapering to the nearly sessile base, more or less glabrous on the upper surface, pubescent on the lower surface, up to 5.5 cm long, up to 1 cm broad; flowers to 2 cm across, in dichotomous cymes, with bracts scarious along the margins, with pedicels up to 1.5 cm long; sepals 5, free, lanceolate, acute, 4–9 mm long; petals 5, free, 2-cleft at the apex, white, (10–) 12–20 mm long; stamens usually 10; styles 5; capsules cylindrical, 6–16 mm long, curved; seeds obovoid, 0.5–1.3 mm long, reddish brown, warty.

Common Name: Field Mouse-eared Chickweed.
Habitat: Rocky woods; river banks; sandy prairies; open woods; old fields.
Range: Ontario to Idaho, south to Missouri, Illinois, and Virginia.
Illinois Distribution: Scattered throughout the state.

This is the largest flowered *Cerastium* in Illinois and, because of the large white petals, it is a reasonably pretty species.

This plant was known for a long time as *C. velutinum*, but most botanists today believe it to be only one of the varieties of the very variable *C. arvense*.

Cerastium arvense var. *villosum* flowers in May and June.

3. **Cerastium fontanum** Baum. Enum. Stirp. Transs. 1:425. 1818. Fig. 99.
Cerastium triviale Link, Enum. Hort. Ber. 1:433. 1821.
Cerastium viscosum L. var. *glandulosum* Boenn. Prodr. Fl. Monast. 133. 1824.
Cerastium vulgatum var. *hirsutum* Fries f. *glandulosum* (Boenn.) Druce ex Moss, Cambr. Brit. Fl. 3:50. 1920.

Perennial herb with basal leafy offshoots; stems tufted, ascending to erect, pubescent, rarely with glandular hairs, occasionally branched, up to 45 cm tall; basal and lower leaves oblong to spatulate to oval, obtuse to subacute at the apex, tapering to the base, more or less hirsute on both surfaces, nearly sessile, up to 3 (–4) cm long, up to 1.0 (–1.5) cm broad; upper leaves oblong to ovate-lanceolate, smaller; flowers up to 5 mm across, several-flowered in dichotomously branched cymes, with scarious-margined bracts, with pedicels up to 2 cm long; sepals 5, free, ovate-lanceolate, acute or less commonly obtuse at the apex, pubescent, scarious along the margins, 4–7 mm long; petals 5, free, 2-cleft at the apex, white, 4–8 mm long; stamens usually 10; styles 5; capsules cylindrical, 6.5–11.0 mm long, more or less curved; seeds obovoid, 0.5–0.7 mm long, reddish brown, roughened.

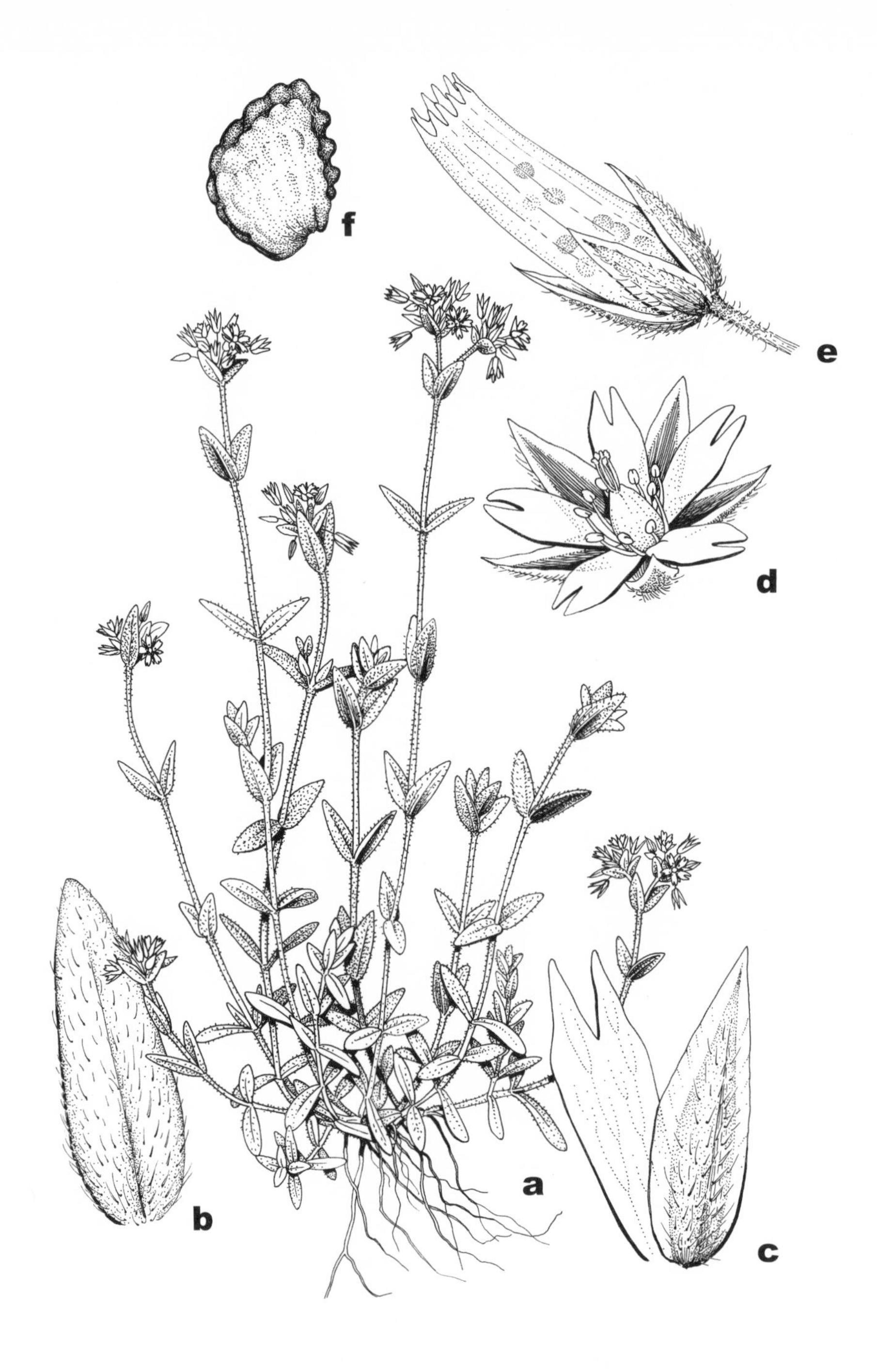

99. **Cerastium fontanum**
a. Habit
b. Leaf
c. Petal *(left),* sepal *(right)*
d. Flower
e. Fruit
f. Seed

Common Name: Common Mouse-eared Chickweed.
Habitat: Disturbed soil, including lawns and gardens.
Range: Native to Europe and Asia; naturalized throughout the United States.
Illinois Distribution: Common throughout the state.

This is the most common species of *Cerastium* throughout the state. It is abundant in lawns and other disturbed areas and even occurs in disturbed woods.

Both glandular-pubescent and eglandular-pubescent specimens occur in Illinois.

Although this plant has been called *C. vulgatum* for decades in the United States, that binomial apparently applies to a different European species. *Cerastium fontanum* seems to be the correct binomial for our species. Link's *C. triviale* appears to be the same species.

Cerastium fontanum flowers from March to November.

4. **Cerastium pumilum** Curtis, Fl. Lond. Fasc. 6, t. 30. 1777. Fig. 100.

Annual without basal offshoots; stems tufted, ascending to erect, usually glandular-pubescent, occasionally branched, up to 20 cm tall; leaves oblong to elliptic to ovate, acute or obtuse at the apex, rounded or tapering to the nearly sessile base, pubescent on both surfaces, up to 2 cm long, less than 1 cm broad; flowers to 4 mm across, in cymes, the uppermost bracts with narrow scarious margins, the lower bracts green throughout, with pedicels up to 1.2 cm long; sepals 5, free, lanceolate, acute, glandular-pubescent, narrowly scarious along the margins, 3–5 mm long; petals 5, free, deeply 2-cleft at the apex, white, with branched veins, 3–5 mm long; stamens usually 5; styles 5; capsules cylindrical, 4–7 mm long, slightly curved to nearly straight, on erect pedicels; seeds obovoid, 0.4–0.5 mm long, brown, minutely warty.

Common Name: Mouse-eared Chickweed.
Habitat: Grassy areas.
Range: Native to Europe; sparingly adventive in the eastern United States.
Illinois Distribution: Scattered in the state; probably more common than our records indicate.

This species has only five stamens, a character shared only with *C. diffusum* and *C. semidecandrum* in the genus. *Cerastium diffusum*, however, does not have scarious bracts and sepals, and *C. semidecandrum* has broad scarious margins of the bracts.

Cerastium pumilum flowers from April to June.

5. **Cerastium semidecandrum** L. Sp. Pl. 488. 1753. Fig. 101.

Annual without basal offshoots; stems tufted, ascending to erect, usually glandular-pubescent, sometimes branched, up to 20 cm tall; leaves oblong to elliptic, acute to

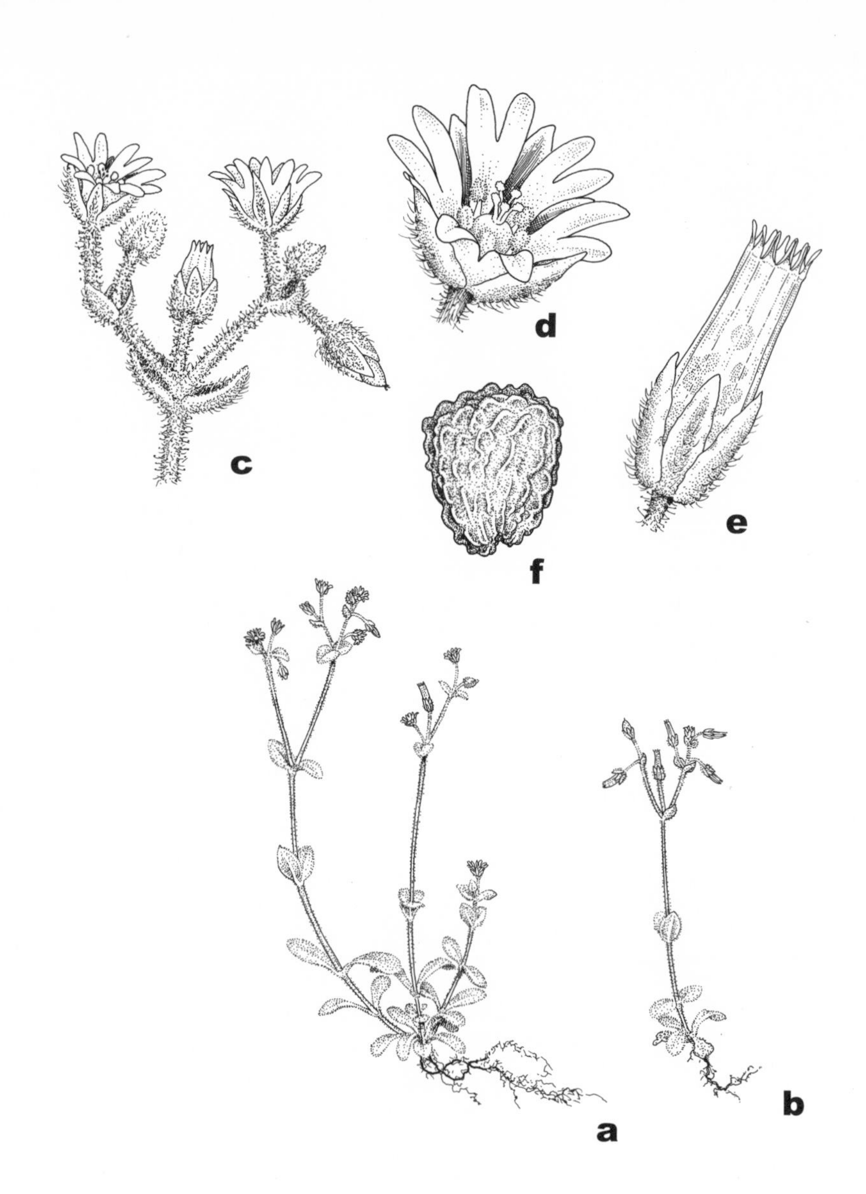

100. **Cerastium pumilum**
a. Habit
b. Habit
c. Flowering branch
d. Flower
e. Fruit
f. Seed

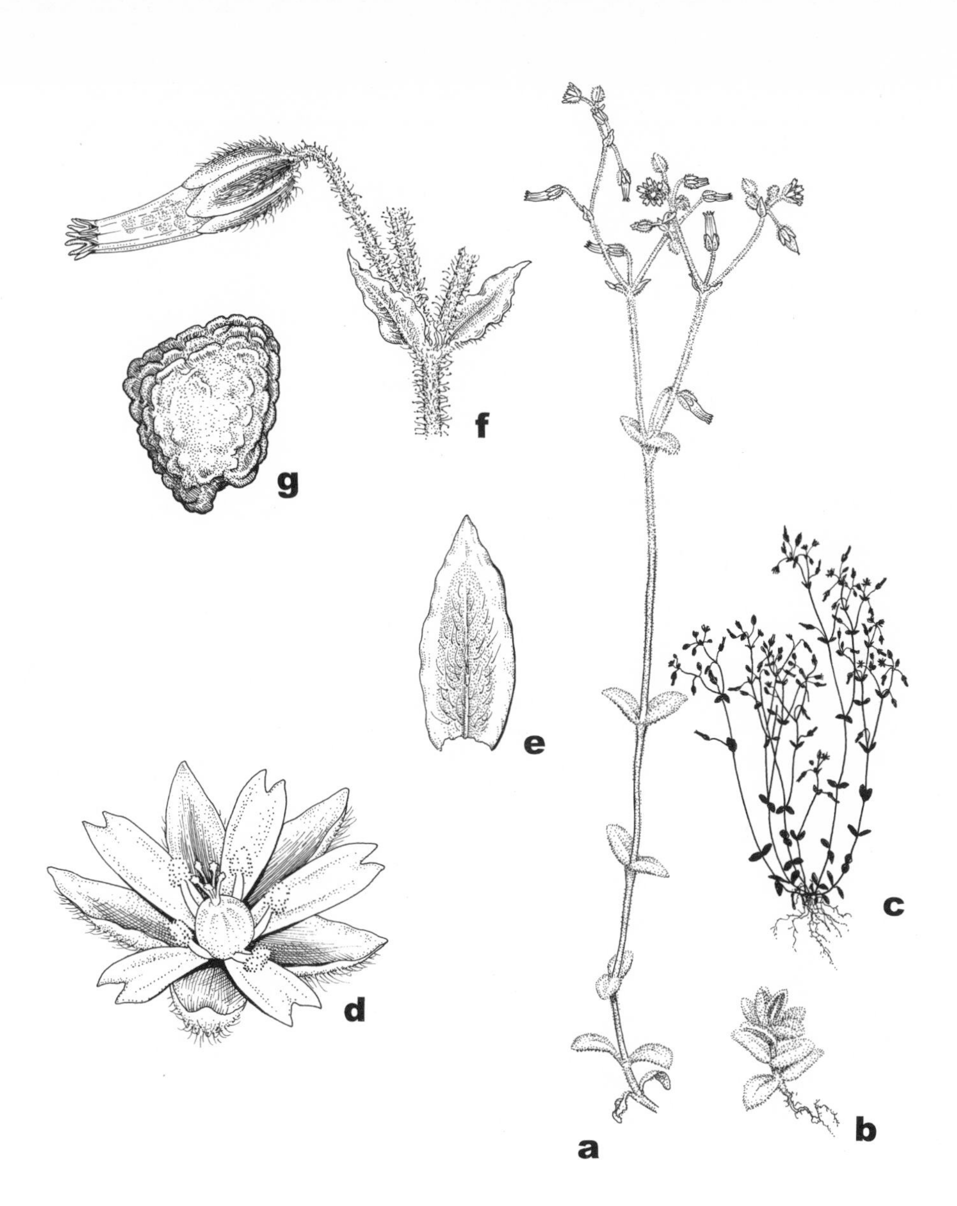

101. **Cerastium semidecandrum**
a. Habit
b. Habit
c. Habit
d. Flower
e. Sepal
f. Fruiting branch
g. Seed

obtuse at the apex, rounded or tapering to the nearly sessile base, usually pubescent on both surfaces, up to 1.5 (–2.0) cm long, usually less than 1 cm broad; flowers to 5 mm across, in cymes, the uppermost bracts as much as half composed of scarious tissue, with pedicels up to 1.2 cm long; sepals 5, free, lanceolate, acute, glandular-pubescent, scarious along the margins, 3–5 mm long; petals 5, free, shallowly 2-cleft at the apex, white, with unbranched veins, 3–5 mm long; stamens usually 5; styles 5; capsules cylindrical, 4–7 mm long, usually slightly curved, on deflexed pedicels; seeds obovoid, 0.4–0.6 mm long, brown, smooth or minutely warty.

Common Name: Mouse-eared Chickweed.
Habitat: Disturbed, usually sandy soil.
Range: Native to Europe; scattered as an adventive across the United States.
Illinois Distribution: Known from Cook, Effingham, Iroquois, Kankakee, Lake, and Will counties.

This species is very similar to *C. pumilum* from which it differs in its more broadly scarious-margined upper bracts, its deflexed fruiting pedicels, and the unbranched veins of its petals.

Close study of the species of *Cerastium* in Illinois will probably reveal a wider distribution of this species.

Cerastium semidecandrum flowers from late March to mid-June.

6. Cerastium diffusum Pers. Syn. 1:520. 1805. Fig. 102.
Cerastium tetrandrum Curtis, Fl. Loud. fasc. 6, t. 31. 1777.

Annual from slender roots; stems ascending to erect, puberulent with viscid-glandular hairs, sometimes branched, up to 20 cm tall; median and upper leaves ovate to ovate-lanceolate, acute to obtuse at the apex, tapering or rounded at the sessile base, puberulent with glandular or eglandular hairs, up to 17 mm long, up to 6 mm broad; lower leaves elliptic to obovate, acute at the apex, tapering to the base, otherwise similar to the other leaves; flowers up to 5 mm across, in dichotomous cymes, with nonscarious bracts, with pedicels glandular-pubescent, up to 13 mm long; sepals 5, free, green, broadly lanceolate, pubescent, with a broad, scarious margin; petals 4 or 5, free, acute, 2-cleft at the apex, white; stamens 4 (–5), the filaments glabrous; styles 5; capsules cylindrical, 7–8 mm long, slightly curved, the pedicel at least twice as long as the capsule; seeds obovoid, 0.4–0.6 mm long, pale brown, minutely warty.

Common Name: Mouse-eared Chickweed.
Habitat: Disturbed soil, particularly along roads.
Range: Native to Europe; rarely adventive in the United States.
Illinois Distribution: Scattered in the central and southern parts of Illinois.

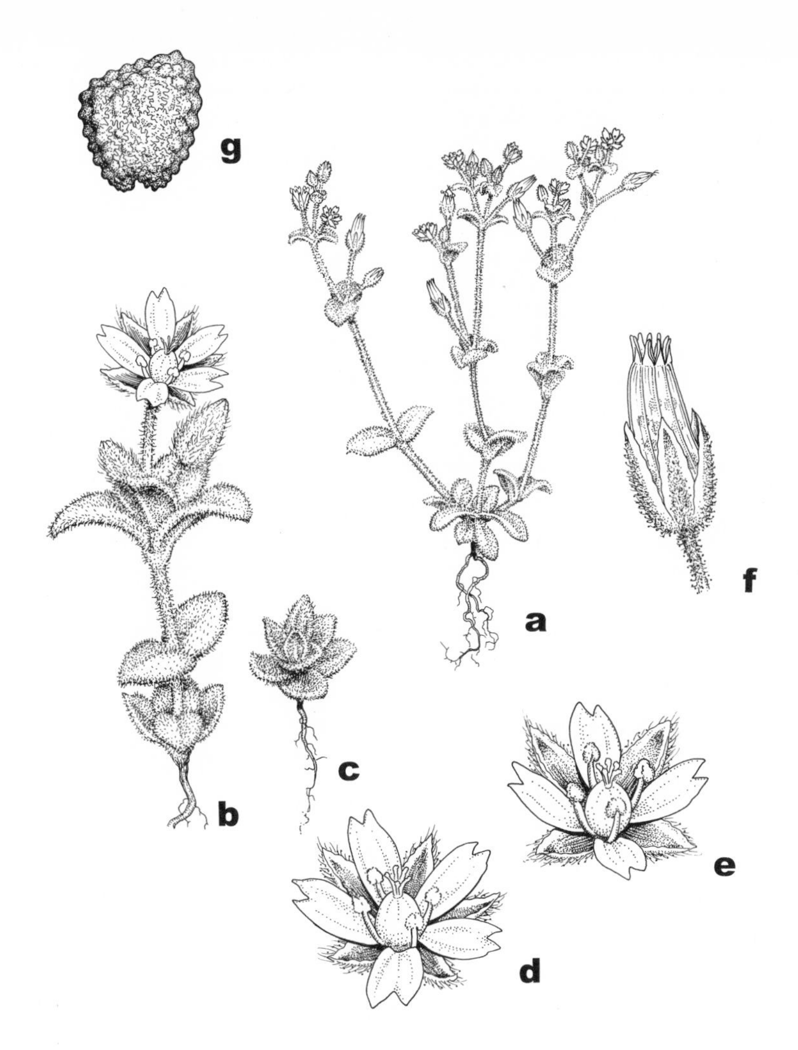

102. **Cerastium diffusum**
a. Habit
b. Habit
c. Habit
d. Flower with 5 petals
e. Flower with 4 petals
f. Fruit
g. Seed

This species is similar in appearance to *C. glomeratum* except that it is generally smaller in most characteristics. It is further distinguished by its usually four stamens.

Cerastium tetrandrum, by which name this species was first reported from Illinois, is a synonym for *C. diffusum.*

Cerastium diffusum flowers in March and April.

7. **Cerastium nutans** Raf. Prec. Decouv. 36. 1814. Fig. 103.
Cerastium longepedunculatum Muhl. Cat. 46. 1813, *nomen nudum.*

Annual from slender roots; stems spreading to ascending, weak, finely glandular-pubescent to sometimes nearly glabrous, branched, up to 50 cm tall; median and upper leaves oblong to lanceolate, acute at the apex, tapering to a short-petiolate base, otherwise similar to the other leaves; flowers up to 7 mm across, in dichotomous cymes, with nonscarious bracts, with pedicels spreading to ascending, becoming deflexed by fruiting time, to 1.2 cm long, at least 2 times longer than the sepals; sepals 5, free, lanceolate, acute to obtuse, green with scarious margins, pubescent, 3–5 mm long; petals 5, free, 2-cleft at the apex, white, 3–6 mm long; stamens usually 10, the filaments glabrous; styles 3; capsules cylindrical, to 15 mm long, curved, the pedicels deflexed; seeds obovoid, 0.5–0.6 mm long, reddish brown, warty.

Common Name: Nodding Mouse-eared Chickweed.
Habitat: Open, disturbed soil.
Range: Quebec to British Columbia, south to Arizona, Texas, and Florida.
Illinois Distribution: Throughout the state, although apparently less common in the northwestern counties.

This is one of the native species of *Cerastium* in Illinois, although it is usually found in areas of disturbed soil.

It is similar in appearance to *C. brachypodum,* from which it differs mainly by its longer pedicels, and to *C. brachypetalum,* from which it differs by its glabrous filaments and its uniformly pubescent sepals.

Although the binomial *C. longepedunculatum* predates *C. nutans* by one year, it is illegitimate because it is a *nomen nudum,* that is, it was published without a description.

Nodding mouse-eared chickweed blooms from March to June.

8. **Cerastium brachypetalum** Pers. Syn. 1:520. 1805. Fig. 104.

Annual from slender roots; stems ascending to erect, puberulent but the hairs glandular and viscid only on the upper part of the stem, sometimes branched, up to 25 cm tall; median and upper leaves broadly lanceolate, mostly acute at the apex, tapering to the sessile base, puberulent with eglandular hairs, up to 20 mm long, up to 7 mm broad; lower leaves elliptic to spatulate, otherwise similar to the other leaves; flowers up to 6 mm across, in dichotomous cymes, with nonscarious bracts bearing a tuft of hairs at the apex, with spreading to ascending pedicels up to 15 mm long; sepals 5, free, lanceolate, acute, with a tuft of hairs at the apex, 4–5 mm long;

103. **Cerastium nutans**
a. Habit
b. Habit
c. Flower
d. Sepal
e. Petal
f. Fruit
g. Seed

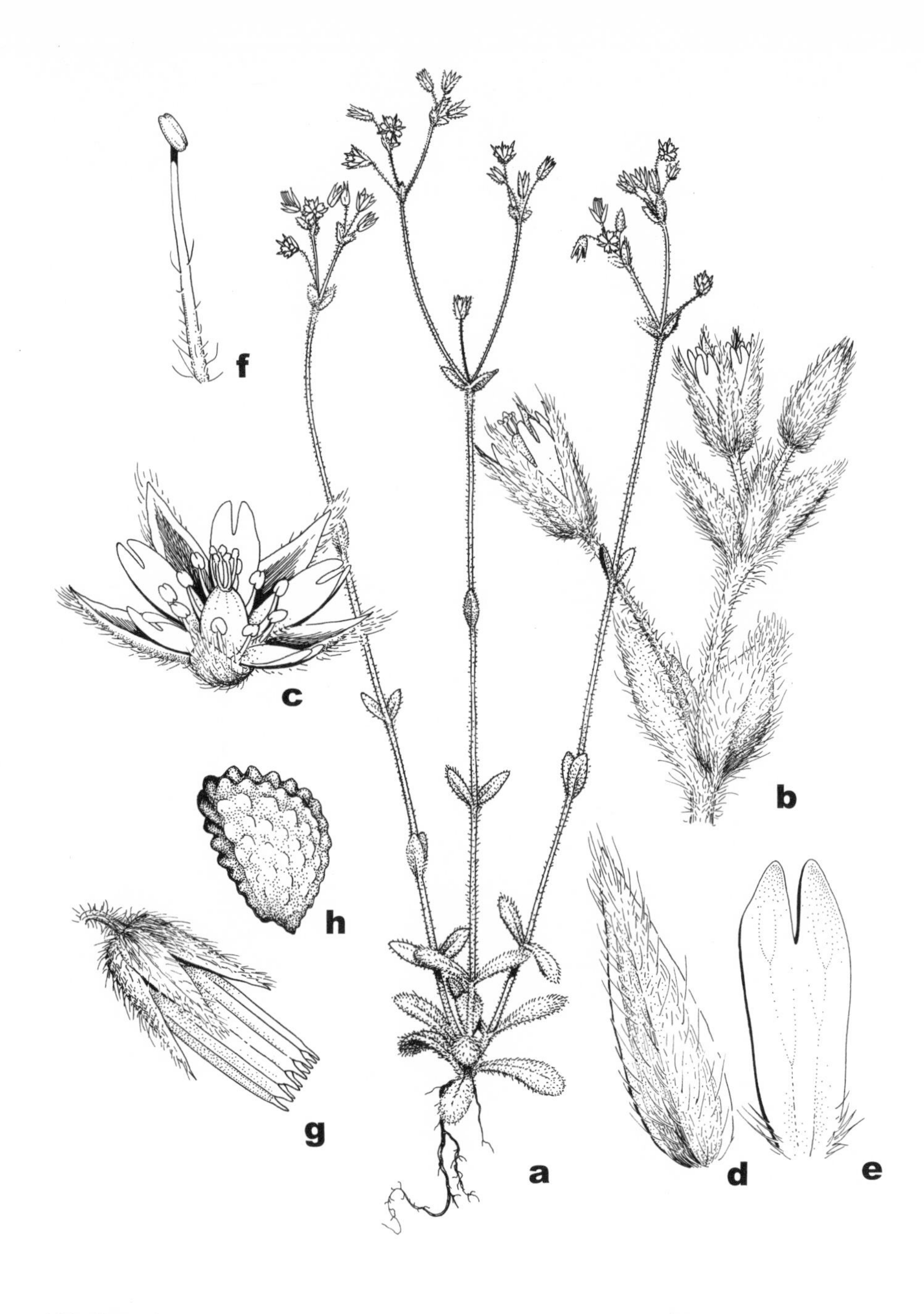

104. **Cerastium brachypetalum**
a. Habit
b. Flowering branch
c. Flower
d. Sepal
e. Petal
f. Stamen
g. Fruit
h. Seed

petals 5, free, 2-cleft at the apex, white; stamens 10, the filaments ciliate; styles 5; capsules cylindrical, 7–8 mm long, slightly curved, the pedicel less than twice as long as the capsule; seeds obovoid, 0.4–0.6 mm long, pale brown, minutely warty.

Common Name: Mouse-eared Chickweed.
Habitat: Along roads.
Range: Native to Europe; rarely adventive in the United States.
Illinois Distribution: Known from Jackson, Montgomery, Pulaski, Shelby, Tazewell, and Union counties.

This species appears gray when seen growing in dense patches along roads.

Although this species is similar to other *Cerastium* species in Illinois, it differs in its conspicuous tuft of hairs on the bracts and sepals and in its pubescent filaments of the stamens.

Cerastium brachypetalum flowers during April and May.

9. Cerastium glomeratum Thuill. Fl. Paris, ed. 2, 226. 1799. Fig. 105.
Cerastium viscosum L. Sp. Pl. 437. 1753, *nomen ambiguum.*

Annual from slender roots; stems tufted, spreading to ascending to erect, viscid-pubescent, branched, up to 40 cm tall; leaves ovate to elliptic to obovate to spatulate to lanceolate, usually obtuse at the apex, tapering or somewhat rounded at the sessile base, viscid-pubescent on both surfaces, up to 3.5 cm long, up to 1 cm broad; flowers up to 7 mm across, in congested cymes, the bracts nonscarious, the pedicels up to 5 mm long; sepals 5, free, lanceolate, acute, pubescent, sometimes glandular-pubescent, green with scarious margins, 3.5–5.0 mm long; petals 5, free, 2-cleft at the apex, white, 3.0–3.5 mm long, or petals absent in flowers on some of the lateral branches; stamens usually 10, the filaments glabrous; styles 5; capsules cylindrical, to 8 (–10) mm long, curved, nearly twice as long as the sepals; seeds obovoid, 0.4–0.6 mm long, pale brown, minutely warty.

Common Name: Clammy Mouse-eared Chickweed.
Habitat: Disturbed soil, including lawns.
Range: Native to Europe and Asia; naturalized in most of the United States.
Illinois Distribution: Scattered throughout the state.

This mouse-eared chickweed is the most viscid-pubescent species of *Cerastium* in Illinois. The inflorescence, although cymose, is more congested than that of any other species of *Cerastium.* It differs from *C. brachypodum* in its acute sepals and its capsules that are only twice as long as the sepals.

For years this plant has been known usually as *C. viscosum,* but that Linnaean binomial has been determined to be a *nomen ambiguum.*

Cerastium glomeratum flowers during April and May.

105. **Cerastium glomeratum**
a. Habit
b. Habit
c. Flower
d. Sepal
e. Petal
f. Fruit
g. Seed

10. Cerastium brachypodum (Engelm. ex Gray) B. L. Robins. Proc. Am. Acad. Arts 29:277. 1894. Fig. 106.
Cerastium nutans Raf. var. *brachypodum* Engelm. ex Gray, Man. Bot. ed. 5, 94. 1867.

Annual from slender roots; stems spreading to ascending, weak, finely glandular-pubescent to eglandular, branched, up to 45 cm tall; median and upper leaves oblong to linear-oblong, acute at the apex, tapering to the sessile base, more or less puberulent, up to 2 cm long, up to 1 cm broad; lower leaves oblanceolate to spatulate, obtuse at the apex, tapering to the short-petiolate base, up to 2.5 cm long, up to 1.2 cm broad; flowers up to 7 mm across, in dichotomous cymes, with nonscarious bracts, with pedicels spreading to ascending, not becoming deflexed during fruiting, up to 1.5 cm long; sepals 5, free, lanceolate, obtuse to subacute, pubescent, green with scarious margins, 3–5 mm long; petals 5, free, 2-cleft at the apex, white, 3–6 mm long; stamens usually 10, the filaments glabrous; styles 5; capsules cylindrical, up to 15 mm long, curved, about three times as long as the capsules; seeds obovoid, 0.5–0.8 mm long, reddish brown, warty.

Common Name: Mouse-eared Chickweed.
Habitat: Disturbed, usually moist soils; open woods; railroad ballast.
Range: Virginia to Illinois to North Dakota and Washington, south to Arizona and Georgia; Alberta; Mexico.
Illinois Distribution: Occasional throughout the state.

At one time this plant was considered to be a variety of *C. nutans* from which it differs primarily in its pedicels three times as long as the capsules. *Cerastium brachypodum* is similar to *C. glomeratum*, but *C. brachypodum* has obtuse to subacute sepals and a capsule at least three times longer than the sepals.

This is a species primarily of waste ground.

Cerastium brachypodum flowers from late March to June.

10. **Moenchia** L.—Moenchia

Annuals; leaves opposite, entire, without stipules; inflorescence a few-flowered, terminal cyme, or the lowest flowers solitary in the leaf axils; flowers perfect, actinomorphic; sepals 4, free; petals 0 or 4, free, entire at the apex; stamens 4 or 8; styles 4, opposite the sepals; ovary superior, 1-locular, with many ovules; fruit a capsule, 8-toothed; seeds numerous.

This genus is unique in the Caryophyllaceae in its 4 sepals, 4 or 8 stamens, 4 styles, and 8-toothed capsules.

Six species comprise this European genus. Only the following adventive species occurs in Illinois.

1. Moenchia erecta P. Gaertn., Meyer, & Schreb. Fl. Wett. 1:219. 1800. Fig. 107.

Annual with a slender taproot; stems erect, glabrous, glaucous, to 15 cm tall; basal leaves oblanceolate, obtuse or subacute at the apex, tapering to a more or less petiolate base, glabrous, glaucous, to 1.5 cm long; cauline leaves stiff, linear to

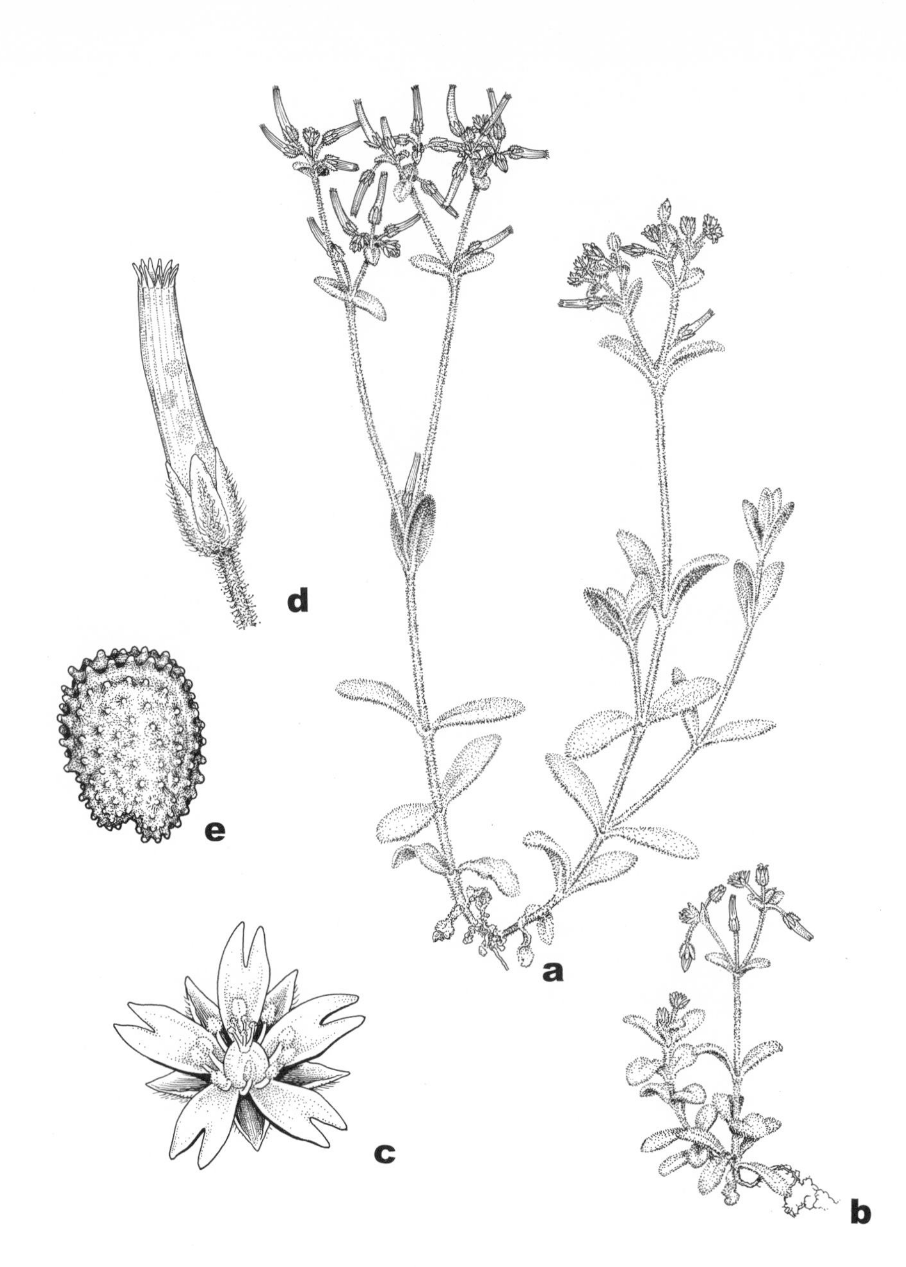

106. **Cerastium brachypodum**
a. Habit
b. Habit
c. Flower
d. Fruit
e. Seed

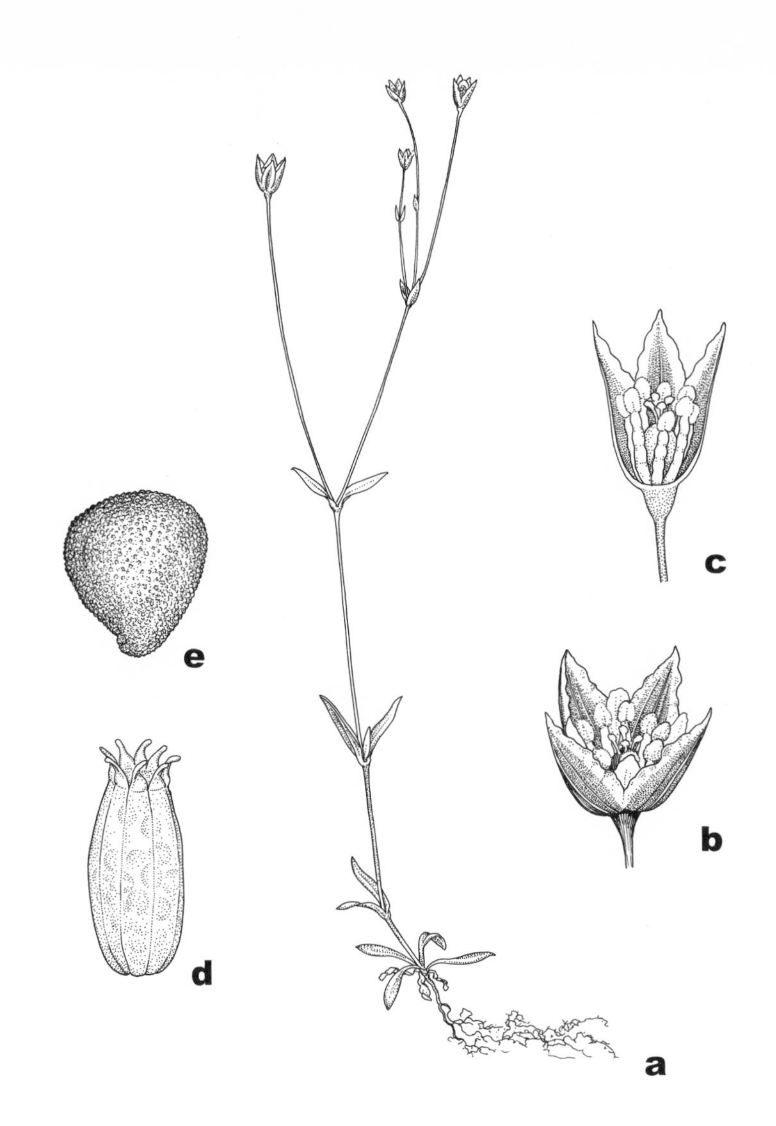

107. **Moenchia erecta**
a. Habit
b. Flower
c. Flower with petal removed
d. Fruit
e. Seed

linear-lanceolate, subacute at the apex, sessile, to 1.2 cm long; flowers solitary or few in a cyme, on glabrous pedicels up to 7 cm long; sepals 4, free, lanceolate, glabrous, 3-veined, 4–7 mm long; petals 0 or 4, free, white, lanceolate, entire at the apex, 3–6 mm long; stamens 4 or 8; styles 4, opposite the sepals; capsules cylindrical, more or less straight, with 8 teeth at the tip; seeds numerous, 0.4–0.6 mm long, reddish brown.

Common Name: Moenchia.
Habitat: Disturbed ground in a field.
Range: Native to Europe; rarely adventive in the United States.
Illinois Distribution: Known only from Clay County.

This species was discovered in an old field in Clay County in 1992 by David Ketzner. Clay County is one of the few localities in North America for this species.

Moenchia erecta flowers during May in Illinois.

11. **Minuartia** L.—Sandwort

Annual or perennial herbs; leaves subulate, setaceous, filiform, or linear; stipules absent; inflorescence cymose, terminal or axillary, or flower occasionally solitary; flowers with a very short hypanthium, perfect, actinomorphic; sepals 5, free or nearly so; petals 5, free, entire at the apex; stamens 10; styles 3; ovary superior, 1-locular, with many ovules; fruit a capsule, 3-toothed; seeds usually numerous, often with thickened margins.

Although Linnaeus described *Minuartia, Arenaria,* and *Moehringia* as three distinct genera, most botanists in the twentieth century combined the three, using the name *Arenaria.* However, McNeill (1980) resurrected all three genera, a decision I have chosen to follow in Illinois. *Minuartia* differs from *Arenaria* and *Moehringia* in its capsules that have three teeth at the apex. In addition, the leaves of species of *Minuartia* are extremely narrow, while those of *Arenaria* and *Moehringia* in Illinois are lanceolate to ovate.

There are about 120 species of *Minuartia* found from the Arctic to Mexico, as well as in north Africa and south Asia. Two species, identifiable by the following key, occur in Illinois.

1. Leaves often in fascicles, setaceous to linear, flat; seeds 0.8–1.2 mm long; branches of the inflorescence glabrous . . . . 1. *M. michauxii*
1. Leaves opposite, not in fascicles, terete, more or less fleshy; seeds 0.5–0.8 mm long; branches of the inflorescence usually glandular-pubescent . . . . 2. *M. patula*

1. **Minuartia michauxii** (Fenzl) Farw. Rep. Mich. Acad. Sci. 20:177. 1918. Fig. 108.
Arenaria stricta Michx. Fl. Bor. Am. 1:274. 1803, non Sw.
Alsine michauxii Fenzl, Verbr. Alsin. 18. 1833.
Arenaria michauxii (Fenzl) Hook. f. Trans. Linn. Soc. 23:297. 1867.

Perennial herb from a long, tough rootstock; stems branching from the base, wiry, very slender, glabrous or rarely puberulent, erect or nearly so, up to 15 cm tall;

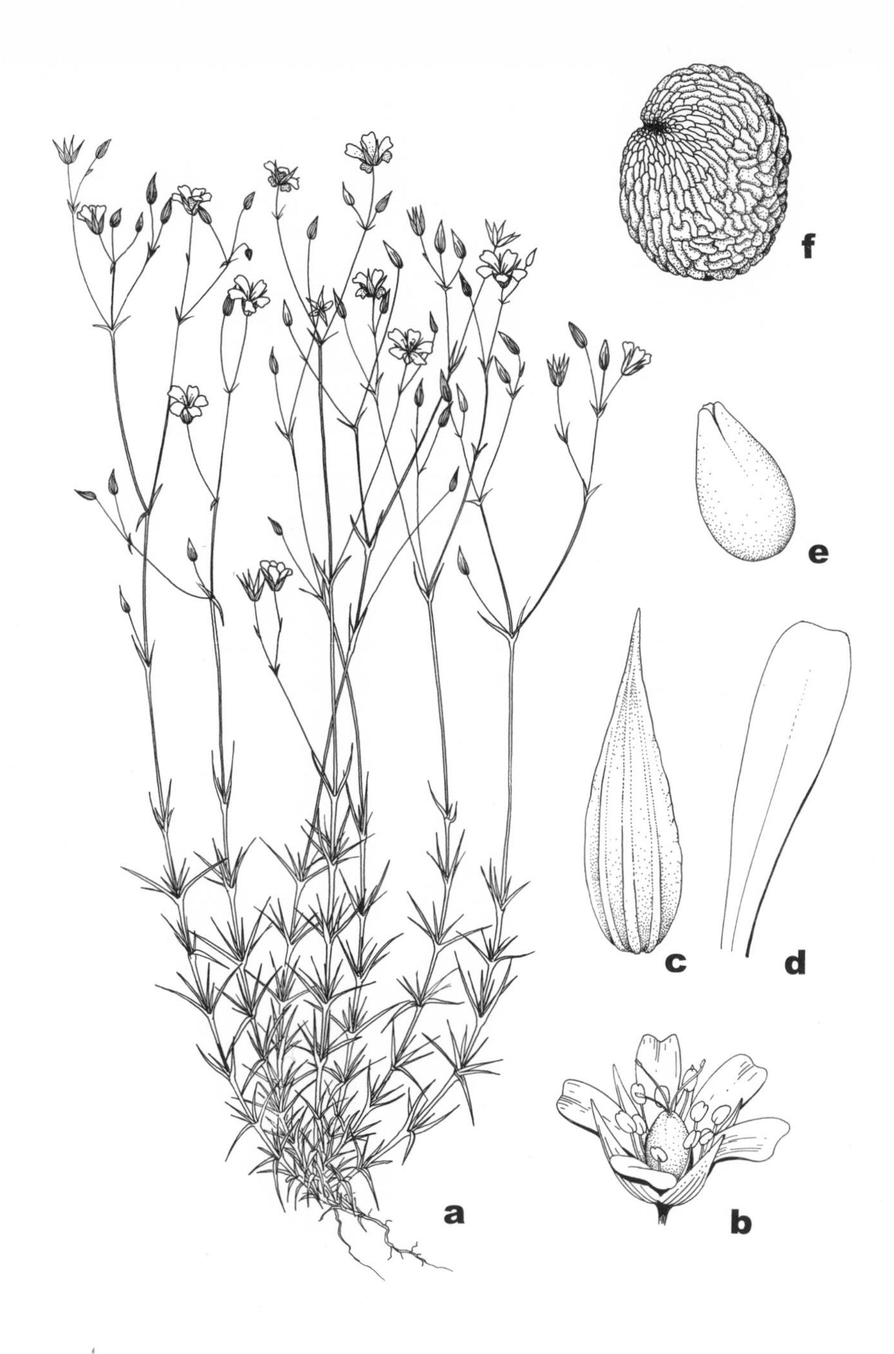

108. **Minuartia michauxii**
a. Habit
b. Flower
c. Sepal
d. Petal
e. Fruit
f. Seed

leaves confined to the lower part of the plant, fascicled, subulate to linear, stiff, flat, connate at the base, glabrous, up to 20 mm long, flowers up to 1 cm across, in dichotomous cymes, with glabrous pedicels up to 20 mm long; sepals 5, lanceolate, acute to acuminate, scarious-margined, glabrous, 4–5 mm long; petals 5, free, 5–8 mm long, white; stamens 10; styles 3; capsules ovoid, to 5 mm long, with 3 teeth at the apex; seeds reniform, 0.8–1.2 mm long, dark brown to nearly black, with low tubercles.

Common Name: Stiff Sandwort.
Habitat: Sandy ridges near Lake Michigan; gravelly limestone glades.
Range: Ontario to South Dakota, south to Texas and Virginia.
Illinois Distribution: Occasional in the northern one-third of Illinois; also St. Clair Co.

Minuartia michauxii is a highly variable species of intergrading characters, including length of leaves and length of sepals.

Michaux first described this species as *Arenaria stricta* and, if this species is retained in the genus *Arenaria*, that would be the correct binomial. However, if transferred to the genus *Minuartia*, this species becomes *M. michauxii* since Swartz had already used the binomial *M. stricta* for a different species.

Many of the areas in Illinois where this species used to be more common have been destroyed.

Minuartia michauxii flowers from May to July.

2. Minuartia patula (Michx.) Mattf. Bot. Jahrb. 57, beibl. 126:28. 1921. Fig. 109.
Arenaria patula Michx. Fl. Bor. Am. 1:273. 1803.
Arenaria pitcheri Nutt. in Torr. & Gray, Fl. N. Am. 1:180. 1838.
Arenaria patula Michx. f. *pitcheri* (Nutt.) Steyerm. Rhodora 43:329. 1941.
Arenaria patula Michx. f. *media* Steyerm. Rhodora 43:331. 1941.

Annual from slender roots; stems branching from the base, slender, glabrous or pubescent with glandular hairs, ascending to erect, up to 40 cm tall; leaves more or less terete, somewhat fleshy, setaceous, filiform, or linear, subulate at the apex, glabrous or nearly so, up to 2 cm long, up to 1.5 mm broad; flowers up to 1 cm across, in terminal cymes, on usually glandular-pubescent pedicels up to 3 cm long; sepals 5, lanceolate, acute to acuminate, glabrous or glandular-pubescent, 4–6 mm long; petals 5, white, free, 5–8 mm long; stamens 10; styles 3; capsules oblongoid, to 6 mm long, with 3 teeth at the apex; seeds reniform, 0.5–0.8 mm long, gray-brown, with low tubercles.

Common Name: Slender Sandwort.
Habitat: Wooded slopes.
Range: Ohio to Minnesota, south to Texas and Alabama.
Illinois Distribution: Confined to a few northeastern counties and St. Clair County.

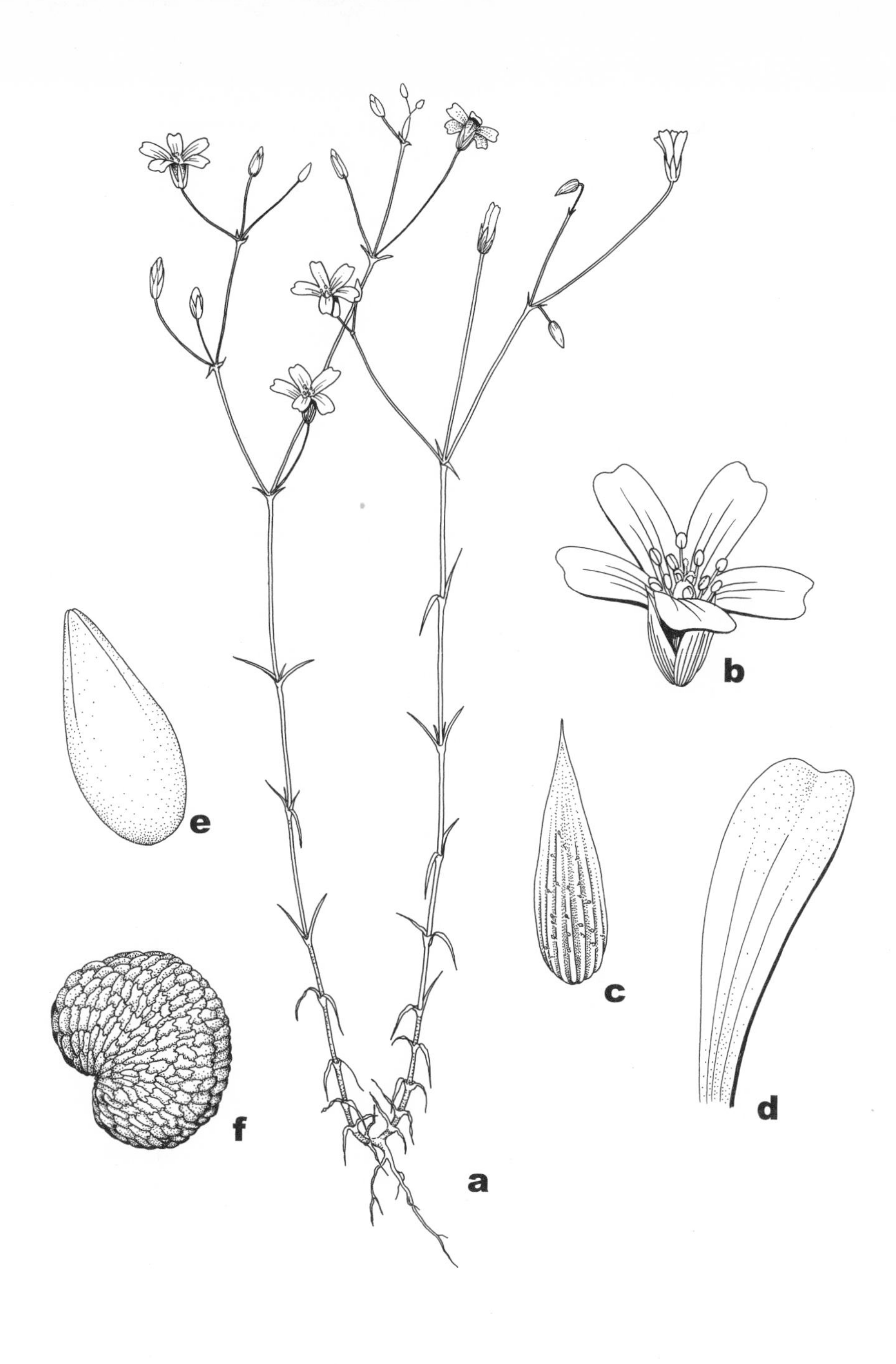

109. **Minuartia patula**
a. Habit
b. Flower
c. Sepal
d. Petal
e. Fruit
f. Seed

Minuaria patula may be distinguished from *M. michauxii* in its terete and often fleshy, nonfascicled leaves; its smaller seeds; and its usually glandular-pubescent branches of the inflorescence.

This very slender species is infrequently found in Illinois, although it is sometimes rather common where it does occur. It prefers wooded slopes with a limestone substrate.

Slender sandwort flowers from May to July.

12. **Moehringia** L.—Sandwort

Perennial herbs with rhizomes; leaves opposite, simple, lanceolate to elliptic, entire; stipules absent; inflorescence cymose, terminal or axillary; flowers with a very short hypanthium, perfect, actinomorphic; sepals 5, free or nearly so; petals 5, free, white, entire at the apex; stamens 10; styles 3; ovary superior, 1-locular, with few ovules; fruit a capsule, 6-toothed; seeds few, each with a spongy appendage.

Twenty-five species comprise this genus. It is often merged with *Arenaria* from which it differs in its 6-toothed capsules, its spongy-appendaged seeds, and its broader leaves.

Only the following species occurs in Illinois.

1. **Moehringia lateriflora** (L.) Fenzl, Verbr. Alsin. 18. 1833. Fig. 110.
Arenaria lateriflora L. Sp. Pl. 423. 1753.
Moehringia lateriflora (L.) Fenzl var. *typica* Regel, Bull. Soc. Nat. Moscow 35:377. 1862.
Arenaria lateriflora L. var. *typica* (Regel) St. John, Rhodora 19:260. 1917.

Perennial herbs from slender rhizomes; stems ascending to erect, branched or unbranched, puberulent, up to 30 cm tall; leaves oval to oblong to elliptic, obtuse to subacute at the apex, puberulent, up to 2.6 cm long, up to 1 cm broad; flowers up to 9 mm across, in few-flowered terminal or lateral cymes, on slender pedicels up to 2.5 cm long; sepals 5, oblong to ovate, obtuse, glabrous or puberulent, 2–3 mm long; petals 5, white, 4–6 mm long; stamens 10; styles 3; capsules ovoid, 3–7 mm long; seeds reniform, 1.0–1.3 mm long, black, with an elongated spongy appendage.

Common Name: Sandwort.
Habitat: Moist or dry woods; prairie remnants.
Range: Labrador to Alaska, south to New Mexico, northern Missouri, northern Illinois, and Maryland.
Illinois Distribution: Occasional in the northern half of the state; also St. Clair and Wabash counties.

The seeds of this species are unique because of an elongated spongy appendage that is attached to each of them. The 6-toothed capsule also distinguishes *Moehringia* from *Minuartia.*

Although *M. lateriflora* is primarily a species of woodlands, it sometimes occurs in prairie remnants.

This species flowers from April to July.

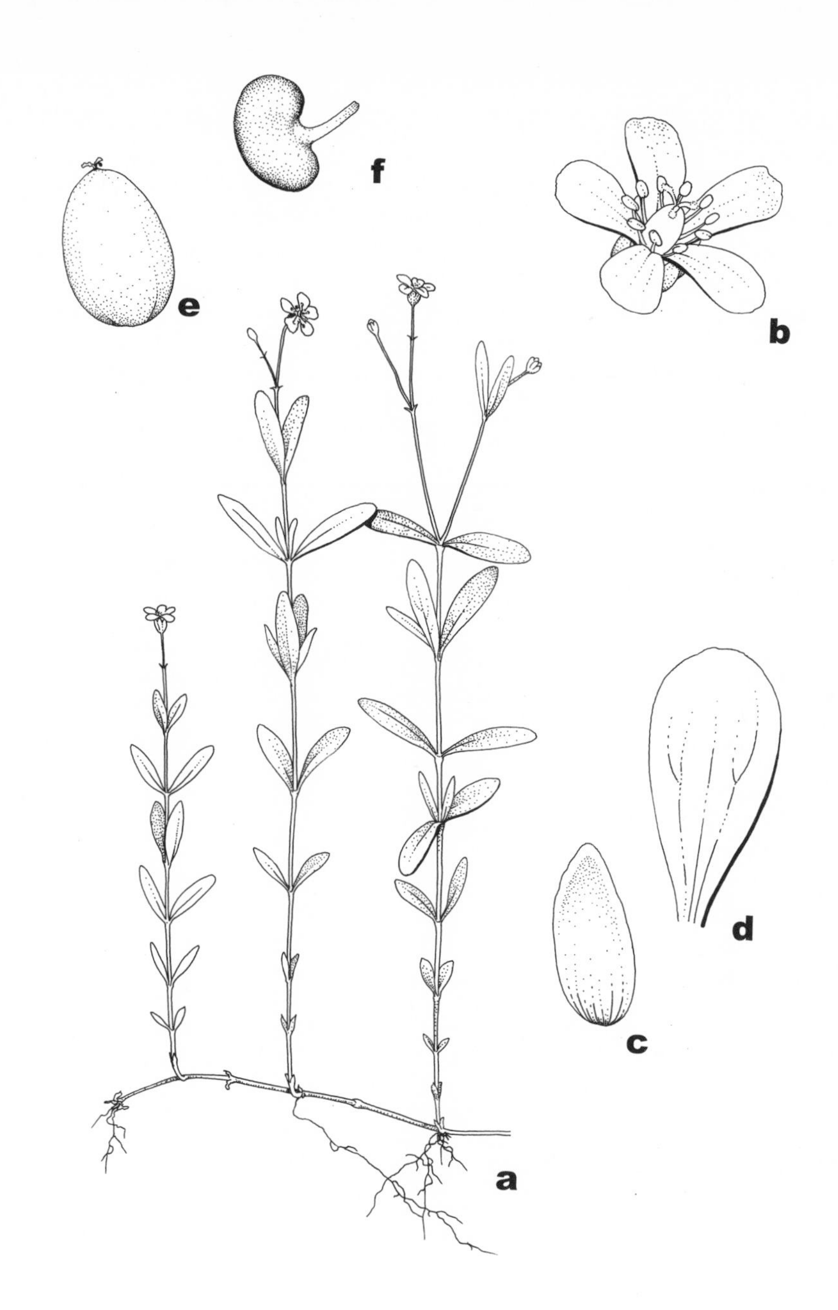

110. **Moehringia lateriflora**
a. Habit
b. Flower
c. Sepal
d. Petal
e. Fruit
f. Seed

13. **Arenaria** L.—Sandwort

Annual or perennial herbs; leaves simple, opposite, entire, mostly ovate, sometimes threadlike (but not in Illinois); stipules absent; inflorescence cymose, terminal or axillary; flowers with an extremely short hypanthium, perfect, actinomorphic; sepals 5, free or nearly so; petals 5, free, white, entire at the apex; stamens 10; styles 3; ovary superior, 1-locular, with few to many ovules; fruit a capsule, 6-toothed; seeds few to many, without a spongy appendage.

In the past, most Illinois botanists have included within *Arenaria* plants that in this work are segregated into *Minuartia* and *Moehringia. Minuartia* differs in its 3-toothed capsules, while *Moehringia* differs in having a spongy appendage attached to the seed.

About 150 species comprise *Arenaria.* Nearly all species occur in the northern hemisphere.

Only the following species is found in Illinois:

1. Arenaria serpyllifolia L. Sp. Pl. 423. 1753. Fig. 111.

Annual from slender roots; stems prostrate to ascending, wiry, puberulent, branched, up to 25 cm tall; leaves ovate, acute to acuminate at the apex, tapering or rounded to the sessile base, minutely rough-hairy, up to 5 (–7) mm long, up to 4 mm broad; flowers up to 5 mm across, subtended by bracts resembling the leaves, borne in reduced terminal and axillary cymes, with pedicels up to 2 cm long; sepals 5, ovate to ovate-lanceolate, acute to acuminate, minutely rough-hairy, sometimes glandular, 2.5–4.0 mm long; petals 5, free, white, 1.0–2.5 mm long; stamens 10; styles 3; capsules ovoid, 4–5 mm long, dehiscing by 6 teeth; seeds few, reniform, 0.4–0.6 mm long, gray-black, with low tubercles.

Common Name: Thyme-leaved Sandwort.
Habitat: Disturbed sandy soil, often along roads and paths.
Range: Native to Europe and Asia; naturalized throughout Canada and the United States.
Illinois Distribution: Scattered throughout the state.

This naturalized species has become widespread in sandy soils throughout most of temperate North America.

The small, acute to acuminate, ovate leaves and the reduced inflorescence help to distinguish this species.

Arenaria serpyllifolia flowers from April to August.

14. **Dianthus** L.—Pink; Carnation

Annual, biennial, or perennial herbs; leaves opposite, simple, entire, without stipules; inflorescence cymose or reduced to solitary flowers; flowers perfect, actinomorphic, usually showy, bracteate; sepals 5, united into a tube, weakly 10- to 40-ribbed; petals 5, free, clawed, usually irregularly toothed at the apex; stamens 10; styles 2; ovary superior, 1-locular, with many ovules, stipitate; fruit a capsule, more or less tubular, dehiscing by 4 (–5) teeth, stipitate; seeds many, flattened.

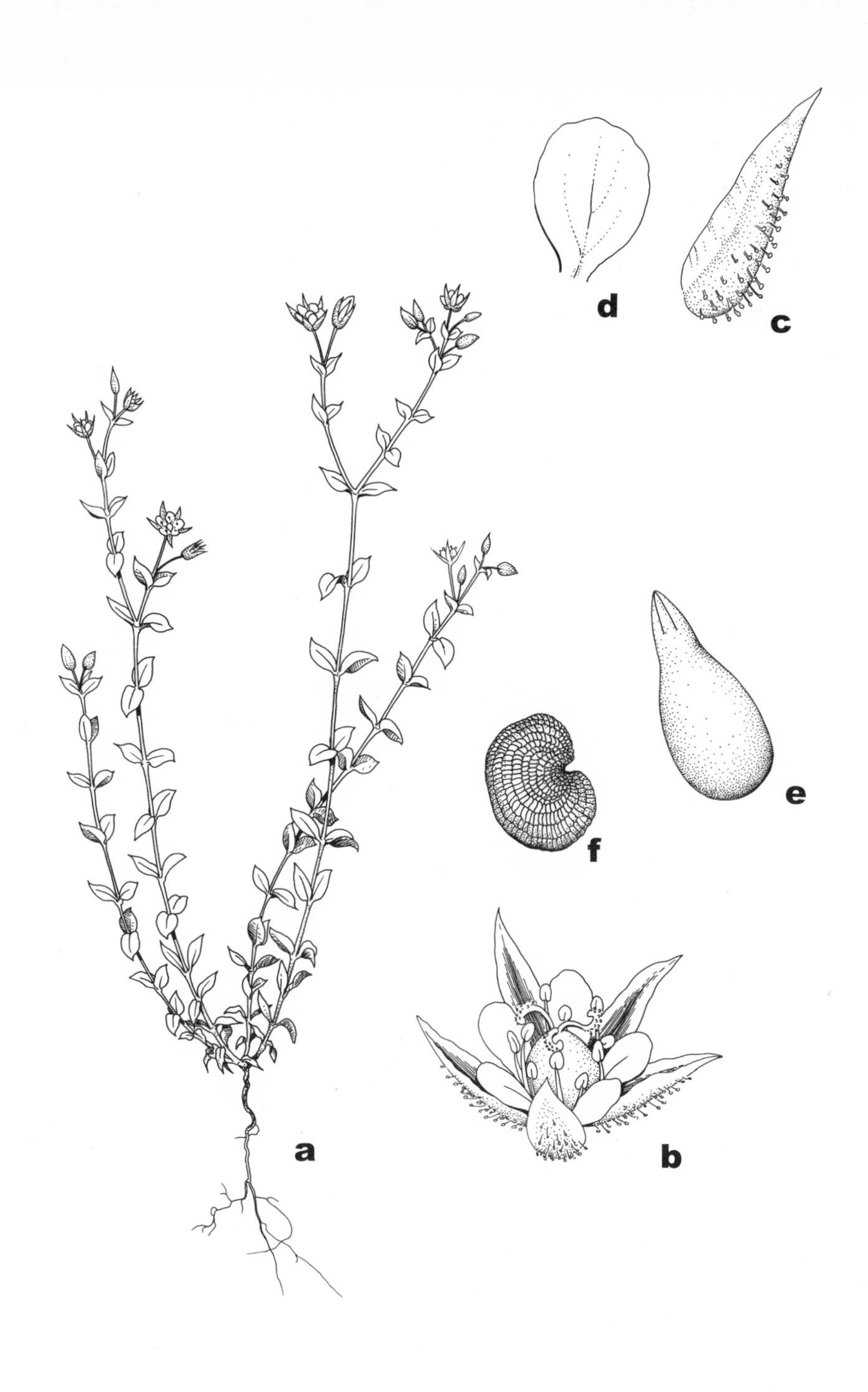

111. **Arenaria serpyllifolia**
a. Habit
b. Flower
c. Sepal
d. Petal
e. Fruit
f. Seed

More than three hundred species, mostly native to Europe and Asia, comprise this genus. Several species are favorites in gardens, including the ones described below. In addition to these, other ornamentals include *D. carophyllus*, the carnation, and *D. chinensis*, the rainbow pink.

Four adventive species are known from Illinois. They may be distinguished by the following key:

1. Leaves elliptic to oblong, 9 mm broad or broader; flowers in dense heads . . . . . 1. *D. barbatus*
1. Leaves linear, up to 9 mm broad; flowers not in dense heads.
  2. Annual or biennial; calyx densely pubescent; bracts about as long as the calyx . . . . 2. *D. armeria*
  2. Perennial; calyx glabrous or sparsely pubescent; bracts up to half as long as the calyx.
    3. Petals lacerate; leaves stiff; flowers strongly clove-scented . . . . 3. *D. plumarius*
    3. Petals sharply toothed; leaves lax; flowers not clove-scented . . . . 4. *D. deltoides*

1. Dianthus barbatus L. Sp. Pl. 409. 1753. Fig. 112.

Perennial herb from stout rhizomes; stems erect, mostly unbranched, glabrous, up to 1 m tall; leaves opposite, entire, lanceolate to ovate-lanceolate, acute at the apex, tapering to the base, up to 10 cm long, up to 1.8 cm broad, the basal leaves a little larger than the cauline and sometimes oblong; flowers borne in dense terminal heads up to 3 cm across, sessile or on pedicels up to 3 mm long, with bracts about as long as the calyx; calyx 5-parted, up to 18 mm long, 34- to 40-nerved, glabrous, the lobes lanceolate, acuminate; petals 5, pink to white, sometimes purple or violet, up to 25 mm long, the apex erose; stamens 10; styles 2; capsules ellipsoid, up to 1 cm long, on a stalk 3–4 mm long.

Common Name: Sweet William.
Habitat: Along roadsides.
Range: Native to Europe; rarely escaped from cultivation in the United States.
Illinois Distribution: Known from Hancock, Jackson, and McLean counties.

Sweet William is a commonly cultivated ornamental. It was found growing along a road in Lake Murphysboro State Park in Jackson County. Kibbe (1952) recorded it from Hancock County, but I have not seen a specimen to verify this report.

Dianthus barbatus flowers from June to August.

2. Dianthus armeria L. Sp. Pl. 410. 1753. Fig. 113.

Annual or biennial herb with slender taproots; stems erect, sparingly branched, puberulent or rarely nearly glabrous, up to 60 cm tall; leaves opposite, entire, linear to linear-lanceolate to oblanceolate, acute at the apex, tapering to the base, up to 8 cm long, up to 4 mm broad; flowers borne in dense or open terminal clusters, up to 3 cm across, with bracts about as long as the calyx; calyx 5-parted, up to 20 mm long, 20- to 25-nerved, densely pubescent, the lobes narrowly lanceolate; petals 5,

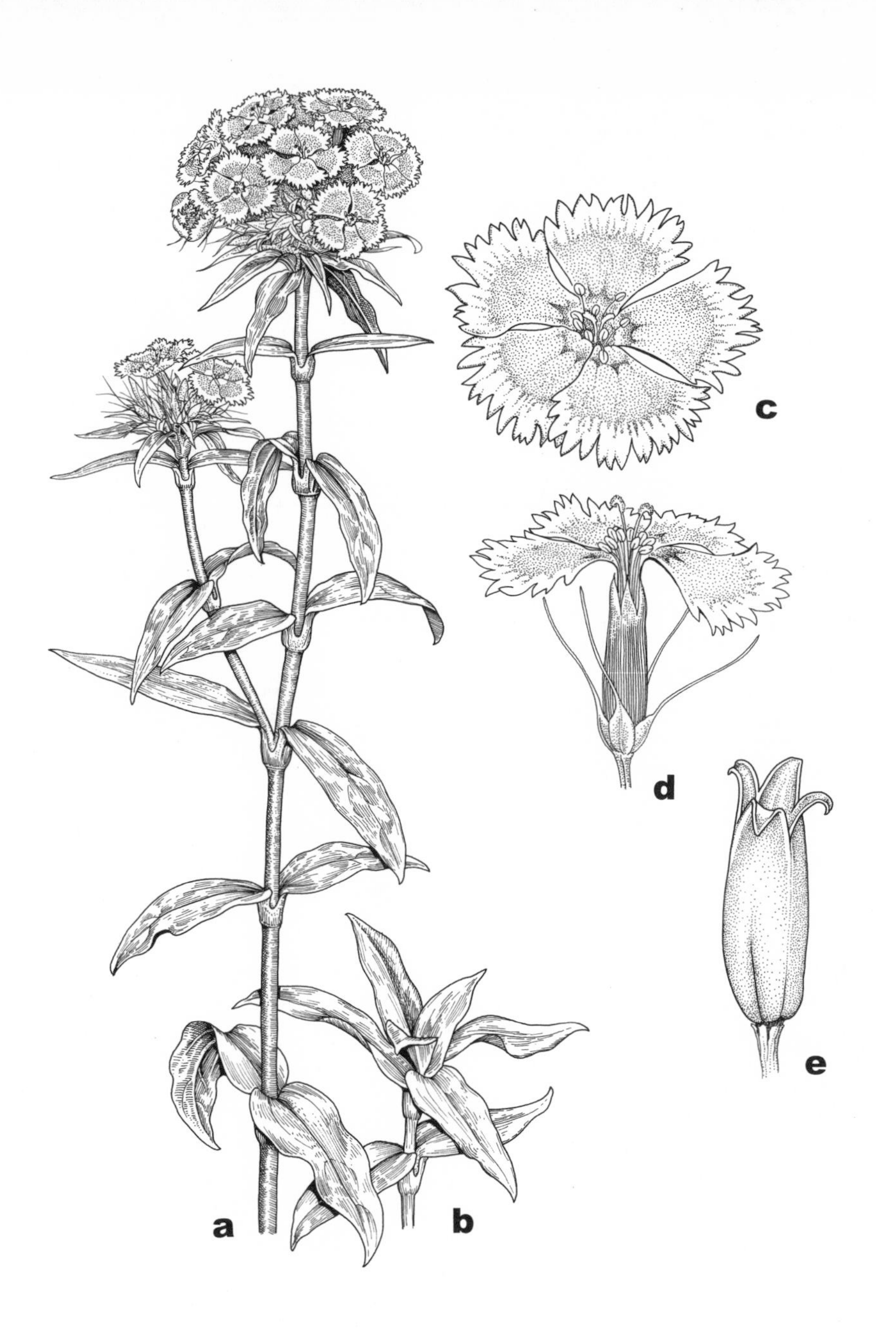

112. **Dianthus barbatus**
a. Habit
b. Leafy shoot
c. Flower, face view
d. Flower with one petal removed
e. Capsule

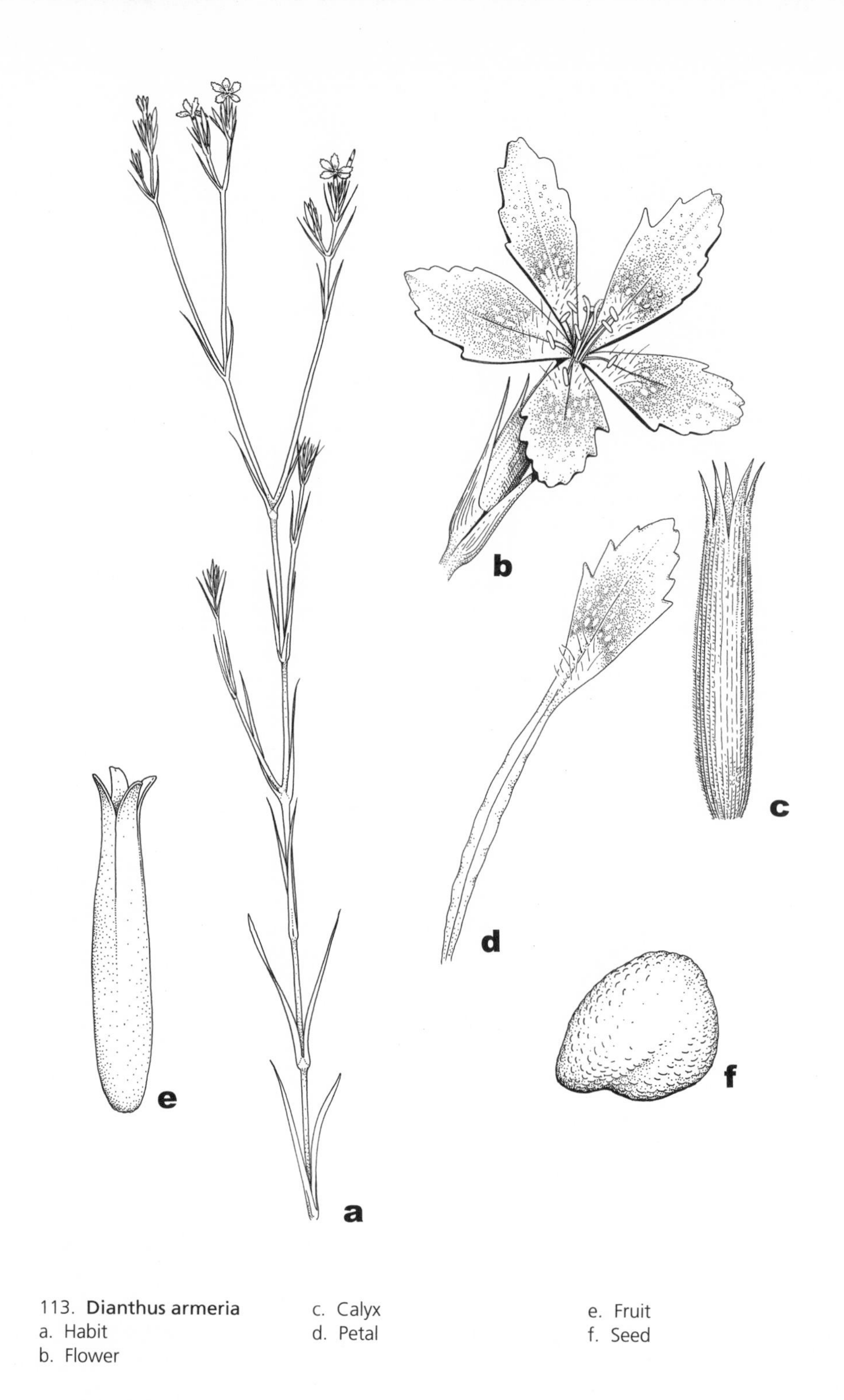

113. **Dianthus armeria**
a. Habit
b. Flower
c. Calyx
d. Petal
e. Fruit
f. Seed

free, rose or pink, speckled with white, rarely entirely white, 15–25 mm long, the apex erose; stamens 10; styles 2; capsules ellipsoid, up to 2 cm long, sessile or on a stalk up to 1 mm long.

Common Name: Deptford Pink.
Habitat: Fields, along roads, edges of woods.
Range: Native to Europe; adventive throughout the eastern United States.
Illinois Distribution: Scattered throughout the state.

Deptford pink is a bright-flowered species that has become increasingly common during the last half of the twentieth century. It differs from other species of *Dianthus* in Illinois in its annual or biennial habit and in its densely pubescent calyx tube.

Dianthus armeria flowers from May to September.

3. **Dianthus plumarius** L. Sp. Pl. 411. 1753. Fig. 114.

Perennial herb from rhizomes; stems tufted, erect, mostly unbranched, glabrous, glaucous, up to 70 cm tall; leaves opposite, entire, linear, acute at the apex, tapering to the base, up to 10 cm long, up to 9 mm broad, stiff, finely serrulate, the basal a little longer than the cauline; flowers borne in open or dense clusters, strongly clove-scented, with bracts up to half as long as the calyx; calyx 5-parted, up to 12 mm long, several-nerved, glabrous or sparsely pubescent, the lobes short and mucronate; petals 5, free, pink or purplish, spinulose-toothed along the margins, up to 25 mm long, lacerate; stamens 10; styles 2; capsules ellipsoid, often not developed.

Common Name: Cottage Pink; Garden Pink.
Habitat: Disturbed sandy soil, particularly in cemeteries.
Range: Native to Europe; rarely escaped from cultivation in the United States.
Illinois Distribution: Known from Cook County.

The deeply lacerate petals distinguish this species from any other member of the genus in Illinois.

Dianthus plumarius flowers during June.

4. **Dianthus deltoides** L. Sp. Pl. 411. 1753. Fig. 115.

Perennial herb from slender rhizomes, with trailing basal offshoots; stems ascending, glabrous or puberulent, branched, to 40 cm tall; median leaves linear-lanceolate, acute at the apex, tapering to the sessile base, up to 4 cm long, up to 4 mm broad, glabrous or puberulent; lower leaves oblanceolate, otherwise similar to the other leaves; flower solitary, up to 3 cm across, on a pedicel up to 2 cm long, the bracts acuminate to awned, up to half as long as the calyx; calyx 5-parted, 12–20 mm long, glabrous to sparsely pubescent, 25- to 30-nerved, the lobes linear to lanceolate; petals 5, free, purple to red to white, with a darker zigzag band near the base,

114. **Dianthus plumarius**
a. Habit
b. Node
c. Petal

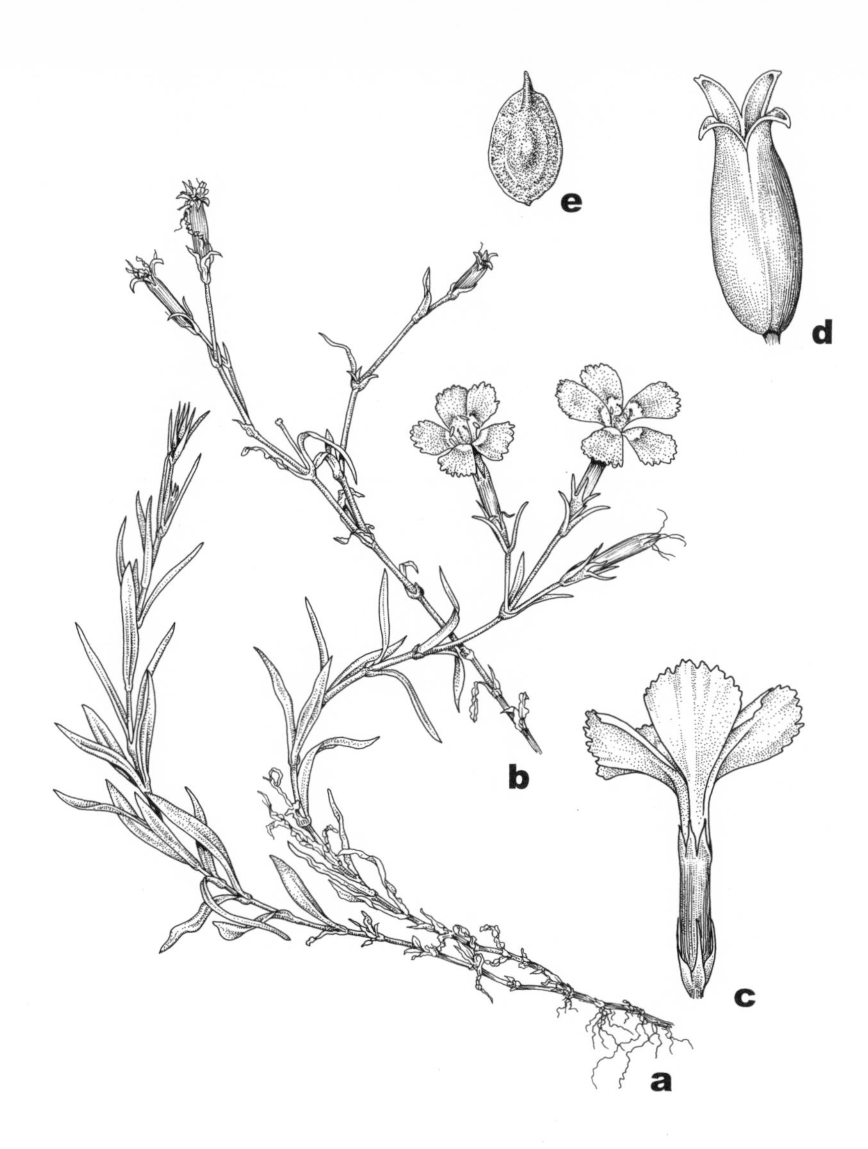

115. **Dianthus deltoides**
a. Habit
b. Fruiting branch
c. Flower
d. Fruit
e. Seed

the apex erose; stamens 10; styles 2; capsules ellipsoid to lanceoloid, up to 20 mm long, the stalk 2–3 mm long.

Common Name: Maiden Pink.
Habitat: Disturbed soil.
Range: Native to Europe; seldom escaped from cultivation.
Illinois Distribution: Known from Champaign, DuPage, Kane, Menard, Piatt, and Vermilion counties.

This is a handsome garden ornamental that only rarely escapes from cultivation. The solitary flower on a long pedicel is a distinctive feature.

The maiden pink flowers from June to August.

15. **Petrorhagia** (Ser. in DC.) Link—Saxifrage Pink

Annual or perennial herbs; leaves opposite, simple, entire, without stipules; flowers solitary or in headlike cymes, perfect, actinomorphic, with 2–6 bracts per flower; sepals 5, united below; petals 5, free; stamens 10; styles 2; ovary superior, 1-locular, with many ovules; capsules with 4 teeth at the apex; seeds flattened, many, brown to black.

This genus of Europe and Asia has about thirty species. Rabeler (1985) has monographed the genus.

Only the following species occurs in Illinois.

1. **Petrorhagia saxifraga** (L.) Link, Handbuch 2:235. 1831. Fig. 116.
Dianthus saxifraga L. Sp. Pl. 413. 1753.
Tunica saxifraga (L.) Scop. Fl. Carn. ed. 2, 300. 1772.

Perennial herb from creeping rhizomes; stems spreading to ascending, up to 35 cm long, much branched, glabrous or nearly so; leaves tufted below, linear to subulate, glabrous but ciliate along the margins, connate at the base, up to 10 mm long, up to 1 mm broad; flowers up to 7 mm across, usually in small panicles or cymes, subtended by small, scarious bracts; calyx campanulate, 5-lobed, 4–5 mm long, green, glabrous, 5-ribbed, the lobes oblong, obtuse; petals 5, free, pink or purple, 5–6 mm long, broadly notched at the apex; stamens 10; styles 2; capsules globose to ovoid, with 4 terminal teeth; seeds flattened, brown.

Common Name: Saxifrage Pink.
Habitat: Roadsides.
Range: Native to Europe; rarely adventive in the United States.
Illinois Distribution: Known from Champaign and Cook counties.

The saxifrage pink is an attractive plant sometimes planted in rock gardens but rarely escaped from cultivation.

Although earlier botanists in Illinois called this plant *Tunica saxifraga*, Rabeler (1985) has shown that the correct binomial should be *Petrorhagia saxifraga*.

This species flowers during June and July.

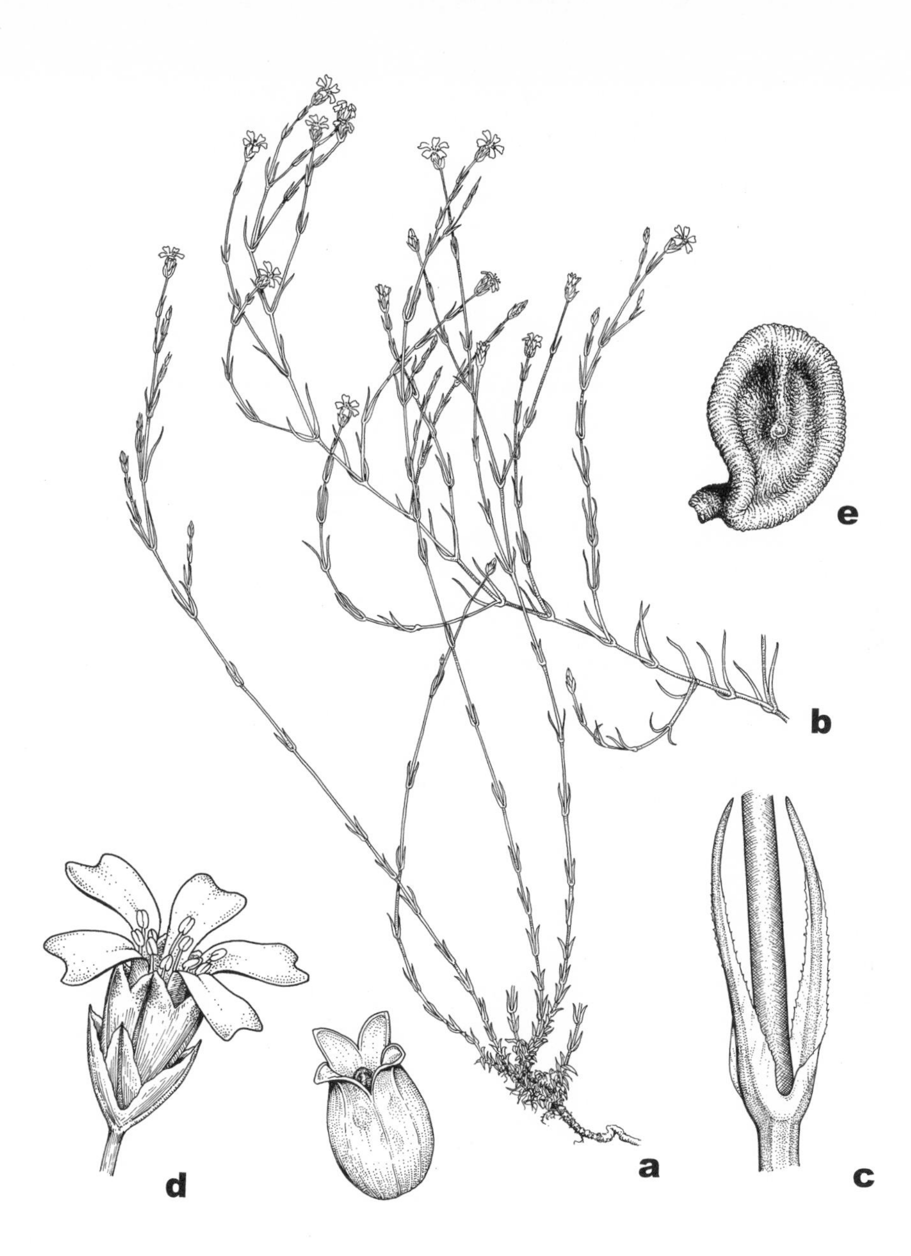

116. **Petrorhagia saxifraga**
a. Habit
b. Habit
c. Node
d. Flower
e. Seed

16. **Lychnis** L.—Campion

Biennial or perennial herbs; leaves opposite, simple, entire, without stipules; inflorescence cymose; flowers perfect, actinomorphic, bracteate; calyx mostly tubular, 5-lobed, the tube hairy and strongly 5-ribbed; petals 5, free, notched at the apex, clawed, with 2 appendages attached above the claw; stamens 10; styles 5; ovary superior, 1-locular, with many ovules; fruit a capsule dehiscing usually by 5 teeth; seeds many, usually warty.

The petals of the genus are unique in having 2 appendages attached above the claw.

About thirty-five species comprise this genus. They are found generally in the cooler parts of the world. Some species traditionally placed in the genus *Lychnis* are now in *Silene*.

Two species of *Lychnis* occur in Illinois, distinguished by the following key:

1. Plants densely white-tomentose; flowers red-purple; calyx lobes twisted . . . . . 1. *L. coronaria*
1. Plants variously pubescent but not white-tomentose; flowers scarlet; calyx lobes straight . . . . 2. *L. chalcedonica*

1. Lychnis coronaria (L.) Desr. in Lam. Encycl. 3:363. 1789. Fig. 117.
Agrostemma coronaria L. Sp. Pl. 436. 1753.

Perennial herb from a thick taproot; stems erect, branched, densely white-woolly, up to 75 cm tall; leaves opposite, entire, densely white-woolly, the upper lanceolate to oblanceolate to oblong, acute at the apex, tapering or rounded at the sessile base, up to 7 cm long, up to 2.5 cm broad, the lower spatulate to elliptic, up to 10 cm long, up to 3 cm broad, petiolate; inflorescence sparsely branched, few-flowered, the flowers perfect, opening in the morning, not fragrant; calyx more or less campanulate, ellipsoid, up to 1.5 cm long, pubescent, 10-nerved, with 5 filiform to linear, twisted lobes up to 7 mm long; petals 5, free, notched at the apex, purple-red, 2–3 cm long, with 2 appendages 2–3 mm long, the appendages awl-like, thickened; stamens 10; styles 5; capsules ovoid, up to 1 cm long, with 5 teeth at the apex; seeds 1.0–1.3 mm in diameter, black, with rounded tubercles.

Common Name: Mullein Pink.
Habitat: Disturbed soil.
Range: Native to Europe; occasionally escaped from cultivation in North America.
Illinois Distribution: Scattered in Illinois.

Mullein pink is a beautiful garden ornamental with its large red-purple flowers and its densely white-woolly pubescence.

This species flowers from July to September.

2. Lychnis chalcedonica L. Sp. Pl. 436. 1753. Fig. 118.

Perennial herb from a thick taproot; stems erect, stout, unbranched or sparingly branched, finely pubescent to hirsute, not white-tomentose, up to 80 cm tall; leaves opposite, entire, finely pubescent, the upper lanceolate, the lower ovate to

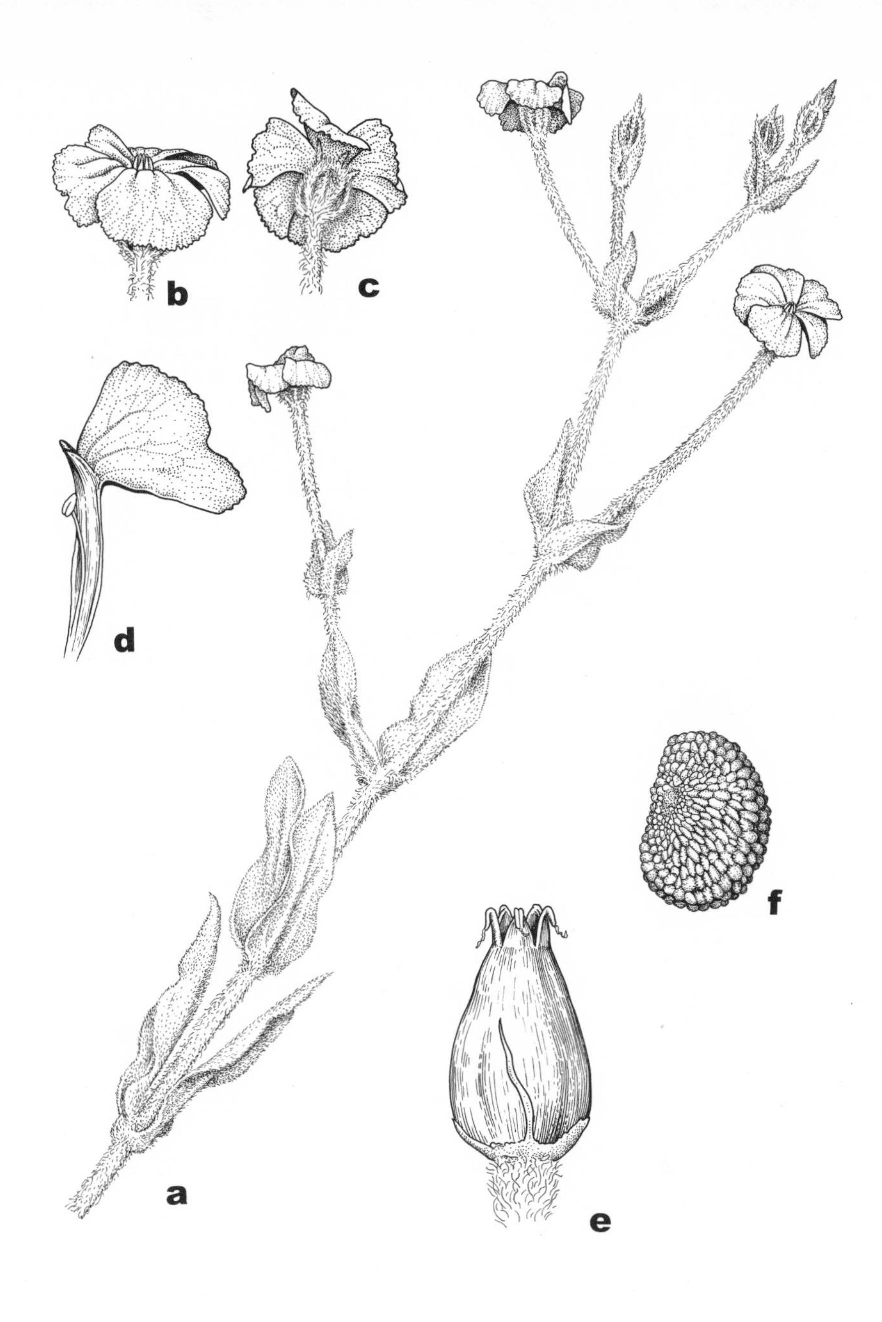

117. **Lychnis coronaria**
a. Flowering branch
b. Flower
c. Flower from below
d. Petal
e. Fruit
f. Seed

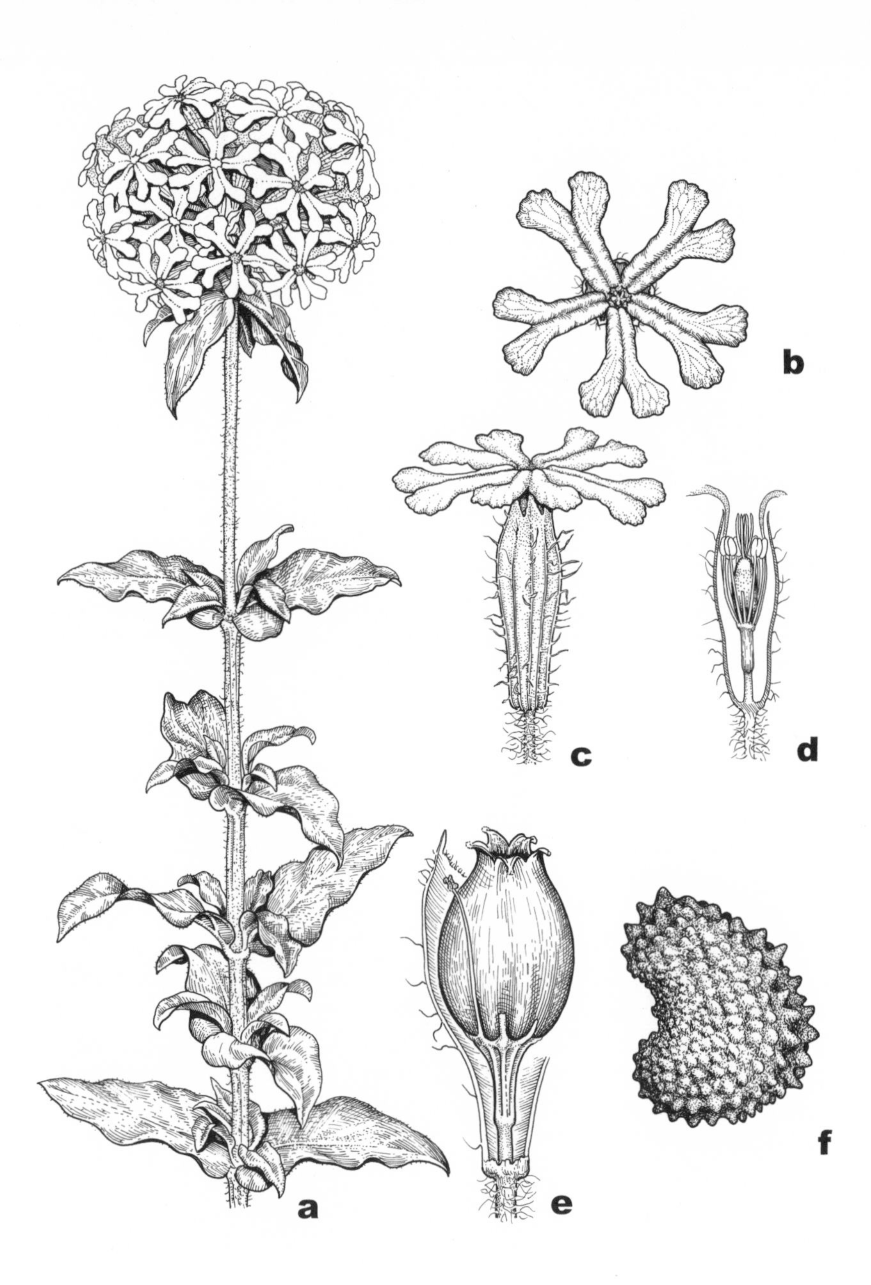

118. **Lychnis chalcedonica**
a. Flowering branch
b. Flower
c. Flower, side view
d. Longitudinal section through flower
e. Fruit
f. Seed

ovate-lanceolate, acute or acuminate at the apex, rounded or subcordate at the sessile or clasping base, up to 12 cm long, up to 7.5 cm broad; inflorescence sparsely branched, many-flowered, the flowers perfect; calyx more or less oblong, up to 2 cm long, finely pubescent, 10-nerved, with 5 very short lobes; petals 5, free, deeply notched at the apex, scarlet, 2–3 cm long, with 2 linear appendages; stamens 10; styles 5; capsules ovoid, up to 1 cm long, with 5 teeth at the apex; seeds 1.0–1.2 mm in diameter, black, with rounded tubercles.

Common Name: Maltese Cross.
Habitat: Disturbed soil.
Range: Native to Russia and Siberia; rarely escaped from cultivation in the United States.
Illinois Distribution: This species was collected once in DuPage County in 1938 and again in an old orchard in Vermilion County; also Jefferson County.

Lychnis chalcedonica is a handsome ornamental with its deeply notched scarlet petals. It flowers from June to September.

17. **Silene** L.—Catchfly

Annual, biennial, or perennial herbs; leaves opposite or whorled, simple, entire, without stipules; inflorescence cymose or reduced to a solitary flower; flowers perfect or less commonly unisexual, actinomorphic, bracteate; calyx mostly tubular, sometimes inflated, 5-lobed, usually with 10 or more veins; petals 5, free, sometimes appendaged, entire or notched or erose at the apex; stamens 10; styles 3, rarely 5; ovary superior, 1-locular, with many ovules; fruit a capsule dehiscing usually by 6 teeth; seeds usually tuberculate or minutely spiny.

Silene is similar to *Lychnis* but differs in its usually 3 styles and usually 6 teeth at the apex of the capsule. Two species at one time placed in the genus *Lychnis* but now in *Silene* have 5 styles.

There are about five hundred species in the genus *Silene,* confined almost entirely to the northern hemisphere. Because of the beauty of many of the species, some are frequently grown as ornamentals.

Key to the Species of **Silene** in Illinois

1. Styles 5.
   2. Flowers perfect or unisexual, white, opening at dusk; capsules ovoid .... 1. *S. pratensis*
   2. Flowers unisexual, pink or red, opening in the morning; capsules globose .... 2. *S. dioica*
1. Styles 3.
   3. Petals deep red, crimson, or scarlet.
      4. Stem with 15 or more pairs of leaves; petals entire or emarginate ...... 3. *S. regia*
      4. Stem with up to 8 pairs of leaves; petals bifid .... 4. *S. virginica*
   3. Petals white or pink or purplish.
      5. Leaves whorled .... 5. *S. stellata*

5. Leaves opposite.
   6. Flowers up to 4 mm across, or petals absent .... 6. *S. antirrhina*
   6. Flowers 1 cm or more across.
      7. Calyx with about 30 ribs .... 7. *S. conica*
      7. Calyx with 5–10 ribs.
         8. Stems glutinous below each node; flowers pink or purple .... 8. *S. armeria*
         8. Stems not glutinous below each node; flowers white or only pink-based.
            9. Stems viscid-pubescent or hirsute.
               10. Flowers nodding, opening at dusk; viscid-pubescent annual .... 9. *S. noctiflora*
               10. Flowers ascending, opening during the day; hirsute or pubescent biennial but not viscid.
                  11. Calyx 9–14 mm long at anthesis; pubescence of calyx never glandular; lobes of calyx 2–4 mm long .... 10. *S. dichotoma*
                  11. Calyx up to 9 mm long at anthesis; pubescence of calyx sometimes glandular on the nerves; lobes of calyx 1.0–1.5 mm long .... 11. *S. gallica*
            9. Stems glabrous.
               12. Calyx glabrous, more or less inflated; petals 2-lobed.
                  13. Plants green; leaves acuminate; flower solitary in the upper axils .... 12. *S. nivea*
                  13. Plants glaucous; leaves acute; flowers in cymose panicles.
                     14. Calyx up to 15 mm long, scarcely inflated; vertical nerves of calyx sparsely or not at all connected by cross-veins .... 13. *S. cserei*
                     14. Calyx up to 20 mm long, inflated; vertical nerves of calyx regularly connected by cross-veins .... 14. *S. vulgaris*
               12. Calyx pubescent, not inflated; petals fringed ...... 15. *S. ovata*

1. Silene pratense (Spreng.) Godron & Gren. in Gren. & Godron, Fl. Fr. 1:216. 1847. Fig. 119.
Lychnis alba Mill. Gard. Dict. ed. 8, 4. 1768, non E. H. L. Krausse.
Lychnis pratensis Spreng. Fl. Hal. 138. 1806.

Biennial or perennial herb from a thick taproot; stems erect, branched, pubescent, at least the uppermost hairs glandular-viscid, to nearly 1 m tall; leaves opposite, entire, the lowest oblanceolate to ovate-lanceolate, the uppermost lanceolate, acute at the apex, tapering to the usually sessile base, pubescent, 3- to 5-nerved, up to 10 cm long, up to 4 cm broad, the lower leaves short-petiolate; inflorescence sparsely branched, few-flowered, the flowers staminate, pistillate, or more uncommonly perfect, opening in the evening and closing the next morning, somewhat fragrant; calyx tubular, becoming ovoid, up to 2 cm long, pubescent, 10-nerved in the staminate flower, more or less 20-nerved in the pistillate and perfect flowers, with 5 triangular-lanceolate, acute lobes 3–6 mm long; petals 5, free, notched at the apex, white, 2.5–4.0 mm long, with appendages 1.0–1.5 mm long; stamens 10; styles 5; capsules ovoid, up to 1.5 cm long, with 10 or less erect teeth at the apex; seeds 1–2 mm long, brown, low-warty.

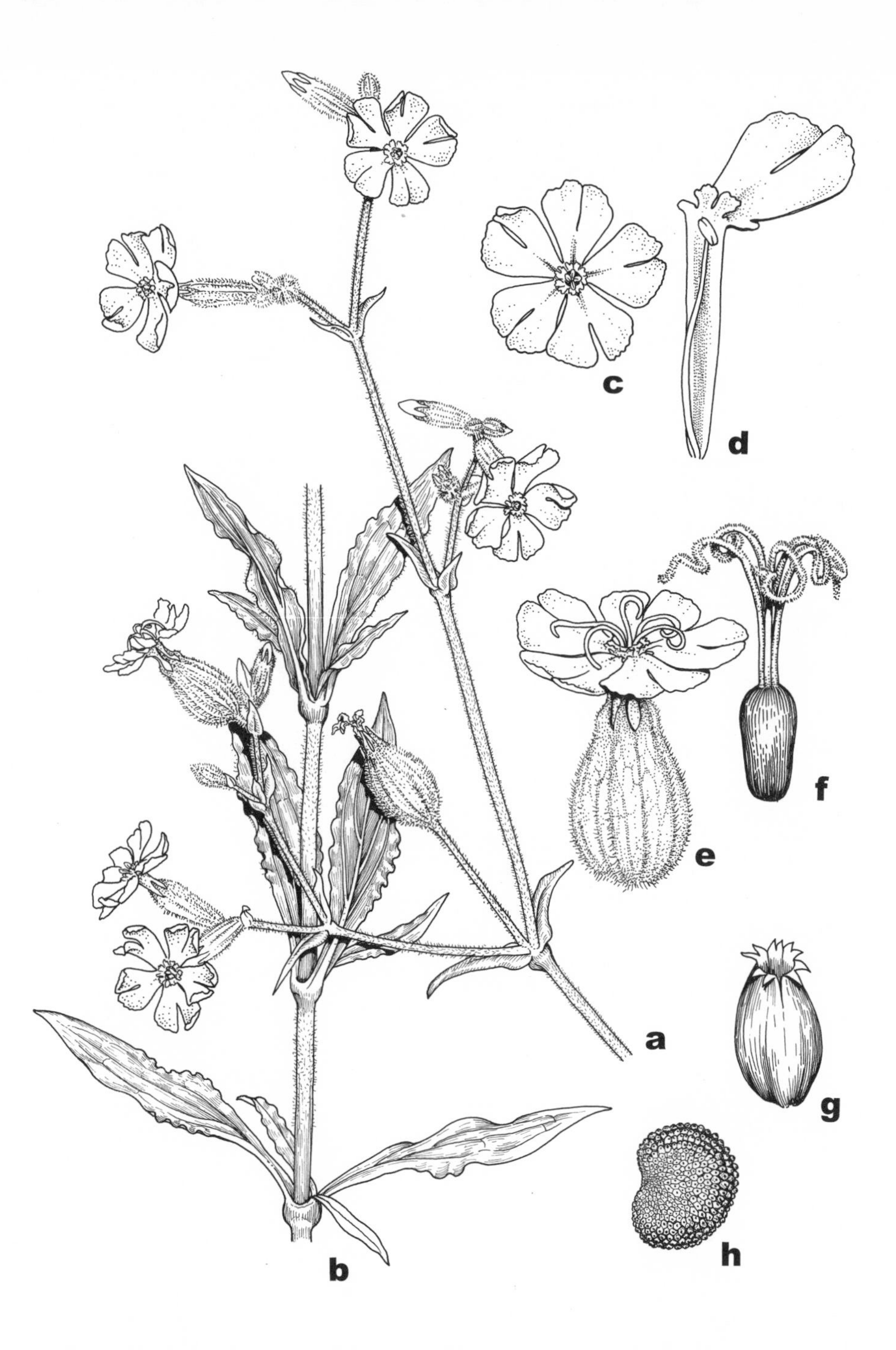

119. **Silene pratense**
a. Flowering branch
b. Stem with leaves
c. Staminate flower
d. Portion of staminate flower
e. Pistillate flower
f. Young ovary
g. Fruit
h. Seed

Common Name: White Campion; Evening Campion.
Habitat: Waste ground, particularly along railroads and in cultivated fields.
Range: Native to Europe and Asia; adventive throughout most of the United States.
Illinois Distribution: Very common in the northern counties, becoming much rarer southward.

This species in the past has been considered to be a species of *Lychnis,* but McNeill (1978) gives reasons for transferring it to *Silene.* It differs from the other dioecious species of *Silene* in Illinois in its white flowers that open in the evening.

This attractive, night-blooming species is very similar in appearance to, and often confused with, *Silene noctiflora. Silene pratensis* differs, however, in being a biennial or perennial, in having most of its flowers unisexual and larger, in having 5 styles, and in having usually 10 apical teeth on the capsule.

The calyx of the staminate flowers is ellipsoid and 10-nerved, while the calyx of the pistillate and perfect flowers is ovoid and more or less 20-nerved.

White campion flowers from May to October.

2. Silene dioica (L.) Clairv. Man. Herb. Suisse Valais 146. 1811. Fig. 120.
Lychnis dioica L. Sp. Pl. 437. 1753.

Biennial or perennial herb from a thick root; stems erect, branched, pubescent but not glandular-viscid, to nearly 1 m tall; leaves opposite, entire, oblong to lance-oblong, acute at the apex, tapering to the usually sessile base, pubescent, 3- to 5-nerved, up to 10 cm long, up to 3.5 cm broad, the lower leaves short-petiolate; inflorescence sparsely branched, few-flowered, the flowers unisexual, opening in the morning, not fragrant; calyx tubular, becoming inflated in fruit, up to 1.8 cm long, pubescent, 10-nerved in the staminate flowers, 10-nerved in the pistillate flowers, with 5 lance-elliptic, acute lobes 2–4 mm long; petals 5, free, notched at the apex, purple, red, or white, 2.0–3.5 cm long, with appendages about 1 mm long; stamens 10; styles 5; capsules globose, up to 1 cm in diameter, with 10 slightly recurved teeth at the apex; seeds low-warty.

Common Name: Red Campion.
Habitat: Waste ground.
Range: Native to Europe and Asia; occasionally adventive in the eastern half of the United States.
Illinois Distribution: Scattered in the northern three-fifths of the state.

Red campion is an attractive species of disturbed soil in the United States. It was at one time a fairly popular garden ornamental.

Although its unisexual flowers relate it to *S. pratensis,* it differs by its usually red flowers that open during the day.

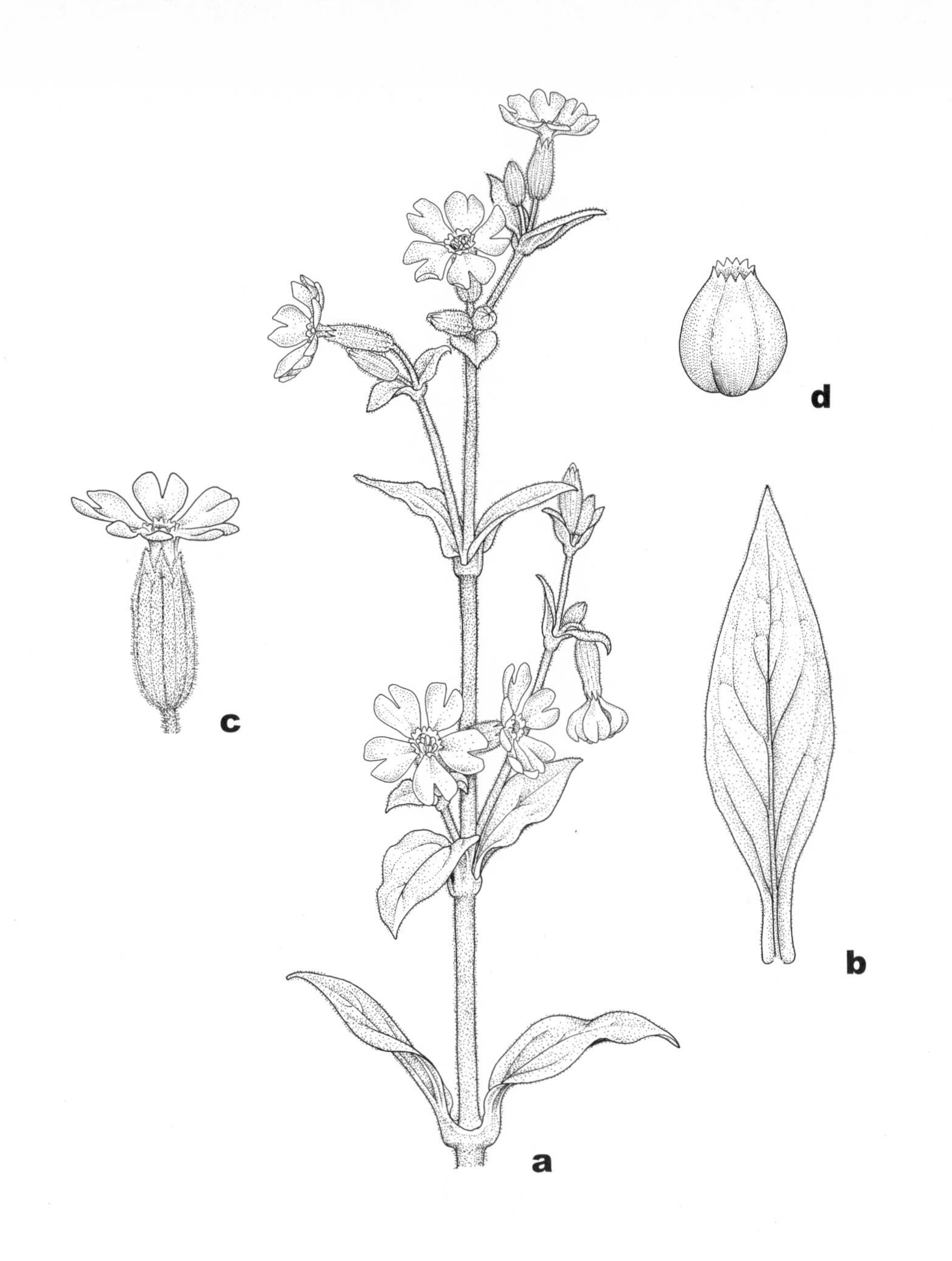

120. **Silene dioica**
a. Habit
b. Leaf
c. Flower
d. Seed

For most of the nineteenth and twentieth centuries, this species was considered to be a species of *Lychnis.*

Silene dioica flowers from May to October.

3. Silene regia Sims, Bot. Mag. 41:pl. 1724. 1815. Fig. 121.
Melandrium illinoense Rohrb. Linnaea 3636:250. 1870.
Silene illinoensis (Rohrb.) Kellerm. Rept. Geol. Surv. Ohio 7:178. 1893.

Perennial herb from a thickened root; stems erect, stout, unbranched to sparsely branched, pubescent, sometimes viscid, up to 1.5 m tall; leaves opposite, entire, broadly lanceolate to ovate, acute at the apex, rounded at the base, up to 12 cm long, up to 7 cm broad, pubescent, the upper leaves sessile, the lowermost short-petiolate; inflorescence narrowly paniculate, several-flowered, with leafy bracts; flowers up to 2.5 cm across, the bracts similar to the leaves but smaller; calyx tubular, oblongoid, up to 2.5 cm long, mostly 10-nerved, glandular-pubescent; petals 5, free, crimson, the limb elliptic to narrowly oblong, up to 2 cm long, with appendages up to 4 mm long, rounded to emarginate to irregularly toothed at the apex; stamens 10; styles 3; capsules fusiform, up to 2 cm long, with 6 apical teeth; seeds low-warty.

Common Name: Royal Catchfly.
Habitat: Dry soil, usually along roads; prairies.
Range: Ohio to Missouri, south to Oklahoma, Louisiana, and Georgia.
Illinois Distribution: Known from Clark, Cook, Lawrence, St. Clair, Wabash, White, Will, and Winnebago counties.

Royal catchfly is one of the most beautiful as well as one of the rarest wildflowers in Illinois.

Many of the Illinois collections were made during the first half of the twentieth century.

The large scarlet flower with petals that are entire or merely emarginate at the apex readily distinguishes this species. The only other *Silene* with red flowers that are all perfect is *S. virginica. Silene regia* differs from *S. virginica* by having at least 15 pairs of leaves on the stem.

Silene regia flowers during July and August.

4. Silene virginica L. Sp. Pl. 419. 1753. Fig. 122.

Perennial herb from a thickened root; stems erect, slender, somewhat branched, viscid-pubescent, up to 75 cm tall; upper leaves broadly lanceolate, obtuse to acute at the apex, tapering to the sessile base, pubescent or at least ciliate; lower leaves spatulate, obtuse to acute at the apex, tapering to the petiolate base, pubescent or at least ciliate, up to 15 cm long, up to 2.5 cm broad; inflorescence cymose to paniculate, several-flowered, with reduced leafy bracts; flowers up to 3.5 cm across; calyx tubular, cylindrical, green, up to 2.5 cm long, mostly 10-nerved, pubescent; petals 5, free, crimson, the limb linear to oblong, up to 2.5 cm long, with appendages up to

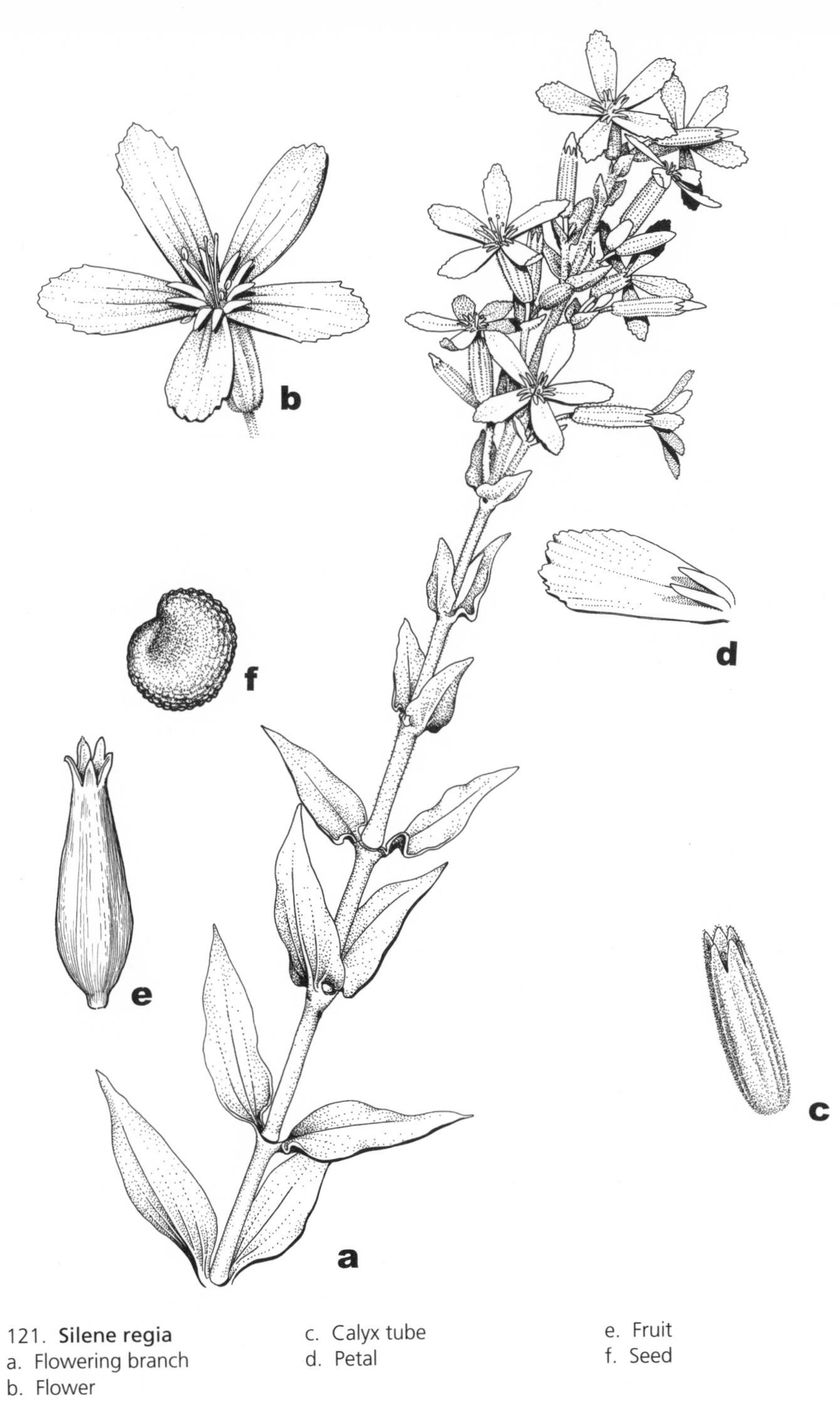

121. **Silene regia**
a. Flowering branch
b. Flower
c. Calyx tube
d. Petal
e. Fruit
f. Seed

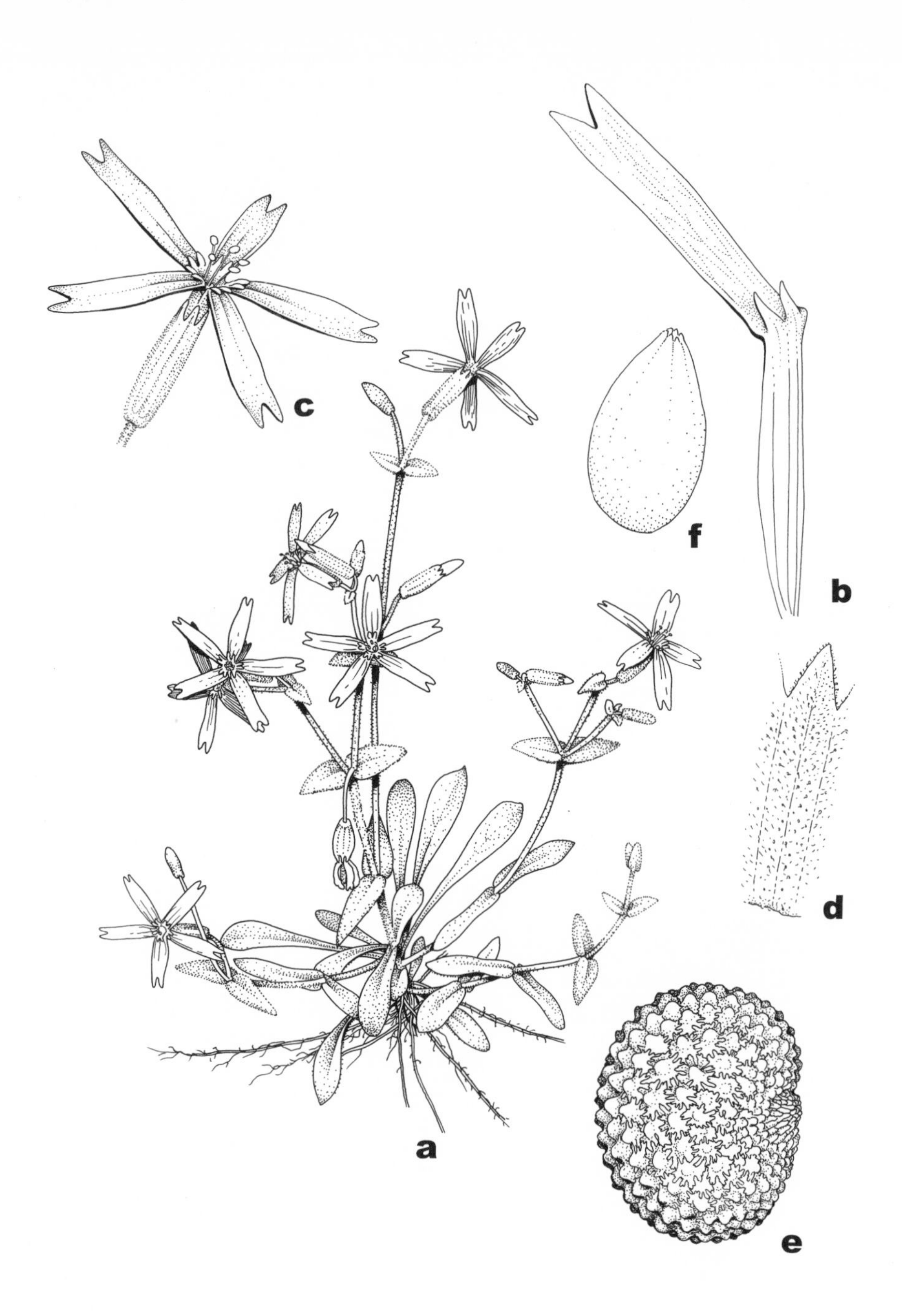

122. **Silene virginica**
a. Habit
b. Node
c. Flower
d. Portion of calyx
e. Seed
f. Fruit

3 mm long, deeply notched at the apex; stamens 10; styles 3; capsules broadly ellipsoid, up to 2 cm long, with 6 apical teeth; seeds low-warty.

Common Name: Firepink.
Habitat: Open woods, both rich and dry.
Range: New York to southern Ontario and Minnesota, south to Oklahoma, Arkansas, and Georgia.
Illinois Distribution: Scattered throughout the state but not common.

This handsome wildflower has strikingly crimson flowers like that found in *S. regia.* It differs from *S. regia* in having 8 or fewer pairs of leaves on the stem, whereas *S. regia* has 15 or more pairs. In addition, the petals of *S. virginica* are strongly bifid.

Firepink may occur in relatively rich woods as it does in the northeastern counties or in dry woods as it does in southern Illinois.

Silene virginica flowers from April to July.

5. Silene stellata (L.) Ait. f. Hort. Kew. ed. 2, 3:84. 1811. Fig. 123.
Cucubalus stellatus L. Sp. Pl. 414. 1753.
Evactoma stellata (L.) Raf. var. *scabrella* Nieuwl. Am. Midl. Nat. 3:58. 1913.
Silene stellata (L.) Ait. f. var. *scabrella* (Nieuwl.) Palmer & Steyerm. Rhodora 42:99. 1940.
Silene scabrella (Nieuwl.) G. N. Jones, Trans. Ill. Acad. Sci. 35:71. 1942.

Perennial herb from a stout crown; stems erect, slender, sparsely branched, puberulent, up to nearly 1 m tall; leaves mostly in whorls of 4, or the uppermost or lowermost merely opposite, narrowly to broadly lanceolate, acuminate at the apex, tapering to the sessile base, mostly glabrous or nearly so, rarely densely pubescent, up to 10 cm long, up to 4 cm broad; inflorescence paniculate, several-flowered, with reduced leafy bracts; flowers up to 2 cm across, showy; calyx campanulate, green, up to 12 mm long, glabrous or puberulent, inconspicuously nerved; petals 5, free, white, up to 12 mm long, woolly at the base, fringed along the outer edge; stamens 10; styles 3; capsules spherical to ovoid, up to 12 mm long, with usually 6 apical teeth; seeds low-warty, about 1 mm long.

Common Name: Starry Campion.
Habitat: Open woods, prairies.
Range: Massachusetts to Minnesota, south to Texas and Georgia.
Illinois Distribution: Occasional throughout the state.

Starry campion is one of the most beautiful of our summer wildflowers. Its white, fringed petals are most attractive.

There is variation in the amount of pubescence. Some plants are nearly glabrous, while others are densely puberulent on the stems, leaves, and calyces. Those extremely hairy plants have been segregated as var. *scabrella.*

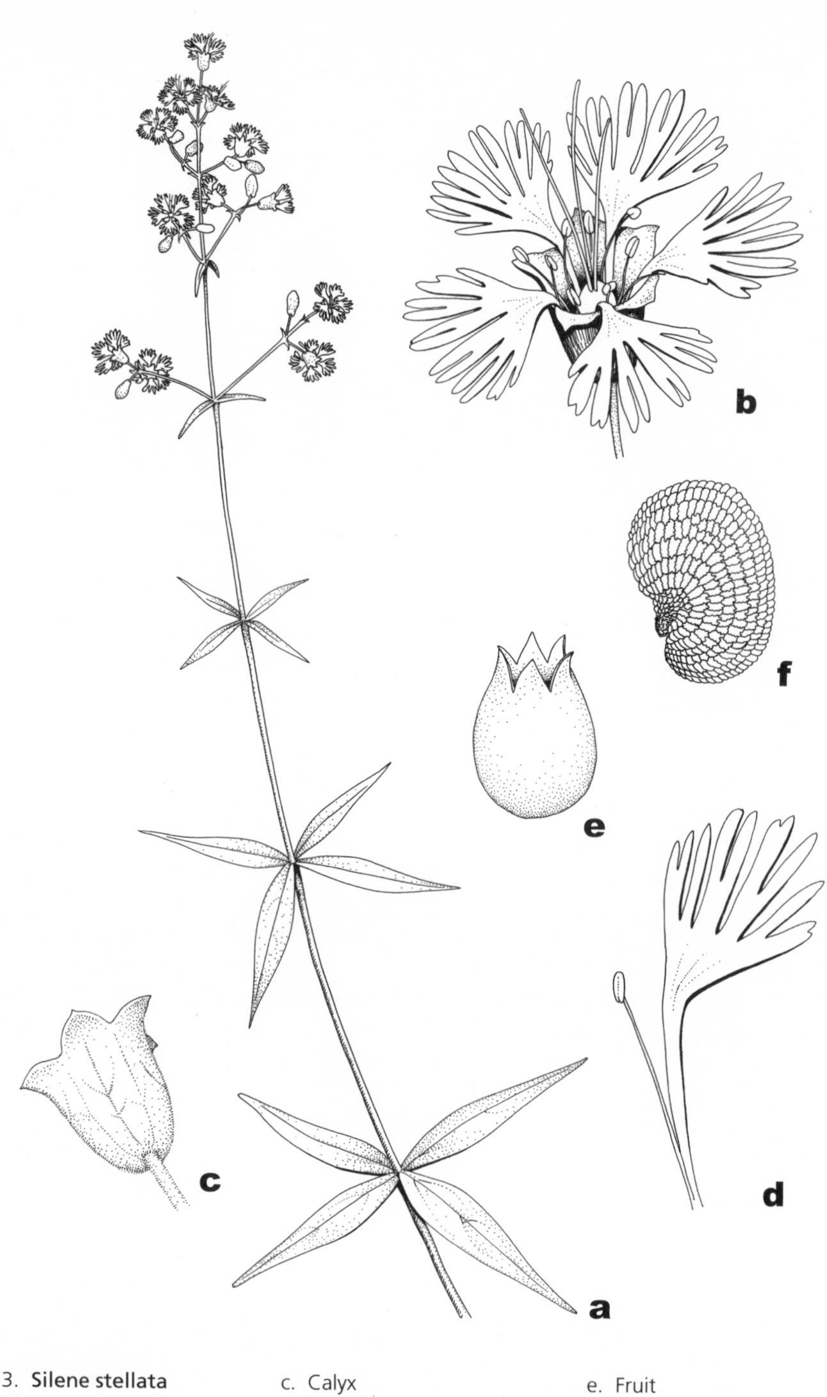

123. **Silene stellata**
a. Flowering branch
b. Flower
c. Calyx
d. Petal with stamen
e. Fruit
f. Seed

Silene stellata is commonly found in rather dry, open woods. It may also occur in prairies, particularly in the northern half of Illinois.

This species flowers from June to October.

6. Silene antirrhina L. Sp. Pl. 416. 1753. Fig. 124.
Silene antirrhina L. var. *divaricata* Robins. Proc. Am. Acad. 28:132. 1893.
Silene antirrhina L. f. *deaneana* Fern. Rhodora 17:96. 1915.
Silene antirrhina L. f. *apetala* Farw. Papers Mich. Acad. Sci. 3:97. 1924.
Silene antirrhina L. f. *bicolor* Farw. Am. Midl. Nat. 11:55. 1928.

Annual from slender roots; stems erect, slender, branched or unbranched, glabrous to puberulent, usually with sticky areas below the nodes, up to 75 cm tall; leaves opposite, entire, glabrous to puberulent, ciliate near the base, all but the lowest leaves linear to lanceolate to oblanceolate, acute to subacute at the apex, up to 4 cm long, up to 8 mm broad, the lowermost leaves spatulate to oblanceolate, acute to obtuse at the apex, up to 6 cm long, up to 1.2 cm broad; inflorescence cymose to paniculate, several-flowered, with much reduced foliar bracts; flowers up to 4 mm across; calyx fusiform, becoming ovoid in fruit, up to 1 cm long, 10-nerved, glabrous; petals 5, free, or absent, white to pink or rose, up to 10 mm long, notched at the apex; stamens 10; styles 3; capsules ovoid, up to 1 cm long, usually with 6 apical teeth; seeds with rows of warty projections, less than 1 mm long.

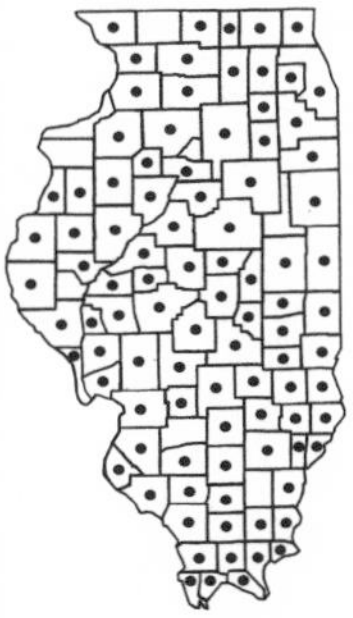

Common Name: Sleepy Catchfly.
Habitat: Disturbed soil, often along railroads.
Range: Quebec to British Columbia, south to California, Texas, and Florida.
Illinois Distribution: Occasional to common throughout the state.

The presence of a sticky area below many of the nodes serves to trap ants, gnats, and other small insects, although the plant is not able to utilize the nutrients from its "catch." Specimens that lack such a sticky area are referred to as f. *deaneana.*

Specimens with spreading branches and apetalous flowers, called var. *divaricata,* were originally described from a Bebb collection near Rockford. Other apetalous plants have been called f. *apetala.*

The normal petal color is white to pink or rose. Rare specimens that have the petals white above and pink or rose below may be known as f. *bicolor.*

Silene antirrhina flowers from May to July.

7. Silene conica L. Sp. Pl. 418. 1753. Fig. 125.

Annual from slender roots; stems erect, solitary but usually forked above, puberulent or even hirsutulous on the upper internodes, to 60 cm tall; leaves linear-lanceolate, acute at the apex, cuneate to the sessile base, up to 3 (–4) mm broad, puberulent on both surfaces; flowers few in cymes, the pedicels 1–3 cm long, hirsutulous; calyx ovoid, rounded or truncate at the base, the body 10–17 mm long,

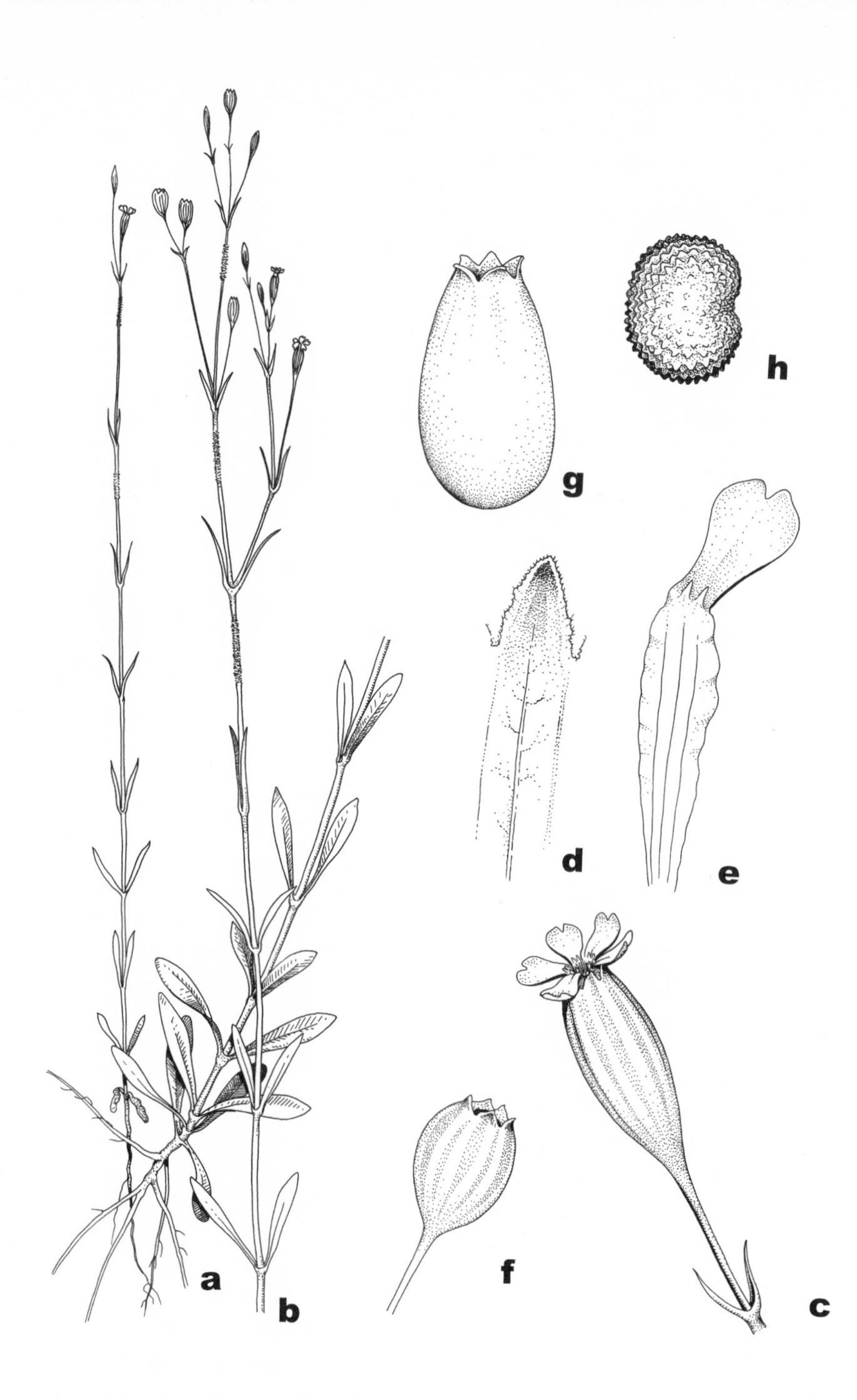

124. **Silene antirrhina**
a. Habit
b. Habit
c. Flower
d. Portion of calyx
e. Petal
f. Fruit
g. Fruit
h. Seed

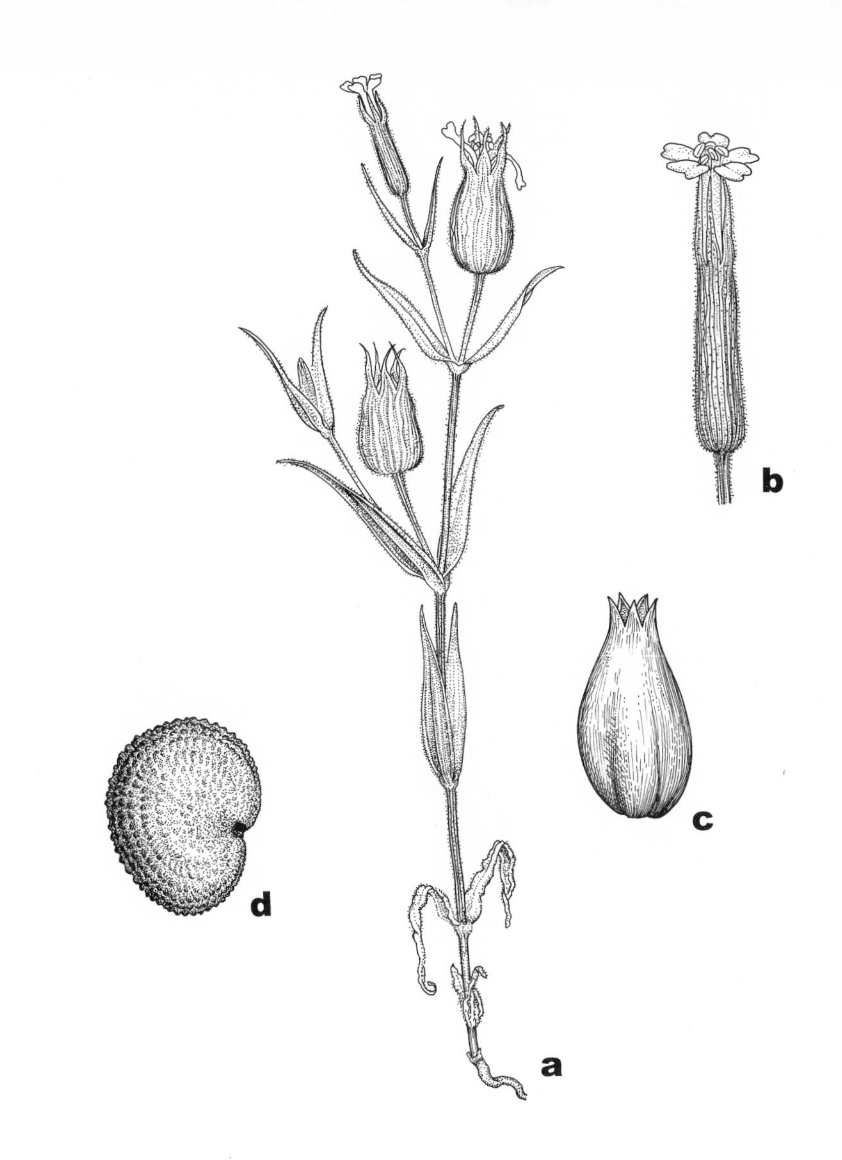

125. **Silene conica**
a. Habit
b. Flower
c. Fruit
d. Seed

with about 30 strong, green ribs, hirsutulous, the teeth long-attenuate, to 7 mm long; petals pink, entire or shallowly 2-cleft, up to 2 cm long, a little longer than the calyx; capsules oblongoid to ovoid, to 10 mm long, the seeds 0.6–1.0 mm long.

Common Name: Striate Catchfly.
Habitat: In cinders and sand along a railroad (in Illinois).
Range: Native to Europe and Asia; sparingly adventive in the eastern United States.
Illinois Distribution: Known from Cass County.

This is the only annual species of *Silene* in Illinois with a hirsutulous calyx tube that is coarsely striated with about thirty ribs. It is similar to another weedy species, *S. conoidea,* which is not known from Illinois. *Silene conoidea* has larger calyces, petals, capsules, and seeds.

The only Illinois collection was made by R. T. Rexroat along a railroad in Cass County on June 16, 1966. This specimen was originally determined as *S. conoidea.*

Silene conica flowers during June and July.

8. Silene armeria L. Sp. Pl. 420. 1753. Fig. 126.

Annual from slender roots; stems erect, branched, glaucous, glabrous or sparsely puberulent, up to 30 cm tall; leaves glabrous to minutely puberulent, the middle and upper leaves elliptic to ovate, acute to obtuse at the apex, clasping or nearly so at the base, up to 6 cm long, up to 2 cm broad, the lowest leaves spatulate to oblanceolate, obtuse at the apex, tapering to the sessile base, up to 6 cm long, up to 2 cm broad; inflorescence a crowded cyme, with much reduced bracts; flowers up to 7 mm across; calyx narrowly clavate, up to 1.7 cm long, 10-ribbed, usually glabrous; petals 5, free, pink or purple, up to 2 cm long, emarginate at the apex, with linear appendages up to 3 mm long; stamens 10; styles 3; capsules narrowly ovoid, long-stipitate, usually with 6 apical teeth; seeds low-warty, 0.5–0.8 mm long.

Common Name: Sweet William Catchfly.
Habitat: Disturbed soil in cities.
Range: Native to Europe and Asia; occasionally escaped from cultivation throughout the United States.
Illinois Distribution: Scattered in the northern three-fourths of the state.

This species was once commonly grown in flower gardens. In addition to the name sweet William catchfly, it is also known as none-so-pretty.

The pink to purple flowers are crowded together into dense cymes.

Silene armeria flowers are borne from June to August.

9. Silene noctiflora L. Sp. Pl. 419. 1753. Fig. 127.

Annual from thickened taproots; stems erect, branched or unbranched, viscid-pubescent, up to 75 cm tall; leaves opposite, entire, pubescent, the middle and upper

126. **Silene armeria**
a. Habit
b. Habit
c. Flower
d. Longitudinal section of flower
e. Fruit

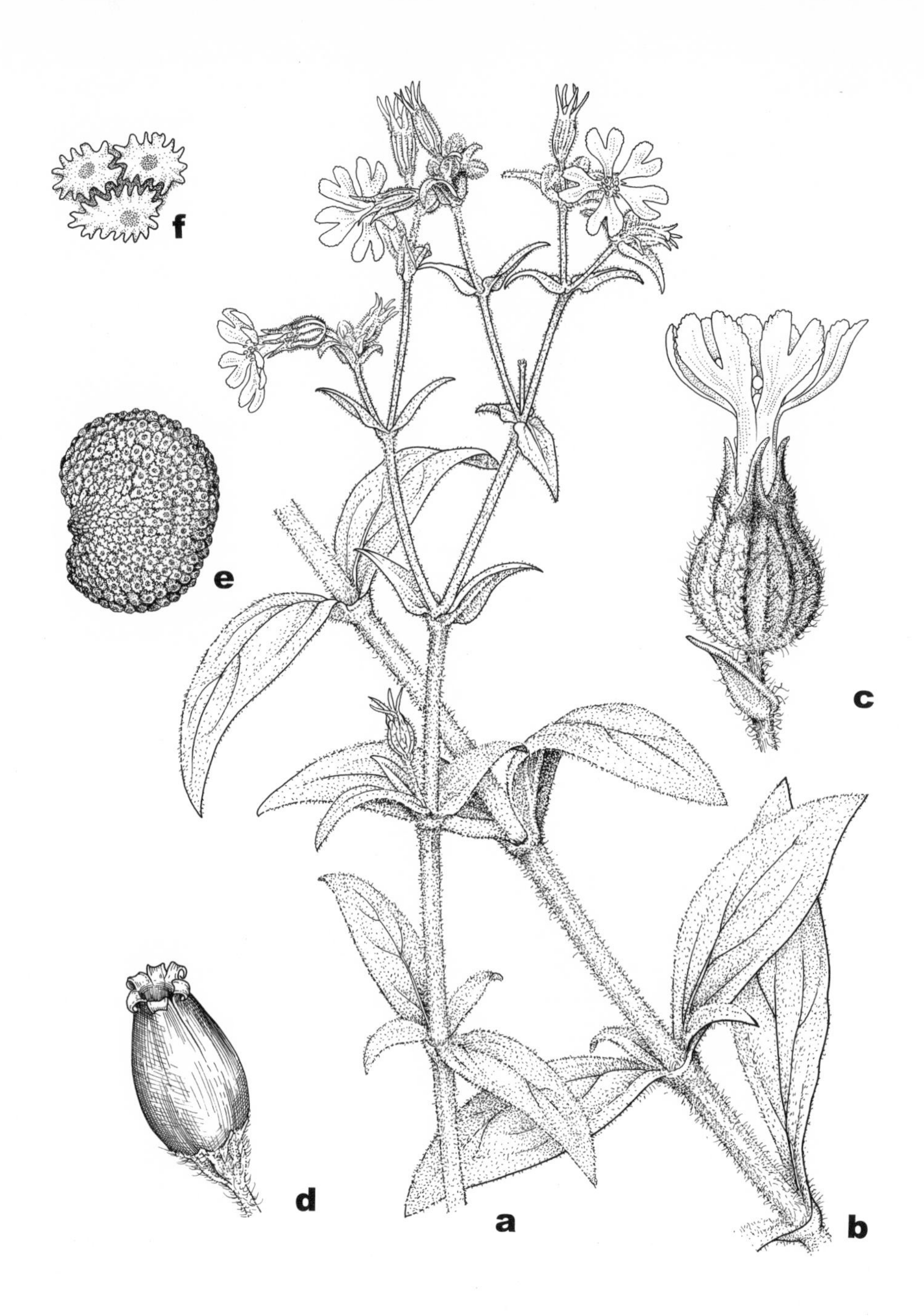

127. **Silene noctiflora**
a. Flowering branch
b. Stem with leaves
c. Flower
d. Fruit
e. Seed
f. Surface of seed

leaves lanceolate to ovate-lanceolate, acute to acuminate at the apex, tapering to the sessile base, up to 8 cm long, up to 3 cm broad, the lowermost leaves spatulate, more or less obtuse at the apex, tapering to a short-petiolate base, up to 10 cm long, up to 4 cm broad; inflorescence a few-flowered cyme, the bracts reduced but foliaceous; flowers opening at night, fragrant, up to 2 cm across; calyx tubular, ovoid, 10-ribbed, the tube up to 1.5 cm long in flower, becoming up to 2.5 cm long in fruit, usually glandular-pubescent, the lobes up to 8 (–13) mm long, linear-lanceolate; petals 5, free, white with usually a pink base, up to 1 cm long, bifid, with appendages up to 1.5 mm long; stamens 10; styles 3; capsules ovoid, with 6 apical teeth, up to 2.5 cm long; seeds low-warty, about 1 mm long, black.

Common Name: Night-flowering Catchfly.
Habitat: Waste ground.
Range: Native to Europe; adventive in much of the United States and Canada.
Illinois Distribution: Known from the northern three-fifths of the state.

This species is remarkably similar in appearance to *Silene pratensis*, another night-blooming species. *Silene noctiflora* differs by having 3 styles, a 10-ribbed calyx, an annual habit, and a capsule with 6 apical teeth.

Silene noctiflora flowers from June to September.

10. Silene dichotoma Ehrh. Beitr. 7:143. 1792. Fig. 128.

Annual or biennial herb from thickened roots; stems erect, branched or unbranched, densely hirsute, up to 1 m or more tall; leaves opposite, entire, densely hirsute, the middle and upper ones linear to lanceolate, acute at the apex, tapering to the sessile base, up to 6 cm long, up to 2 cm broad, the lowest leaves oblanceolate to obovate, more or less obtuse at the apex, up to 8 cm long, up to 3.5 cm broad; inflorescence a series of dichotomous racemes, some of the bracts translucent; flowers white, nodding, short-pedicellate, up to 2 cm across; calyx narrowly tubular, 10-ribbed, hirsute, up to 1.5 cm long, becoming inflated in fruit; petals 5, free, white, bifid, up to 1 cm long, the appendages minute; stamens 10; styles 3; capsules ellipsoid to ovoid, with 6 apical teeth, up to 2 cm long, the teeth 2–4 mm long; seeds rugulate, 1.0–1.5 mm long, dark brown.

Common Name: Forked Catchfly.
Habitat: Disturbed soil.
Range: Native to Europe and Asia; escaped from cultivation throughout most of the United States.
Illinois Distribution: Scattered in the northern half of the state.

This strongly hirsute species at one time was fairly popular as a garden ornamental, but it is rarely seen in gardens today.

The flowers, which bloom from June to August, tend to nod at anthesis.

128. **Silene dichotoma**
a. Flowering branch
b. Stem with leaves
c. Flower
d. Fruit
e. Seed

11. Silene gallica L. Sp. Pl. 1:417. 1753. Fig. 129.

Annual herb from fibrous roots; stems erect to decumbent, branched or unbranched, hirsute, or the uppermost branchlets bristly-pubescent, up to 50 cm tall; leaves opposite, entire, densely pubescent, linear to spatulate, acute at the apex, tapering to the sessile base, up to 3 cm long, up to 1 cm broad, the lowest leaves broadly lanceolate and larger; inflorescence simple or branched, with opaque bracts; flowers erect or ascending, borne singly, short-pedicellate, up to 1.5 cm across; calyx ovoid, 10-ribbed, hirsute, with the hairs on the veins often glandular, 6–9 mm long at anthesis, becoming inflated in fruit; petals 5, free, white (rarely pink), entire or emarginate, up to 1 cm long, the appendages minute; stamens 10; styles 3; capsules ovoid, with 6 apical teeth, the teeth 1.0–1.5 mm long; seeds rugose, 1.0–1.8 mm long, pale brown.

Common Name: Catchfly.
Habitat: Low ground.
Range: Native to Europe and Asia; occasionally adventive in the United States, particularly the Pacific States.
Illinois Distribution: Known from St. Clair County.

This adventive species is similar to *S. dichotoma,* but it has more bristly-pubescent branchlets, solitary flowers with smaller calyces at anthesis, occasional glandular-pubescence on the veins of the calyx, and opaque bracts.

The only Illinois collection was made by Julian O. Neill in the vicinity of Dupo, St. Clair County, on September 22, 1950. It was originally determined as *S. dichotoma,* and indeed the branching pattern is dichotomous. However, all the other diagnostic characteristics of *S. gallica* are present on the specimen.

12. Silene nivea (Nutt.) Otth in DC. Prodr. 1:377. 1824. Fig. 130.
Silene alba Muhl. Cat. 45. 1813, *nomen subnudum.*
Cucubalus niveus Nutt. Gen. 1:287. 1818.

Perennial herb from an elongated rhizome; stems sprawling to ascending to erect, branched or unbranched, usually glabrous, up to 30 cm tall; leaves opposite, entire, lanceolate to oblong-lanceolate, acute to acuminate at the apex, tapering to the sessile or short-petiolate base, glabrous, progressively smaller from the base of the plant to the top, up to 10 cm long, up to 3.5 cm broad; inflorescence usually reduced to solitary, axillary flowers; flowers up to 1.8 cm across, on pedicels up to 2 cm long; calyx tubular-campanulate, thin, obscurely ribbed, puberulent, up to 1.5 cm long; petals 5, free, notched, up to 8 mm long, with oblong appendages up to 1.5 mm long; stamens 10; styles 3; capsules subcylindric, up to 1.5 cm long, with 6 apical teeth; seeds low-warty.

Common Name: Snowy Campion.
Habitat: Wooded ravines; calcareous fens; moist stream banks.

129. **Silene gallica**
a. Flowering branch
b. Flower and bud
c. Fruit

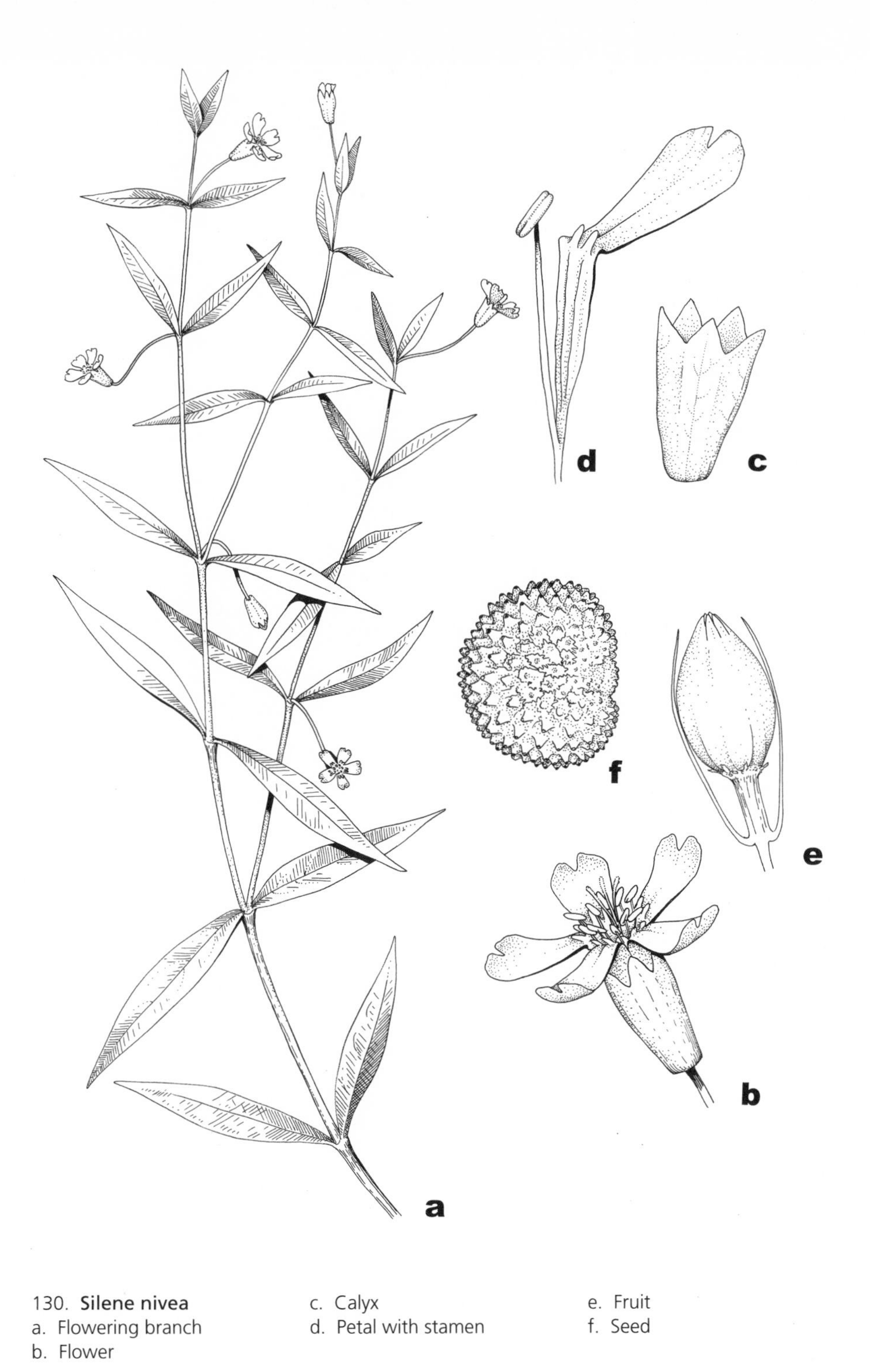

130. **Silene nivea**
a. Flowering branch
b. Flower
c. Calyx
d. Petal with stamen
e. Fruit
f. Seed

Range: Pennsylvania to South Dakota, south to Nebraska, Missouri, and Virginia.
Illinois Distribution: Uncommon in the northern two-thirds of the state.

Snowy campion is an attractive wildflower apparently confined to the northern two-thirds of Illinois where it occurs on moist stream banks, in wooded ravines, and in calcareous fens.

Silene nivea differs from the similar *S. cserei* and *S. vulgaris* in its usually solitary flower and in its acuminate leaves.

The flowers of this species bloom from June to August.

13. **Silene cserei** Baumg. Enum. Stirp. Trans. 3:345. 1818. Fig. 131.

Biennial herb from a thickened crown; stems stout, erect, glabrous, glaucous, branched or unbranched, up to 75 cm tall; leaves opposite, entire, thick, glaucous, glabrous, the middle and upper leaves oblong to oblong-ovate, acute to subacute at the apex, rounded at the more or less clasping base, up to 6 cm long, up to 3 cm broad, the lowest leaves spatulate to oblanceolate, obtuse at the apex, tapering to the base, up to 8 cm long, up to 4 cm broad; inflorescence a narrow raceme, few-flowered; flowers up to 1.7 cm across, subtended by small, narrow bracts; calyx ellipsoid, 10-ribbed above, 20-ribbed below, glabrous, up to 9 mm long, slightly constricted just below the apex; petals 5, free, white, bifid, up to 5 mm long; appendages 0; stamens 10; styles 3; capsules ellipsoid, up to 1.2 cm long; seeds papillose, up to 1 mm long.

Common Name: Glaucous Campion.
Habitat: Waste ground, particularly along railroads.
Range: Native to Europe; becoming increasingly abundant in the United States.
Illinois Distribution: Occasional to common in the northern one-third of the state, rarer elsewhere.

Since Fell and Fell's first report of this species from Illinois in 1949, it has become increasingly common in the northern counties of the state.

This species is similar to *S. vulgaris,* differing in its smaller, less inflated calyx; its less reticulate calyx; and the constriction just below the apex of the calyx.

Silene cserei flowers from May to October.

14. **Silene vulgaris** (Moench) Garcke, Fl. Deutsch. ed. 9, 64. 1882. Fig. 132.
Cucubalus latifolius Mill. Gard. Dict. ed. 8, no. 2. 1768.
Behen vulgaris Moench, Meth. Pl. Hort. Bot. 709. 1794.
Cucubalus inflatus Salisb. Prodr. 302. 1796, *nomen illeg.*
Silene cucubalus Wibel, Prim. Fl. Werth. 241. 1799.
Silene inflata (Salisb.) Sm. Fl. Brit. 2:292. 1800.
Silene latifolia (Mill.) Britten & Rendle, List Brit. Seed Pl. 5. 1907, non Poir. 1789.

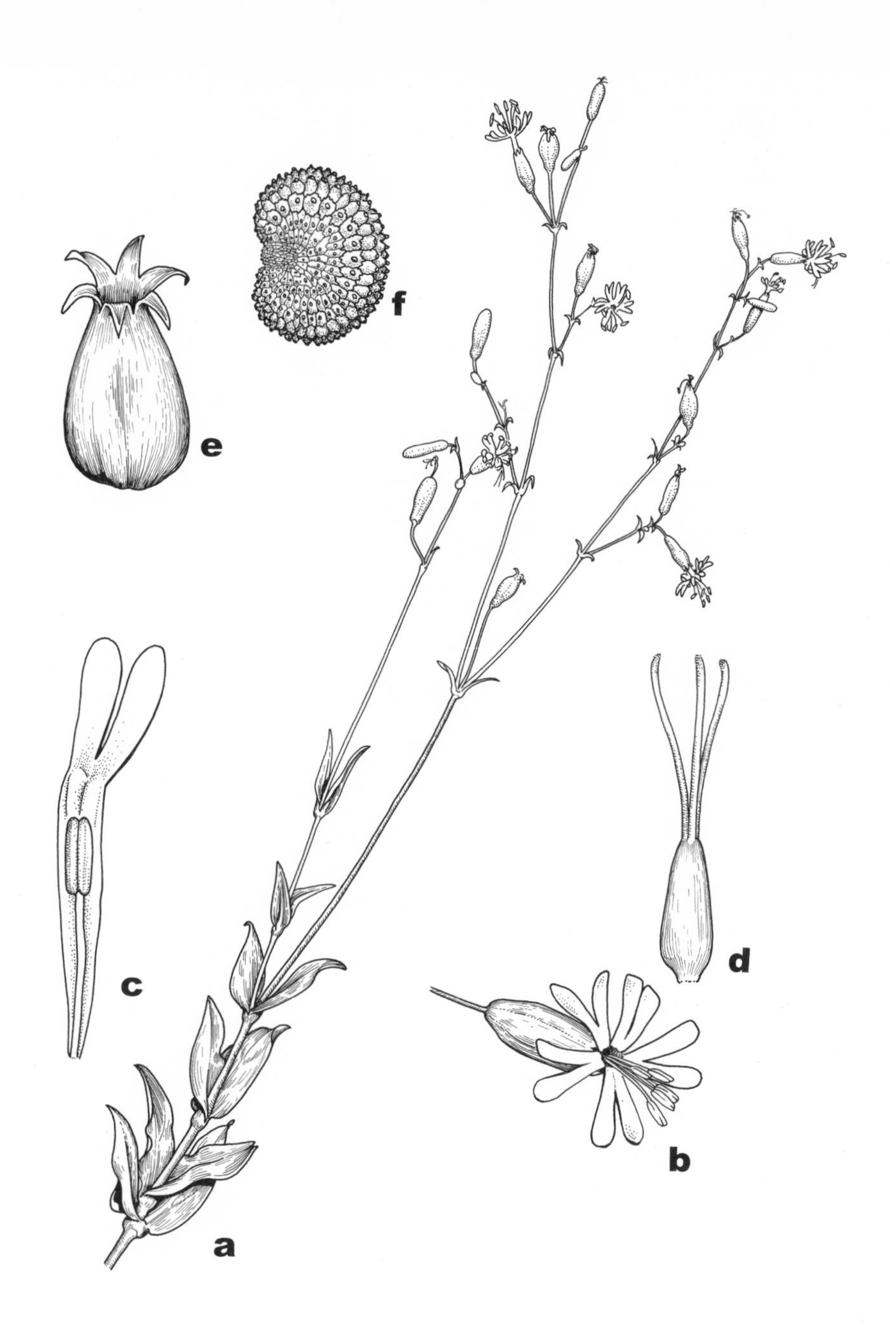

131. **Silene cserei**
a. Flowering branch
b. Flower
c. Petal with stamen
d. Pistil before anthesis
e. Fruit
f. Seed

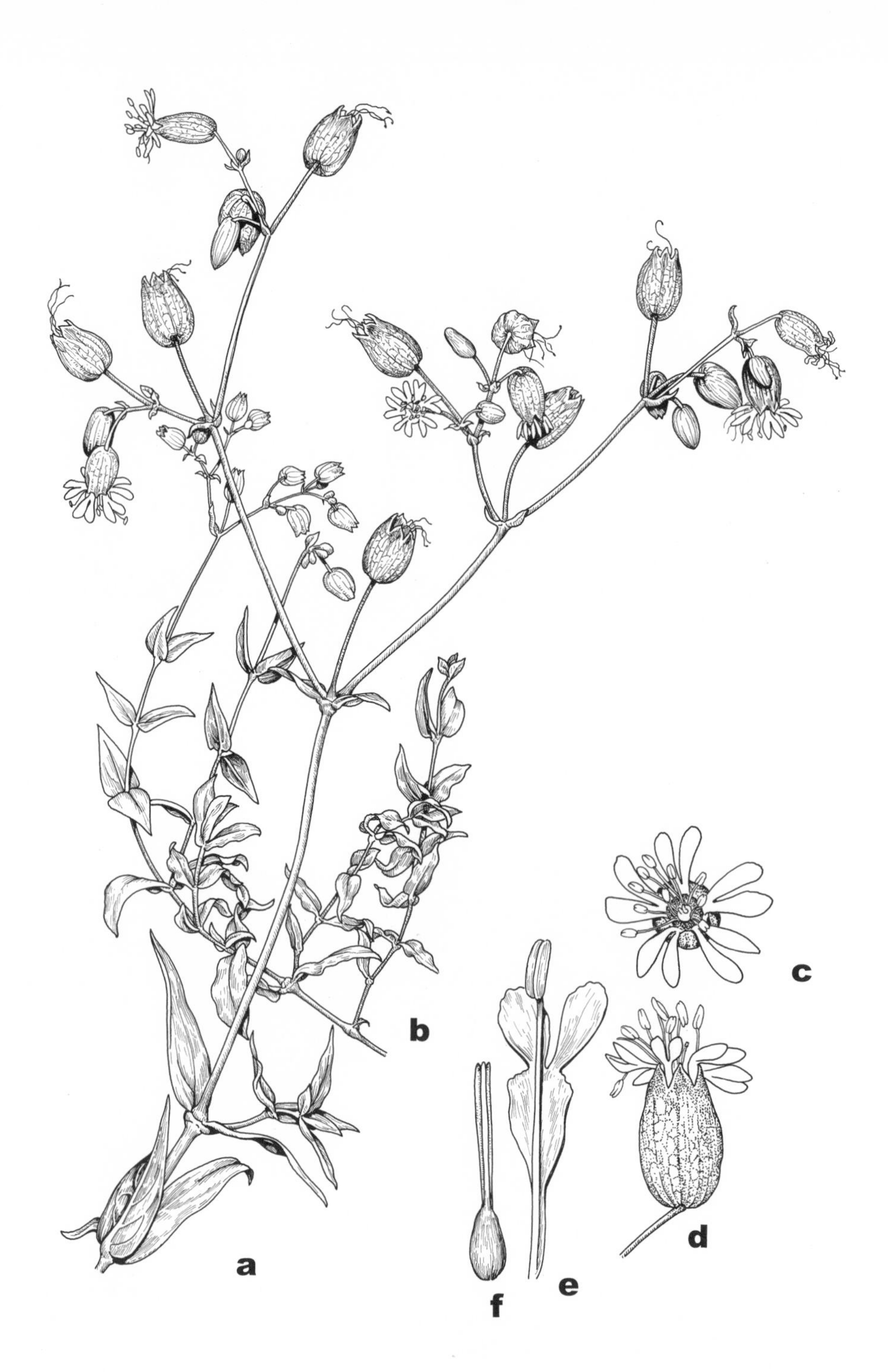

132. **Silene vulgaris**
a. Habit
b. Habit
c. Flower
d. Flower, side view
e. Petal with stamen
f. Pistil

Perennial herb from a short rhizome; stems stout, erect or decumbent, glabrous or rarely hirsutulous, glaucous, branched or unbranched, up to 80 cm tall; leaves opposite, entire, thick, glaucous, usually glabrous, the middle and upper leaves ovate-lanceolate to oblong, acute at the apex, rounded at the sessile base, up to 6 cm long, up to 2.5 cm broad, the lowest leaves spatulate, obtuse to subacute at the apex, tapering to the short-petiolate base, up to 8 cm long, up to 3 cm broad; inflorescence a loose cyme or panicle, several-flowered; flowers up to 2 cm across, subtended by small, narrow bracts; calyx ovoid, becoming inflated, 20-ribbed with conspicuous cross-veins, glabrous, up to 1 cm long; petals 5, free, white, bifid, up to 6 mm long, the appendages minute or none; stamens 10; styles 3; capsules included within the bladdery calyx, up to 1 cm long; seeds papillose, 1.0–1.5 mm long, brown.

Common Name: Bladder Catchfly.
Habitat: Waste ground.
Range: Native to Europe and Asia; adventive throughout most of the United States.
Illinois Distribution: Occasional in the northern half of the state, less common elsewhere.

Silene vulgaris for years has been known as *S. cucubalus,* but Voss (1985) gives reasons for changing the binomial to *S. vulgaris.*

The bladder catchfly has a very beautiful, inflated, reticulate-veined calyx.

Patterson's report in 1874 was the first for this species in Illinois.

Silene vulgaris flowers from May to August.

15. Silene ovata Pursh, Fl. Am. Sept. 1:316. 1814. Fig. 133.

Perennial herb with slender rhizomes; stems firm, erect, glabrous, branched, up to 1.2 m tall; leaves opposite, entire, 5–10 pairs on the stem, ovate, long-acuminate at the apex, rounded at the base, glabrous, up to 8 cm long, up to 3.5 cm broad, sessile; inflorescence a terminal panicle, several-flowered; flowers up to 3 cm across, subtended by small, scarious bracts; calyx slenderly tubular, puberulent, 10-ribbed, 8–10 mm long, with deltate-acuminate teeth; petals 5, free, white, fringed, up to 1.3 cm long; capsules ellipsoid, 7–9 mm long, with 6 teeth; seeds warty.

Common Name: Ovate-leaved Campion.
Habitat: Rich woods.
Range: Western North Carolina and southeastern Kentucky to Arkansas and Georgia; southern Illinois.
Illinois Distribution: Known from a few stations in Hardin County.

The discovery of this species in Hardin County in 1997 by Jody Shimp was totally unexpected since the range of this species was thought to be considerably southeast of Illinois.

Silene ovata differs from all other species in the genus by the combination of puberulent calyx, opposite leaves, and fringed petals.

This species flowers during August.

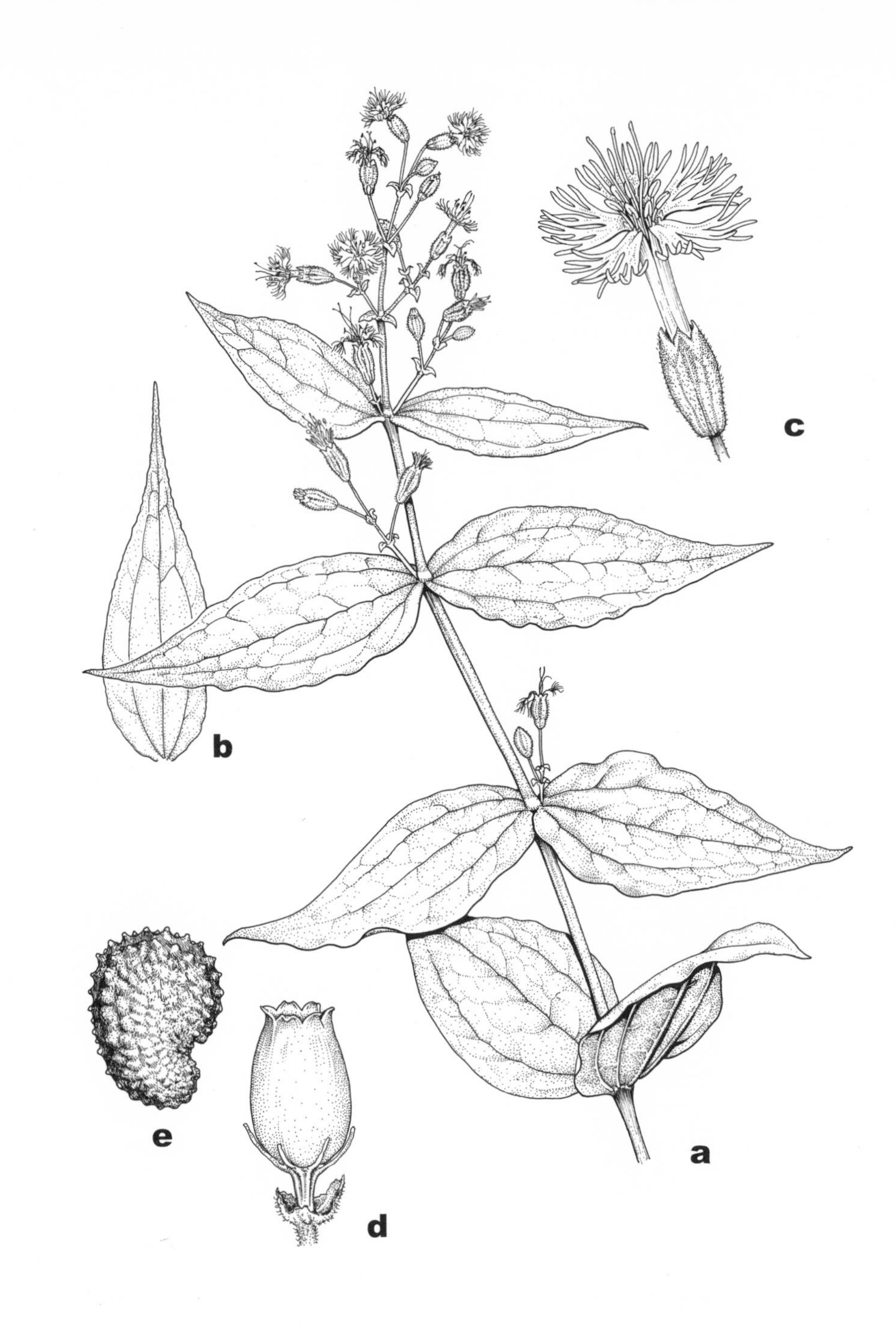

133. **Silene ovata**
a. Flowering branch
b. Leaf
c. Flower
d. Fruit
e. Seed

18. **Agrostemma** L.—Corn Cockle

Annual of biennial herbs; leaves opposite, entire, without stipules; flowers solitary or in terminal cymes; calyx united into a tube, 5-lobed, the lobes longer than the tube, the tube 10-ribbed, pubescent with long hairs; petals 5, free, showy, entire or shallowly notched; stamens 10; styles 5; ovary superior, 1-locular, with many ovules; fruit a capsule, dehiscing by 5 teeth; seeds many, flattened, black.

Two species comprise the genus, with only the following known from Illinois.

1. **Agrostemma githago** L. Sp. Pl. 435. 1753. Fig. 134.
Lychnis githago (L.) Scop. Fl. Carn. ed. 2, 1:310. 1772.

Annual or rarely biennial herb from taproots; stems erect, rather stout, up to 1 m tall, sparingly branched, pubescent; leaves opposite, entire, linear-lanceolate to lanceolate, acute at the apex, tapering to the base, puberulent, up to 15 cm long, up to 1 cm broad; flowers up to 4 cm across, solitary on puberulent pedicels up to 20 cm long; calyx ovoid to ellipsoid, 10-ribbed, puberulent, with 5 foliaceous, linear lobes up to 4 cm long; petals 5, free, purple to purple-red, obovate, obtuse or truncate at the apex, slightly notched at the apex, 2–3 cm long; stamens 10; styles 5; capsules ovoid to oblongoid, up to 20 mm long; seeds 3.0–3.5 mm long, ovate, flattened, black, with low tubercles.

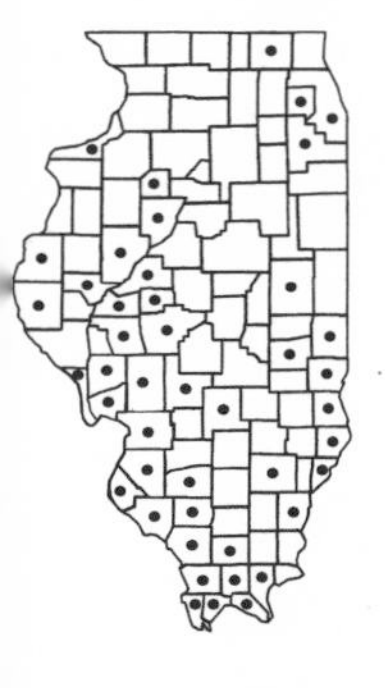

Common Name: Corn Cockle.
Habitat: Fields and other disturbed areas.
Range: Native to Europe and Asia; naturalized throughout the United States.
Illinois Distribution: Occasional throughout the state.

Corn cockle, despite its weedy nature, has large, handsome, purple-red flowers. The five petals are exceeded by the narrow, foliaceous calyx lobes.

This species can be a nuisance in cultivated fields because the seeds are poisonous.

Most of the early Illinois botanists called this plant *Lychnis githago.*

Agrostemma githago flowers from May to July.

19. **Gypsophila** L.—Baby's Breath

Annual, biennial, or perennial herbs; leaves opposite, simple, entire, without stipules; inflorescence cymose to paniculate, or the flower solitary; flowers perfect, actinomorphic, not subtended by bracts; calyx turbinate to campanulate, 5-nerved, white-scarious between the teeth; petals 5, free, entire or notched at the apex; stamens 10; styles 2; ovary superior, 1-locular; fruit a capsule, dehiscing by usually 4 apical teeth; seeds mostly reniform, few to several, black.

Gypsophila is a genus of about 125 species, all native to either Europe, Asia, or northern Africa. Several species are grown as garden ornamentals.

Only the following three adventive species occur in Illinois:

134. **Agrostemma githago**
a. Flowering branch
b. Flower
c. Petal
d. Fruit
e. Calyx
f. Seed

Key to the Species of **Gypsophila** in Illinois

1. Petals at least twice as long as the calyx; calyx 3–5 mm long .... 1. *G. elegans*
1. Petals about as long as the calyx; calyx up to 3 (–4) mm long.
   2. Leaves cuneate, 1-nerved; calyx and pedicels eglandular; petals up to 4 mm long .... 2. *G. paniculata*
   2. Leaves subcordate, 3- to 5-nerved; calyx and pedicels glandular; petals 4–6 mm long .... 3. *G. scorzonerifolia*

1. Gypsophila elegans Bieb. Fl. Taur. Cauc. 1:319. 1808. Fig. 135.

Annual or biennial herbs from a slender taproot; stems erect, glaucous, glabrous, branched, up to 50 cm tall; leaves opposite, entire, glaucous, glabrous, the middle and upper leaves lanceolate, acute to acuminate at the apex, connate at the base, up to 5 cm long, up to 5 mm broad, the lowest leaves spatulate, obtuse to acute at the apex, tapering to the base, up to 8 cm long, up to 2 cm broad; inflorescence an open panicle to corymbiform, several-flowered; flowers up to 1.5 cm across, on glabrous pedicels up to 3.5 cm long; calyx divided nearly to the base, the 5 lobes ovate, obtuse, white-margined, glabrous, 3–5 mm long; petals 5, white to pink, emarginate, 10–15 mm long; stamens 10; styles 2; capsules subglobose, dehiscing by 4 apical teeth; seeds tuberculate.

Common Name: Baby's Breath.
Habitat: Waste ground.
Range: Native to Europe and Asia; uncommonly escaped from cultivation.
Illinois Distribution: Known from Champaign County.

This is the showiest species of baby's breath known from Illinois. The petals exceed the calyx by two or three times.

Gypsophila elegans at one time was a popular garden ornamental.

Time of flowering for this species is May and June.

2. Gypsophila paniculata L. Sp. Pl. 407. 1753. Fig. 136.

Perennial herb from a thickened rhizome; stems erect, glaucous, glabrous, branched, up to 80 cm tall; leaves opposite, entire, lanceolate, acute to acuminate at the apex, narrowed to the base, 1-nerved, glaucous, glabrous, up to 2 cm long, becoming much smaller toward the inflorescence; inflorescence a many-flowered, diffuse panicle, with minute bracts; flowers up to 3 mm across, on glabrous pedicels up to 12 mm long; calyx campanulate, up to 3 mm long, the 5 teeth suborbicular, glabrous, with a narrow green center and a broad scarious margin; petals 5, free, white or pink, usually slightly emarginate, up to 4 mm long; stamens 10; styles 2; capsules globose, up to 3.5 mm in diameter, dehiscing by 4 apical teeth; seeds tuberculate.

Common Name: Baby's Breath.
Habitat: Disturbed soil.
Range: Native to Europe and Asia; occasionally adventive in the United States.

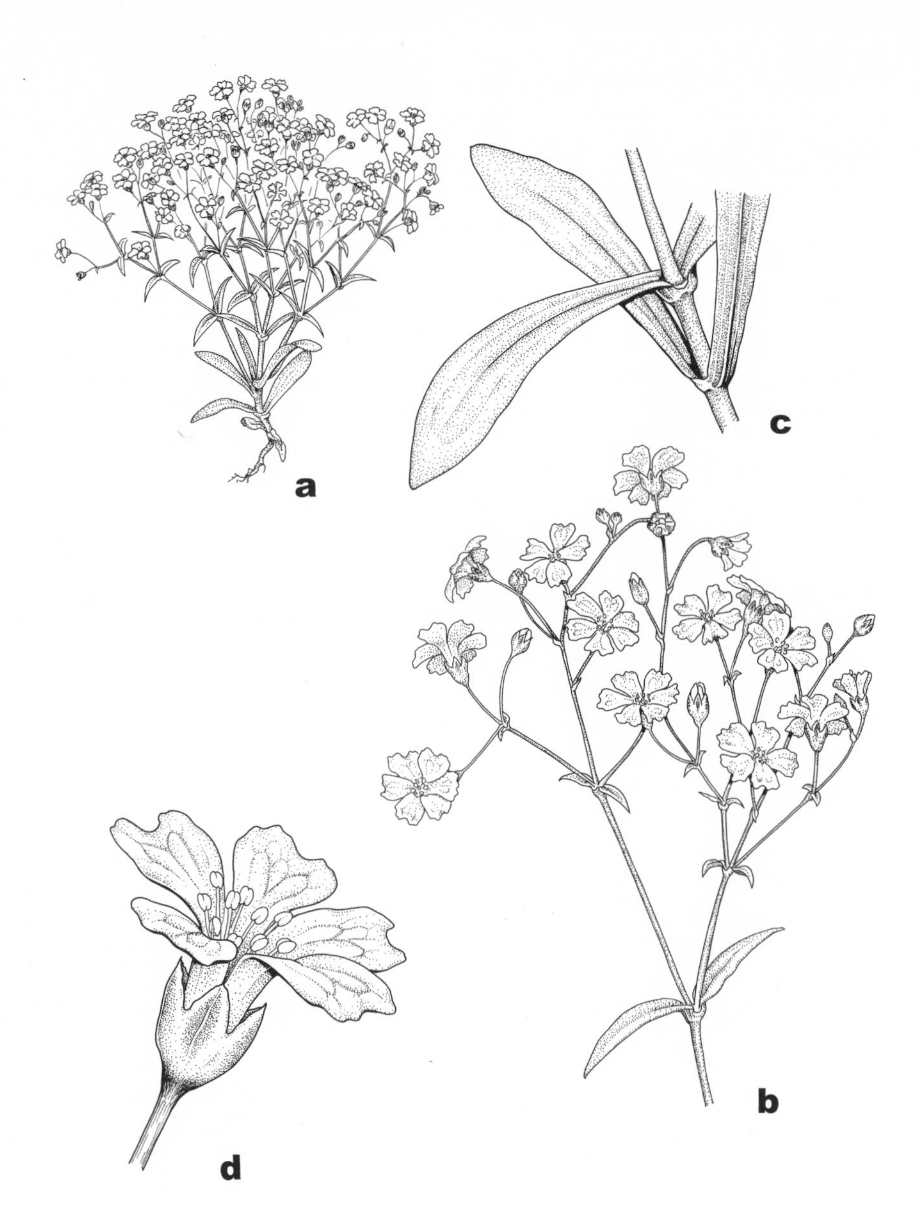

135. **Gypsophila elegans**
a. Habit
b. Flowering branch
c. Lower leaves
d. Flower

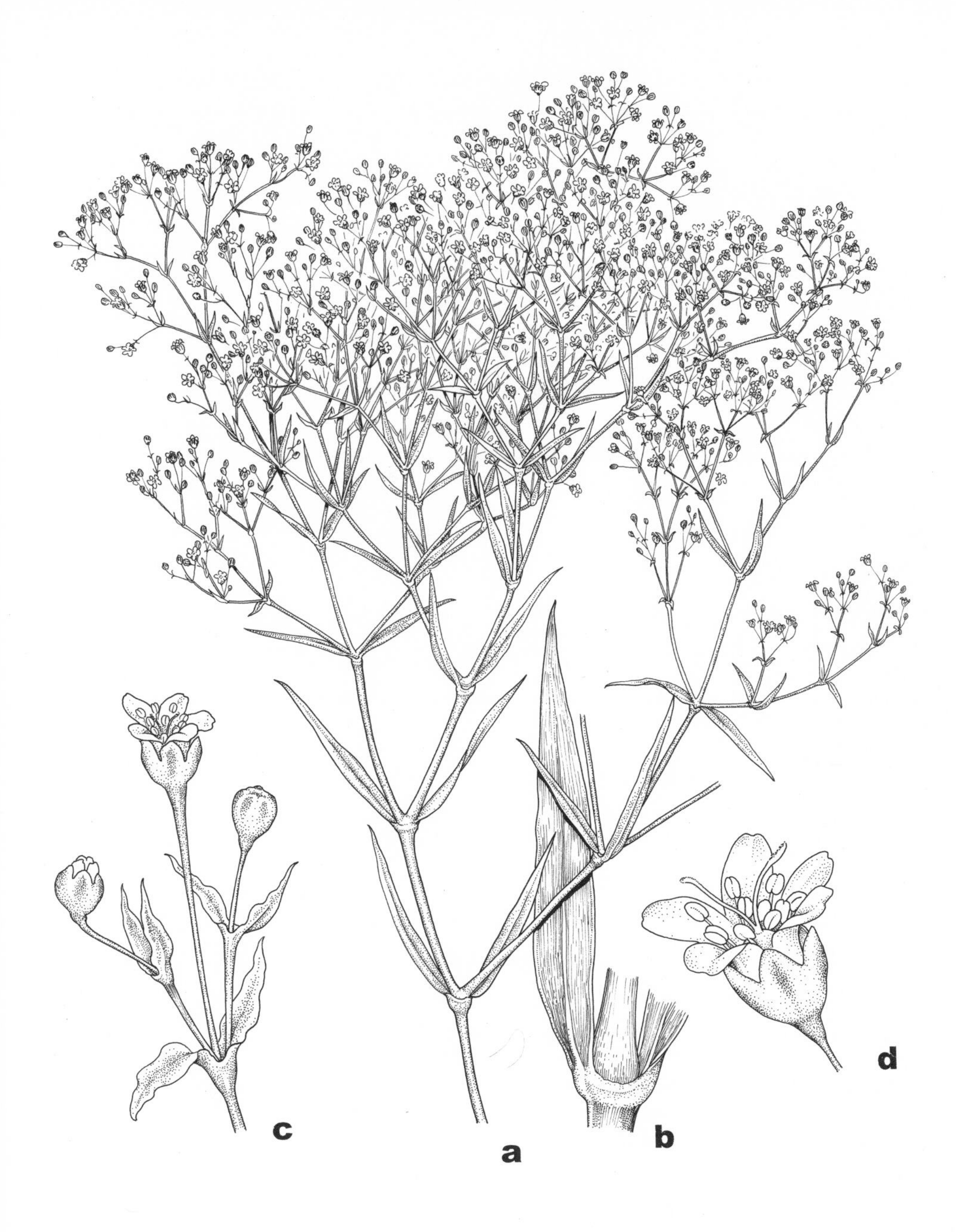

136. **Gypsophila paniculata**
a. Habit
b. Node
c. Flowering branch
d. Flower

Illinois Distribution: Known from Cook, Kane, LaSalle, Mason, Menard and Winnebago counties.

This is an attractive species with many small flowers in a diffuse inflorescence. It is planted occasionally in gardens, but it seldom escapes and persists.

Gypsophila paniculata flowers from late June to August.

3. **Gypsophila scorzonerifolia** Ser. in DC. Prod. 1:352. 1824. Fig. 137.

Perennial herb from short rhizomes; stems erect, sometimes glaucous, glabrous, branched, up to nearly 1 m tall; leaves opposite, entire, lanceolate, obtuse to abruptly acute at the apex, subcordate at the usually clasping base, 3- to 5-nerved, glaucous, glabrous, up to 10 cm long, up to 3 cm broad; inflorescence a many-flowered cyme or panicle, with minute bracts; flowers up to 6 mm across, on glandular pedicels; calyx campanulate, up to 3 (–4) mm long, the 5 teeth obtuse, glandular; petals 5, free, pale pink or white, 4–6 mm long; stamens 10; styles 2; capsules globose, dehiscing by 4 apical teeth; seeds tuberculate.

Common Name: Baby's Breath.
Habitat: Waste ground, particularly along railroads.
Range: Native to Europe; recently adventive near the Great Lakes.
Illinois Distribution: Known from Cook, Kane, Kankakee, and McHenry counties.

This species has recently become more frequent around the Great Lakes. It is known from at least Illinois, Indiana, and Michigan.

Gypsophila scorzonerifolia has been variously misidentified as *G. acutifolia* or *G. perfoliata.* Pringle (1976) has cleared up the identity of this species.

This species is distinguished by its subcordate leaf bases and the glandular calyces and pedicels.

Time for flowering for this species is June to August.

20. **Saponaria** L.—Bouncing Bet

Perennial herbs; leaves opposite, simple, entire, without stipules; inflorescence corymbose or cymose, terminal or axillary, with several flowers; flowers perfect, actinomorphic, bracteate; calyx tubular, 20-nerved; petals 5, free, entire, with 2 appendages at the base; stamens 10; styles 2; ovary superior, 1-locular; fruit a capsule, dehiscing by 4 teeth.

In previous publications on the Illinois flora, I included a second species, *S. vaccaria,* in this genus. I now believe that that species is best treated in the genus *Vaccaria.*

There are thirty species of *Saponaria,* all native to Europe and Asia.

Only the following occurs in Illinois:

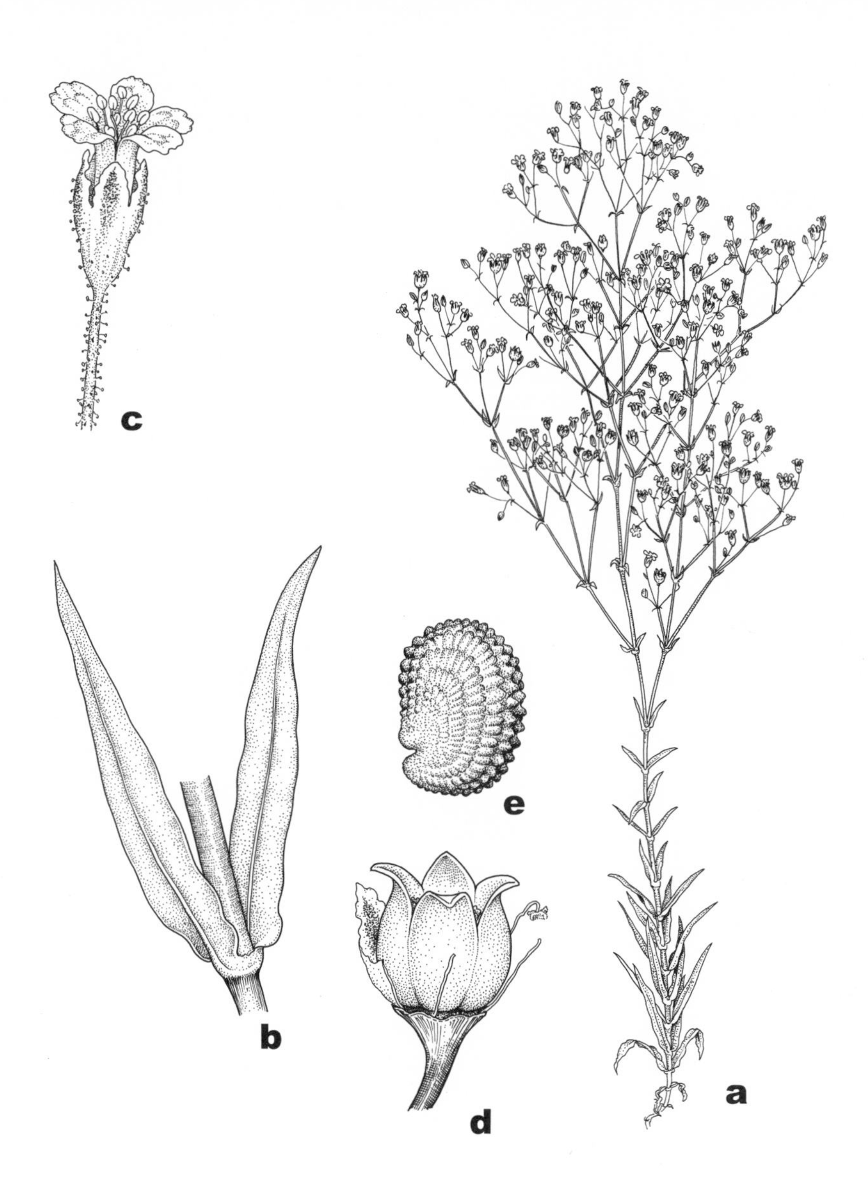

137. **Gypsophila scorzonerifolia**
a. Habit
b. Node
c. Flower
d. Fruit
e. Seed

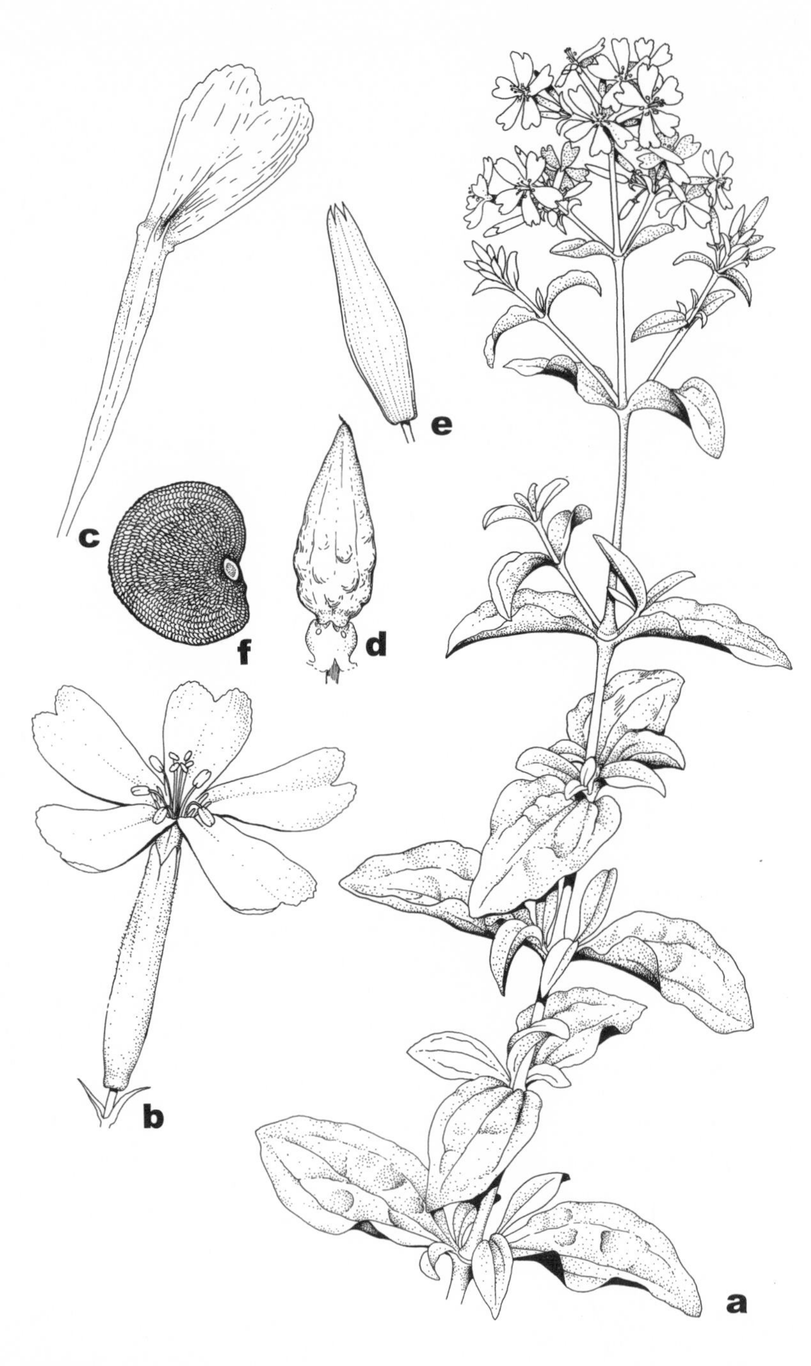

138. **Saponaria officinalis**
a. Flowering branch
b. Flower
c. Petal
d. Fruit
e. Calyx
f. Seed

1. **Saponaria officinalis** L. Sp. Pl. 408. 1753. Fig. 138.

Perennial herb from a creeping rhizome; stems erect, rather stout, glabrous, sparsely branched or unbranched, up to 90 cm tall; leaves opposite, entire, lanceolate to elliptic to oval, acute at the apex, narrowed to a very short petiole, glabrous, up to 10 cm long, up to 4 cm broad; inflorescence a crowded cyme, several-flowered, bracteate; flowers 1.8–2.5 cm across, fragrant, short-pedicellate; calyx tubular-cylindric, 20-nerved, glabrous, up to 2.5 cm long, with 5 small, triangular, usually unequal teeth; petals 5, free, pink or white, obovate, emarginate, up to 1.5 cm long, with 2 awl-shaped appendages at the base of each; stamens 10; styles 2; capsules oblongoid to ovoid, up to 2.5 cm long; seeds reniform, reticulate, 1.8–2.0 mm broad, more or less compressed.

Common Name: Bouncing Bet; Soapwort.
Habitat: Waste ground.
Range: Native to Europe; widely adventive in the United States.
Illinois Distribution: Common throughout the state.

Bouncing Bet is a popular plant of gardens and is readily adapted to adventive habitats where it may persist for years. Double-flowered forms, called var. *caucasica* Hort., are known.

Saponaria officinalis flowers from June to September.

21. **Vaccaria** Moench—Cow Herb

Annual herbs; leaves opposite, simple, entire, without stipules; inflorescence terminal, cymose, with several flowers; flowers perfect, actinomorphic, bracteate; calyx urn-shaped to cylindric, 10-nerved; petals 5, free, obcordate, usually notched, without appendages at the base; stamens 10; styles 2; ovary superior, 1-locular; fruit a capsule, dehiscing by 4 teeth; seeds many, tuberculate.

Only the following species comprises the genus.

1. **Vaccaria hispanica** (Mill.) Rauschert in Wiss. Zeitschr. Martin-Luther-Univ. Math.-Nat. 14:496. 1965. Fig. 139.
Saponaria vaccaria L. Sp. Pl. 409. 1753.
Saponaria hispanica Mill. Gard. Dict. ed. 8, in errata. 1768.
Vaccaria pyramidata Medic. Phil. Bot. 1:96. 1782.
Vaccaria vulgaris Host, Fl. Aust. 1:518. 1827.
Vaccaria segetalis Garcke ex Aschers. Fl. Brandenb. 1:84. 1864.
Vaccaria vaccaria (L.) Britt. in Britt. & Br. Ill. Fl. 2:18. 1897.

Annual from a slender taproot; stems erect, slender, glabrous, glaucous, branched, up to 80 cm tall; leaves opposite, entire, lanceolate to lance-ovate, acute at the apex, tapering or rounded or clasping at the base, glabrous, up to 10 cm long, up to 4 cm broad; inflorescence an open cyme, several-flowered, bracteate; flowers up to 10 mm broad, on long pedicels; calyx ovoid, 10-nerved, glabrous, up to 1.7 cm long, with 5 small, triangular teeth; petals 5, free, red or less commonly pink, emarginate, up

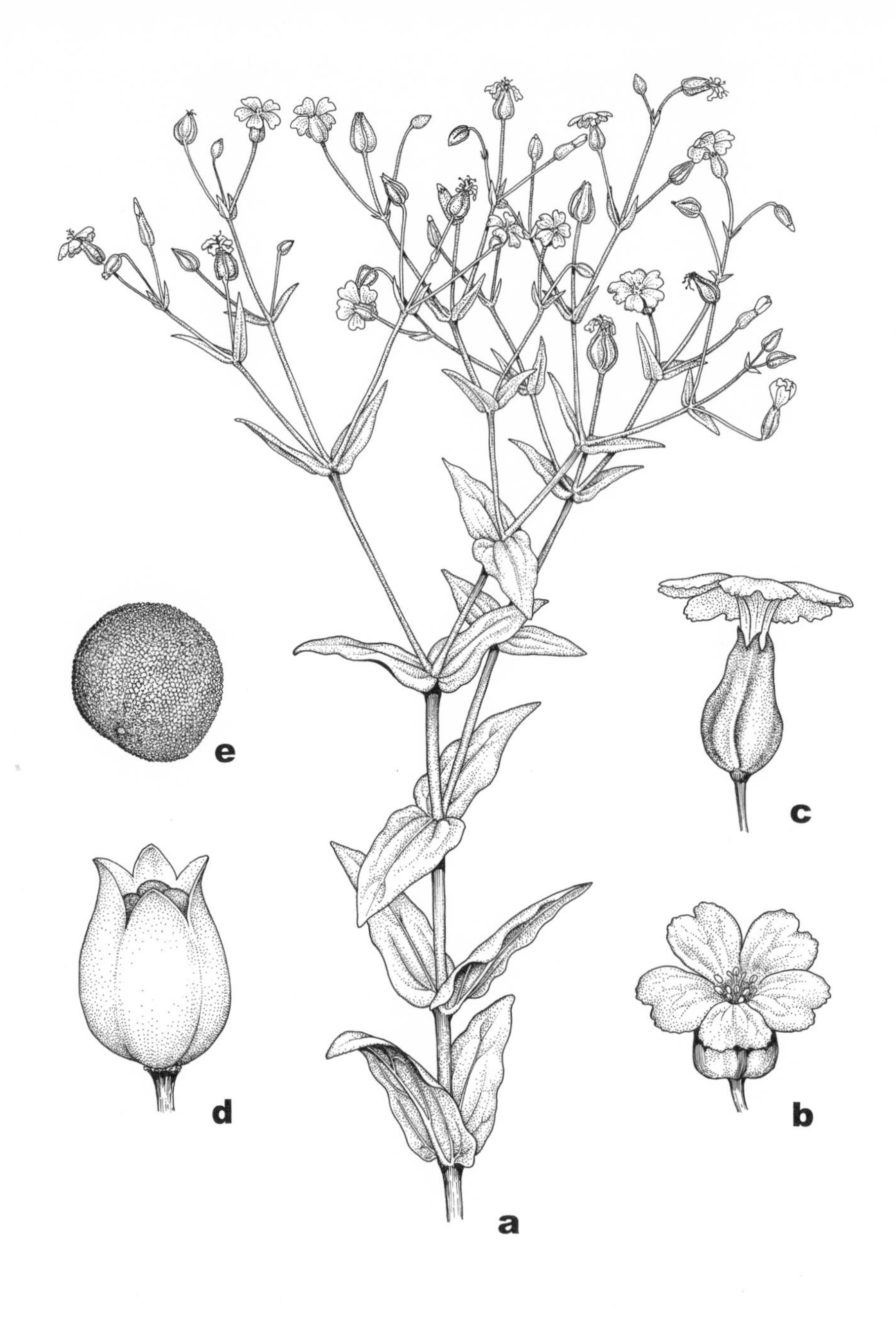

139. **Vaccaria hispanica**
a. Habit
b. Flower, face view
c. Flower, side view
d. Capsule
e. Seed

to 8 mm long, without appendages at the base; stamens 10; styles 2; capsules ovoid, up to 8 mm long; seeds globose, 1.6–1.8 mm in diameter, reddish brown to black, tuberculate.

Common Name: Cow Herb.
Habitat: Waste ground.
Range: Native to Europe; occasionally adventive in the United States.
Illinois Distribution: Scattered throughout the state but very uncommon in the southern one-third of the state.

Cow herb is sometimes placed in the genus *Saponaria,* but it differs in its ovoid calyx and the lack of appendages at the base of each petal.

The flowers, which are usually red, may sometimes be pink.

Vaccaria hispanica flowers from June to August.

Summary of the Taxa Treated in This Volume

Excluded Species

Additional Taxon

Glossary

Literature Cited

Index of Plant Names

Summary of the Taxa Treated in This Volume

Families	Genera	Species	Lesser Taxa
Phytolaccaceae	1	1	
Nyctaginaceae	1	5	
Molluginaceae	1	1	
Cactaceae	1	3	
Portulacaceae	3	6	
Chenopodiaceae	10	42	
Amaranthaceae	6	19	2
Caryophyllaceae	21	62	–
Totals	44	139	2

Excluded Species

Amaranthus pumilus Raf. Huett (1897) attributed this plant erroneously to Illinois. It is a species of sea beaches along the Atlantic coast.

Arenaria groenlandica (Retz.) Spreng. Although Mohlenbrock (1986) reported this plant from Illinois, this was a misidentification of what was actually *Sagina procumbens.*

Claytonia caroliniana Michx. This binomial has been applied by several Illinois botanists and by Fernald (1950) for specimens of *C. virginiana* in Illinois that have very broad leaves. *Claytonia caroliniana* is a different species.

Iresine celosioides L. This binomial was used by Patterson (1876) and Schneck (1876) for *I. rhizomatosa. Iresine celosioides* is a different species unknown from Illinois.

Paronychia dichotoma (L.) Nutt. Although Hyatt (1875) used this binomial for *P. canadensis,* Nuttall's species is not the same.

Paronychia jamesii Torr. & Gray. Mead (1846) gave this binomial to *P. canadensis,* but *P. canadensis* is not the same plant that Torrey and Gray described as *P. jamesii.*

Sagina nodosa (L.) Fenzl. This binomial was used by Brendel (1859) for our *S. decumbens,* and Vasey (1861) used it for our *S. procumbens. Sagina nodosa* is a different species unknown from Illinois.

Sagina subulata Presl. Patterson (1876) mistakenly called our *S. decumbens* by the binomial *S. subulata,* but this latter binomial is a different species not known from Illinois.

Stellaria longipes Goldie. Although this species has been reported from Illinois by Higley and Raddin (1891) and Buhl (1934), the plants on which these reports are based are *S. longifolia.*

Talinum teretifolium Pursh. Several Illinois botanists used this binomial for *T. rugospermum. Talinum teretifolium* is a different species found in the southern United States.

Additional Taxon

A cactus new to Illinois was discovered after this book went to press. It is:

Coryphantha missouriensis (Sweet) Britt. & Rose, Ill. Fl. N. U.S., ed. 2, 570. 1913. *Mammillaria missouriensis* Sweet, Hort. Britt. 171. 1827.

Stems not jointed; areoles with 1–4 central spines and 13–20 radial spines; flowers pale yellow to greenish, the perianth fimbriate; fruit reddish.

Union Co.: among dry leaf litter near waterfall catchpool, vicinity of Water Valley watertower, March 10, 2001, Scott M. Herron.

Glossary

actinomorphic. Having radial symmetry; regular, in reference to a flower.
acuminate. Gradually tapering to a long point.
acute. Sharply tapering to a point.
adnate. Union of unlike parts.
annual. A plant that lives for only one growing season.
anther. The terminal part of a stamen that contains the pollen grains.
anthesis. Flowering time.
anthocarp. The fruit in the genus *Mirabilis* consisting of a utricle enclosed by the calyx tube.
apetalous. Without petals.
apiculate. Abruptly short-pointed at the tip.
appressed. Lying flat against the surface.
areole. An area on the stem of a cactus that often bears spines.
awl. Drawn out into a long slender point, like the end of an ice pick.
awn. A bristle usually terminating a structure.
axillary. Borne from an axil.

berry. A type of fruit where the seeds are surrounded only by fleshy material.
biennial. A plant that completes its life cycle in two years and then perishes.
bifid. Two-cleft.
bract. An accessory structure at the base of many flowers, often appearing leaflike.
bracteole. A secondary bract.

caducous. Falling away early.
calyx. The outermost segments of the perianth of a flower, composed of sepals.
campanulate. Bell-shaped.
capillary. Threadlike.
capitate. Forming a head.
capsule. A dry, dehiscent fruit composed of more than one carpel.
carpel. A simple pistil or one member of a compound pistil.
cauline. Belonging to a stem.
cilia. Marginal hairs.
ciliate. Bearing cilia, or marginal hairs.
ciliolate. Bearing small cilia.
circumscissile. Usually referring to a fruit which dehisces by a horizontal, circular line.
clasping. Referring to a leaf whose base encircles or partially encircles the stem.
clavate. Club-shaped.
clawed. Bearing a narrow, basal stalk, particularly of a petal.
compressed. Flattened.
connate. Union of like parts.
corm. An underground, vertical stem with scaly leaves, differing from a bulb by lacking fleshy leaves.
corymb. A type of inflorescence where the pedicellate flowers are arranged along an elongated axis but with the flowers all attaining about the same height.
corymbiform. Shaped like a corymb.
corymbose. Having the flowers arranged in a corymb.
crested. Provided with an elevated and often flattened ridge at the summit, usually referring to petals.
cucullate. Hood-shaped.
cuneate. Wedge-shaped; tapering to the base.
cuspidate. Terminating in a very short point.
cylindrical. In the form of a cylinder, or rolled-up tube.
cyme. A type of broad and flattened inflorescence in which the central flowers bloom first.
cymose. Having the flowers arranged in a cyme.

deciduous. Falling away.
decumbent. Lying flat but with the tip ascending.
decurrent. Adnate to the petiole or stem and then extending beyond the point of attachment.
deflexed. Turned downward.
dehiscent. Splitting at maturity.
deltate. Referring to a flat object that is triangular in shape.

dentate. With sharp teeth, the tips of which point outward.
denticulate. With small sharp teeth, the tips of which point outward.
dichotomous. With equal branching.
diffuse. Loosely spreading.
dioecious. With staminate flowers on one plant, pistillate on another.

eglandular. Not bearing glands.
ellipsoid. Referring to a solid object that is broadest at the middle, gradually tapering to both ends.
elliptic. Referring to a flat object that is broadest at the middle, gradually tapering to both ends.
emarginate. Having a shallow notch at the tip.
endosperm. Food storage tissue in a seed.
entire. Referring to a leaf margin that has a smooth edge, that is, without teeth.
erose. With an irregularly notched margin.
excurrent. Protruding beyond the tip, usually referring to the midvein extending beyond the tip of a leaf.
exserted. Extended above and beyond.

fasciculate. Clustered.
fetid. Ill-smelling.
fibrous. Referring to roots borne in tufts.
filament. The stalk of a stamen that supports the anther.
filiform. Threadlike.
funnelform. Shaped like a funnel.
fusiform. Spindle-shaped; broadest at the middle, tapering to either end.

glabrate. Becoming smooth.
glabrous. Without pubescence, or hairs.
glandular. Bearing glands.
glaucous. With a whitish covering that can be rubbed off.
globose. Round; globular.
glochidia. Processes bearing barbs.
glomerule. A small compact cluster.
glutinous. Covered with a sticky secretion.
granular. Having a roughened surface.

hastate. Spear-shaped; said of a leaf that is triangular with spreading basal lobes.
hirsute. With stiff hairs.
hoary. Grayish white, usually referring to pubescence.
hyaline. Transparent.
hypanthium. A development of the receptacle beneath the calyx.

indehiscent. Not splitting open at maturity.
indument. A covering; usually referring to the type of pubescence.
inferior. Referring to the position of the ovary when it is surrounded by the adnate portion of the floral tube or is embedded in the receptacle.
inflexed. Turned inward.
inflorescence. A cluster of flowers.
internode. The distance between two nodes or between two joints on the stem.
involucre. A circle of bracts that subtends a flower cluster.

keel. A ridgelike process.

lacerate. Shredded.
laciniate. Divided into narrow, pointed divisions.
lanate. Woolly.
lanceolate. Referring to a flat object that is broadest near base, gradually tapering to the narrower apex; lance-shaped.
lanceoloid. Referring to a solid object that is broadest near base, gradually tapering to the narrower apex.
lenticular. Lens-shaped.
linear. Elongated and narrowly uniform in width throughout.
lobulate. With small lobes.
locular. Referring to the locule, or cavity of the ovary or the anther.
locule. The cavity of the ovary or the anther.

margin. The edge, usually in reference to a leaf.
membranaceous. Thin and membranelike.
monoecious. Bearing both sexes in separate flowers on the same plant.
mucronate. Possessing a short, abrupt tip.
mucronulate. Possessing a very short, abrupt tip.

node. That place on the stem from which leaves and branchlets arise.

obcordate. Reverse heart-shaped.
oblanceolate. Reverse lance-shaped; broadest at apex, gradually tapering to narrow base.
oblong. Referring to a flat object that is broadest at the middle, tapering to both ends, but broader than elliptic.
oblongoid. Referring to a solid object that is broadest at the middle and tapering to both ends.
obovate. Referring to a flat object that is broadly rounded at the apex, becoming narrowed below.
obovoid. Referring to a solid object that is broadly rounded at the apex, becoming narrowed below.
obtuse. Rounded; blunt.
opaque. Incapable of being seen through.
orbicular. Round.
oval. Broadly elliptic.
ovary. The lower swollen part of the pistil that contains the ovules.
ovate. Referring to a flat object that is broadly rounded at the base, becoming narrowed above; broader than lanceolate.
ovoid. Referring to a solid object that is broadly rounded at the base, becoming narrowed above.
ovule. An immature seed.

panicle. A type of inflorescence composed of several racemes.
papilla. A wartlike process.
papillose. Bearing wartlike processes.
pedicel. The stalk of a flower.
pedicellate. Bearing a flower stalk.
peduncle. The stalk of an inflorescence.
pedunculate. Bearing an inflorescence stalk.
pendulous. Nodding; drooping.
perennial. Living more than two years.
perfect. Referring to a flower that has both stamens and pistils.
perfoliate. Referring to a leaf that appears to have the stem pass through it.
perianth. That part of a flower including both the calyx (sepals) and the corolla (petals).
pericarp. The ripened ovary wall.
petal. One segment of the corolla.
petaloid. Being petal-like in color and/or texture.
petiole. The stalk of a leaf.
pilose. Bearing soft hairs.
pinnatifid. Said of a simple leaf or leaf part that is cleft or lobed only partway to its axis.
pistil. The ovule-producing organ of a flower consisting of an ovary, a style, and a stigma.
pistillate. Bearing pistils but not stamens.
placentation. The way in which the ovules are attached within the ovary.
polygamous. Bearing both perfect as well as pistillate flowers.
prostrate. Lying flat.
puberulent. With minute hairs.
pubescent. Bearing some kind of hairs.
punctate. Dotted.
puncticulate. Bearing very tiny dots.

raceme. A type of inflorescence where pedicellate flowers are arranged along an elongated axis.
reflexed. Turned downward.
reniform. Kidney-shaped.
reticulate. Resembling a network.
revolute. Rolled under from the margin.
rhizome. A horizontal, underground stem.
rhombic. Becoming quadrangular, or four-sided.
rudimentary. Reduced to a very small fragment.
rugose. Wrinkled.
rugulate. With small wrinkles.

scapose. Having the flower borne on a leafless stalk.
scarious. Thin and membranous.
scurfy. Bearing scaly particles.
sepal. One part of the calyx.
sericeous. Silky.
serrate. With teeth that project forward.
serrulate. With small teeth that project forward.
sessile. Without a stalk.
setaceous. Bristlelike.
sinuate. Wavy along the margins.
spatulate. Oblong but with the basal end elongated.
spicate. Bearing a spike.
spike. A type of inflorescence where sessile flowers are arranged along an elongated axis.
spinose. Bearing spines.

spinulose. Bearing small spines.
stamen. The pollen producing organ of the flower that consists of an anther and a filament.
staminate. Bearing stamens but not pistils.
staminodia. Sterile stamens.
stellate. Star-shaped.
stigma. The terminal part of a pistil that receives pollen.
stipitate. Bearing a stalk.
stipule. A leaflike or scaly structure found at the point of attachment of a leaf to the stem.
stolon. A slender, prostrate stem.
stoloniferous. Bearing stolons.
stramineous. Straw-colored.
striate. Marked with lines or grooves.
style. The elongated part of the pistil between the ovary and the stigma.
subacute. Nearly pointed at the tip.
subcordate. Nearly heart-shaped.
suborbicular. Nearly spherical.
subulate. With a very short, narrow point.
succulent. Fleshy.
superior. Referring to the position of the ovary when the free floral parts arise below the ovary.

terete. Round in cross-section.
tomentose. Pubescent with matted wool.
tomentulose. Finely pubescent with matted wool.
truncate. Abruptly cut across.
tubercle. A wartlike process.
tuberculate. Warty.
tuberous. Bearing thickened underground stems, or tubers.
tubular. Shaped like a tube.
turbinate. Top-shaped.
turgid. Swollen to the point of bursting.

umbel. A type of inflorescence in which the flower stalks arise from the same level.
undulate. Wavy.
unisexual. Bearing either stamens or pistils in a flower but not both.
utricle. A small, one-seeded, indehiscent fruit with a thin covering.

valve. One part of a capsular fruit.
verrucose. Warty.
villous. With long, soft, slender, unmatted hairs.
viscid. Sticky.

whorl. An arrangement of three or more structures at a point on the stem.

Literature Cited

Aellen, P. 1929. Beitrage zur systematik der *Chenopodium*-arten Amerikas. Fedde, Repert. Spec. Nov. Reg. Veg. 26: 31–64.

Aellen, P., and T. Just. 1943. Key and synopsis of the American species of the genus *Chenopodium*. American Midland Naturalist 30: 47–76.

Bessey, C. E. 1915. The phylogenetic taxonomy of flowering plants. Annals of the Missouri Botanical Garden 2: 109–64.

Blackwell, W. H., M. D. Baechle, and G. Williamson. 1978. Synopsis of *Kochia* (Chenopodiaceae) in North America. Sida 7: 248–54.

Brendel, F. 1859. Additions and annotations to Mr. Lapham's catalogue of Illinois plants. Transactions of the Illinois State Agricultural Society 3: 583–85.

Buhl, C. A. 1934. Supplement to an annotated flora of the Chicago area by H. S. Pepoon. Bulletin of the Chicago Academy of Science 5: 52.

Crawford, D. J. 1975. Systematic relationships in the narrow-leaved species of *Chenopodium* of the western United States. Brittonia 27: 279–88.

Crowe, G. E. 1978. A taxonomic revision of *Sagina* (Caryophyllaceae) in North America. Rhodora 80: 11.

Danin, A., I. Baker, and H. G. Baker. 1978. Cytogeography and taxonomy of the *Portulaca oleracea* L. polyploid complex. Israel Journal of Botany 27: 171–77.

Davis, R. J. 1966. The North American perennial species of *Claytonia*. Brittonia 18: 28–503.

Dorn, R. D. 1988. *Chenopodium simplex*, an older name for *C. gigantospermum* (Chenopodiaceae). Madrono 35: 162.

Engler, A. 1897. Ubersicht uber die Unterabteilungen, Klassen, Reihen, Unterreihen und Familien der Embryophyta Siphonogama. Naturliche Pflanzenfamilien, Nachtr. II–IV.

Engler, A., and H. Prantl. 1897–1915. Die naturlichen Pflanzenfamilien, 20 vols. Leipzig.

Fernald, M. L. 1950. Gray's manual of botany, 8th ed. American Book Company, New York.

Higley, W. K., and C. S. Raddin. 1891. Flora of Cook County, Illinois, and part of Lake County, Indiana. Bulletin of the Chicago Academy of Science 2: 168.

Huett, J. W. 1897. Essay toward a natural history of LaSalle County, Illinois. Flora LaSallensis, Part I. Published by the author.

Hyatt, J. 1875. Western plants observed near Chicago and Peoria. Bulletin of the Torrey Botanical Club 6: 6–68.

Kibbe, A. L. 1952. A botanical study and survey of a typical mid-western county (Hancock County, Illinois). Published by the author at Carthage, Illinois.

Maguire, B. 1951. Studies in the Caryophyllaceae—V. *Arenaria* in America north of Mexico. A conspectus. American Midland Naturalist 46:493–511.

Maihle, N. J., and W. H. Blackwell. 1978. A synopsis of North American *Corispermum* (Chenopodiaceae). Sida 7:382–91.

McNeill, J. 1978. *Silene alba* and *S. dioica* in North America and the generic delimitation of *Lychnis, Melandrium,* and *Silene.* (Caryophyllaceae). Canadian Journal of Botany 56:297–308.

———. 1980. The delimitation of *Arenaria* (Caryophyllaceae) and related genera in North America, with 11 new combinations in *Minuartia.* Rhodora 82: 495–502.

McNeill, N. J., I. J. Bassett, and C. W. Crompton. 1977. *Suaeda calceoliformis* the correct name for *Suaeda depressa* auct. Rhodora 79:133–38.

Mead, S. B. 1846. Catalogue of plants growing spontaneously in the state of Illinois, the principal part near Augusta, Hancock County. Prairie Farmer 6:3–36; 60; 93; 119–22.

Mohlenbrock, R. H. 1986. Guide to the vascular flora of Illinois, revised and enlarged edition. Carbondale: Southern Illinois University Press.

———. 1987. New distribution data for Illinois vascular plants. Erigenia 9:1–6.

———. n.d. Vascular flora of Illinois. Southern Illinois University Press, forthcoming.

Patterson, H. N. 1876. Catalogue of the phaenogamous and vascular cryptogamous plants of Illinois. Published by the author at Oquawka, Illinois.

Pech, F. 1866. Catalogue of the United States plants in the Department of Agriculture. Washington, D.C.

Pepoon, H. S. 1927. An annotated flora of the Chicago region. Chicago: The Chicago Academy of Science.

Pringle, J. S. 1976. *Gypsophila scorzonerifolia* (Caryophyllaceae), a naturalized species in the Great Lakes region. Michigan Botanist 15:215–19.

Rabeler, R. K. 1985. *Petrorhagia* (Caryophyllaceae) of North America. Sida 11:64.

Robertson, K. 1999. Personal communication.

Sauer, J. 1955. Revision of the dioecious amaranths. Madrono 13:56.

———. 1972. The dioecious amaranths: a new species name and major range extensions. Madrono 21:426–34.

Schneck, J. 1876. Catalogue of the flora of the Wabash Valley. Annual Report of the Geological Society of Indiana 7:504–79.

Shinners, L. H. 1965. *Holosteum umbellatum* (Caryophyllaceae) in the United States: a population explosion and fractionated suicide. Sida 2:119–28.

Standley, P. C. 1915. Chenopodiaceae. North American Flora 21:13.

Swink, F., and G. Wilhelm. 1994. Plants of the Chicago region, 4th ed. Indianapolis: Indiana Academy of Science.

Thorne, R. F. 1968. Synopsis of a putatively phylogenetic classification of the flowering plants. Aliso 6:5–76.

Vasey, G. 1861. Additions to the flora of Illinois. Prairie Farmer 22:119.

Voss, E. G. 1985. Michigan Flora. Part II. Dicots (Saururaceae—Cornaceae). Bloomfield Hills, Michigan: Cranbrook Institute of Science.

Wahl, H. A. 1954. A preliminary study of the genus *Chenopodium* in North America. Bartonia 27:16.

Wettstein, R. 1935. Handbuch der systematischen Botanik, 4 ed. 2 vols. Leipzig and Wien.

Index of Plant Names

Names in roman type are accepted names, while those in italics are synonyms and are not considered valid. Page numbers in bold refer to pages that have illustrations.

Robert H. Mohlenbrock taught botany at Southern Illinois University Carbondale for thirty-four years, obtaining the title of Distinguished Professor. After his retirement in 1990, he joined Biotic Consultants as a senior scientist teaching wetland identification classes in twenty-six states to date. Mohlenbrock has been named SIU Outstanding Scholar and has received the SIU Alumnus Teacher of the Year Award, the College of Science Outstanding Teacher Award, the AMOCO Outstanding Teacher Award, and the Meritorious Teacher of the Year Award from the Association of Southeastern Biologists. During his career at Southern Illinois University, ninety graduate students earned degrees under his direction. Since 1984 he has been a monthly columnist for *Natural History* magazine. Among his 45 books and more than 510 publications are *Macmillan's Field Guide to North American Wildflowers, Field Guide to the U.S. National Forests,* and *Where Have All the Wildflowers Gone?*